Innovation Strategies in the Food Industry

Innovation Strategies in the Food Industry

Tools for Implementation

Edited by

Charis M. Galanakis

ELSEVIER

AMSTERDAM • BOSTON • HEIDELBERG • LONDON
NEW YORK • OXFORD • PARIS • SAN DIEGO
SAN FRANCISCO • SINGAPORE • SYDNEY • TOKYO

Academic Press is an imprint of Elsevier

Academic Press is an imprint of Elsevier
125 London Wall, London EC2Y 5AS, UK
525 B Street, Suite 1800, San Diego, CA 92101-4495, USA
50 Hampshire Street, 5th Floor, Cambridge, MA 02139, USA
The Boulevard, Langford Lane, Kidlington, Oxford OX5 1GB, UK

British Library Cataloguing-in-Publication Data
A catalogue record for this book is available from the British Library

Library of Congress Cataloging-in-Publication Data
A catalog record for this book is available from the Library of Congress

ISBN: 978-0-12-803751-5

For information on all Academic Press publications visit our website at https://www.elsevier.com/

Publisher: Nikki Levy
Acquisition Editor: Patricia Osborn
Editorial Project Manager: Jaclyn Truesdell
Production Project Manager: Nicky Carter
Designer: Greg Harris

Typeset by TNQ Books and Journals

Contents

PART B DEVELOPMENT OF INNOVATIONS IN THE FOOD INDUSTRY

PART D CONCLUSIONS AND PERSPECTIVES

Contributors

F.J. Barba
University of Copenhagen, Frederiksberg, Denmark

C. Barrera
Universitat Politècnica de Valencia, Valencia, Spain

B. Batalvi
SB&B Research, Toronto, ON, Canada

E. Betoret
University of Bologna, Cesena, Italy

N. Betoret
Universitat Politècnica de Valencia, Valencia, Spain

B. Bigliardi
University of Parma, Parma, Italy

C.G. Biliaderis
School of Agriculture, Aristotle University, Thessaloniki, Greece

H. Bolini
University of Campinas, San Paolo, Brazil

L. Calabuig-Jiménez
Universitat Politècnica de Valencia, Valencia, Spain

A. Claret
IRTA, Food Research Center, Finca Camps i Armet, Monells, Spain

L. Cottoni
Unitelma-Sapienza, University of Rome, Rome, Italy

J. Cvejic
University of Novi Sad, Novi Sad, Serbia

J. Deiters
University of Bonn, Bonn, Germany

J. Domenech
Universitat Politècnica de València, Valencia, Spain

A.R.H. Fischer
Wageningen University, Wageningen, the Netherlands

P. Fito
Universitat Politècnica de Valencia, Valencia, Spain

C.M. Galanakis
Galanakis Laboratories, Chania, Greece

F. Galati
University of Parma, Parma, Italy

C. García-Viguera
CEBAS-CSIC, Murcia, Spain

A. Gironés-Vilapana
University Miguel Hernández, Orihuela, Spain

M. Gougouli
Perrotis College, American Farm School, Thessaloniki, Greece

L. Guerrero
IRTA, Food Research Center, Finca Camps i Armet, Monells, Spain

M. Hersleth
Nofima AS, Ås, Norway

K. Li
Mind Genomics Associates, White Plains, NY, United States

R.P. Lopes
University of Aveiro, Aveiro, Portugal

C. Malpuech-Brugère
INRA, Clermont Université, Unité de Nutrition Humaine, Clermont-Ferrand, France;
CRNH Auvergne, Clermont-Ferrand, France

V. Martinez-Gomez
Universitat Politècnica de València, Valencia, Spain

F. Mas-Verdú
Universitat Politècnica de València, Valencia, Spain

D.A. Moreno
CEBAS-CSIC, Murcia, Spain

P. Morone
Unitelma-Sapienza, University of Rome, Rome, Italy

H. Moskowitz
Mind Genomics Associates, White Plains, NY, United States

M.J. Mota
University of Aveiro, Aveiro, Portugal

V. Orlien
University of Copenhagen, Frederiksberg, Denmark

S.A. Pereira
University of Aveiro, Aveiro, Portugal

M.J. Reinders
LEI Wageningen UR, Agricultural Economics Research Institute, the Hague, the Netherlands

I.S. Saguy
The Hebrew University of Jerusalem, Rehovot, Israel

J.A. Saraiva
University of Aveiro, Aveiro, Portugal

G. Schiefer
University of Bonn, Bonn, Germany

J.-L. Sébédio
INRA, Clermont Université, Unité de Nutrition Humaine, Clermont-Ferrand, France;
CRNH Auvergne, Clermont-Ferrand, France

L. Segui
Universitat Politècnica de Valencia, Valencia, Spain

A. Segura-Carretero
University of Granada, Granada, Spain

V. Sirotinskaya
The Hebrew University of Jerusalem, Rehovot, Israel

C. Sulmont-Rossé
INRA, UMR1324, CNRS, UMR6265, Université de Bourgogne, Centre des Sciences du Goût et de l'Alimentation, Dijon, France

V. Verardo
University of Almería, Almería, Spain

W. Verbeke
Ghent University, Ghent, Belgium

D. Villaño
CEBAS-CSIC, Murcia, Spain; University Catolica San Antonio de Murcia (UCAM), Murcia, Spain

K.G. Zinoviadou
Perrotis College, American Farm School, Thessaloniki, Greece

Preface

Nowadays the term *innovation* is increasingly used in all science fields. It has developed into a trendy word that is referred to whenever discussions focus on forthcoming developments or future perspectives. However, despite its frequent usage, innovation is neither easy to specify nor to identify. What does innovation truly represent? It is a new idea that when implemented leads to a more effective process, product, service, or technology. Innovation provides better solutions that meet advanced, unaddressed, or existing market needs. It can be considered as a breakthrough that provides a different way of thinking, consuming, or living. However, the key element of innovation is the precondition noted herein—"when implemented." Without implementation, "innovation" turns back to the "idea" status.

Food industry is today facing technical and economic changes in spite of society, manufacture, and food processing. This fact has affected significantly the entire food supply chain (eg, distribution of food to the consumers), forcing companies to pay great attention in food products so that they meet the consumers' demand for a healthy lifestyle. As a consequence, there is an extensive dialogue about the need of food industries to introduce innovations in the market in order to survive competition. Innovation in the food industry has been discussed not just as an opportunity but also as a precondition for assuring the sustainability of the food sector. It is an important instrument to stand out from the competition and satisfy consumer demands. Ultimately, it is the tool to success in a hard economic environment, the carrier to penetrate new markets, and the key to establish new products or processes. On the other hand, food industry is traditionally considered as a sector with low research intensity. Although researchers continuously develop innovative products and technologies, their applications in the food industry face several obstacles. The latest fact concerns more the introduction of the innovations within the food industry, as well as the reaction of related consumers, and less the technological adequacy of the innovative techniques. Indeed, there is a gap between R&D strategy developers and technical R&D associates, as well as a lack of interpretation between the information received by food technologists and consumer scientists.

This book aspires to fill in this gap by providing tools for the implementation of innovations in the food industry. This is conducted by suggesting certain solutions to overcome limitations, meet consumer demands, as well as improve academia/industry interaction and know-how transfer. The book consists of four major sections and 15 chapters. Part A (*Innovation strategies and long-term R&D for the food industry*) includes 4 chapters. Chapter "Food Innovation Dynamics and Network Support" covers aspects of cooperative networks and platforms development. In chapters "Open Innovation and Incorporation Between Academia and Food Industry" and "Open Innovation Opportunities Focusing on Food SMEs" emphasis is given in "open innovation" denoting the incorporation between academia and industry, as well as the particular issues that small and medium enterprises are facing during their innovation efforts. Chapter "Transition to a Sustainable Agro-Food System: The Role of Innovation Policies" focuses on innovation policies and transition to sustainable agri-food systems. Thereafter, *Development of innovations in the food industry* (Part B) is divided into four chapters. Chapter "Innovation in Traditional Food Products: Does It Make Sense?" discusses the implementation of innovations in traditional food products, whereas Chapter "Implementation of Innovation in the Food Industry" discusses the consumer-driven food innovations. The implementation of emerging technologies in the food industry is referred to in Chapter "Implementation of Emerging Technologies" and the characteristics of sustainable innovations are discussed in Chapter "Sustainable Innovation in Food Science and Engineering". Part C concerns the implementation of innovations in cutting-edge food science areas.

These include innovative biobased materials for packaging sustainability (Chapter: Innovative Biobased Materials for Packaging Sustainability), development of functional foods (Chapter: Development of Functional Foods), food use for social innovation by optimizing waste recovery strategies (Chapter: Food Use for Social Innovation by Optimizing Food Waste Recovery Strategies), adoption of ICT innovations in the agri-food sector (Chapter: Adoption of ICT Innovations in the Agri-Food Sector: An Analysis of French and Spanish Industries), and finally implementation of foodomics in the food industry (Chapter: Implementation of Foodomics in the Food Industry). Part D (*Conclusions and perspectives*) concentrates all the notes from previous chapters by discussing consumers' acceptance of novel foods (Chapter: Consumer Acceptance of Novel Foods), as well as challenges, opportunities, and solutions to overcome current implementation problems (Chapter: Challenges and Opportunities).

Conclusively, the ultimate goal of the book is to provide a handbook for anyone who wants to implement innovations in the food industry. It is intended to support researchers, scientists, engineers, and students working in the whole food science area, as well as middle management professionals in the food industry and department chairs in universities. It could be used in graduate and postgraduate levels of food science courses, consumer science and innovation fields, as well as food management programs and business schools. It may even be an appropriate text for a senior seminar course or in non-thesis master's programs at universities that are teaching food science to managers with business degrees. The most important feature of the book is that it covers both strategy designing and food technology issues in order to support new product development specialists in transition from active research to administration (both in academics and industry). This approach allows the reader to come closer and understand professional activities that overlap both strategy designing and technical knowledge.

I would like to take this opportunity to thank all the contributors to this book for their fruitful collaboration and high-quality work in bringing together strategic, theoretical, and technical issues in an integral and comprehensive text. I consider myself fortunate to have had the opportunity to collaborate with so many knowledgeable colleagues from Belgium, Brazil, Canada, Denmark, Germany, Greece, France, Israel, Italy, Norway, Portugal, Serbia, Spain, the Netherlands, and the United States. Their acceptance of editorial guidelines and book concept is highly appreciated. I would also like to thank the acquisition editor Patricia M. Osborn for her honorary invitation to lead this project and the entire Elsevier production team, especially Jaclyn Truesdell and Karen Miller for their assistance during the editing process.

Last but not least, a message for the reader. In a collaborative project of this size it is impossible for it not to contain errors. Thus, if you find errors or have any objections, I would really appreciate it if you would contact me.

Charis M. Galanakis
Research and Innovation Department
Galanakis Laboratories
Chania, Greece
e-mail: cgalanakis@chemlab.gr

CHAPTER

1 FOOD INNOVATION DYNAMICS AND NETWORK SUPPORT

G. Schiefer, J. Deiters

University of Bonn, Bonn, Germany

1.1 INTRODUCTION: SECTOR CHALLENGES AND INNOVATION

The *food sector* as a whole is faced with major challenges that arise from:

1. changes in the sector's economic and noneconomic environments
2. changes in lifestyles
3. global increases in food consumption
4. a diminishing production base (eg, due to the loss of arable land or its divergence for nonfood production alternatives (CIAA, 2007)
5. ultimately changing attitudes of society toward the consequences of the food system's activities for environmental, social, and economic issues, captured in the term of "sustainability" (Fritz and Schiefer, 2008).

For coping with the challenges, innovation has been discussed not just as an opportunity but also as a precondition for success. Innovation support and especially support toward small and medium-sized enterprises (SMEs) has therefore been identified as a core requirement for assuring the sustainability and competitiveness of the food sector. For instance, the mission formulated by the European Commission for its innovation strategy was described as "...with an ageing population and strong competitive pressures from globalization, Europe's future economic growth and jobs will increasingly have to come from innovation in products, services and business models. This is why innovation has been placed at the heart of the Europe 2020 strategy for growth and jobs..." (EU, 2011).

This statement makes it clear that the need for innovation covers a broad spectrum of developments, not just in products and processes but also in services, organization, and management. The term *innovation* has developed into a buzzword that is used whenever discussions focus on future developments. However, despite its frequent reference, innovation is not easy to specify and identify.

Discoveries or inventions without any uptake by industry and impact on the competitive advantage as well as the sustainability of the sector cannot be classified as innovation. In his overview publication on *Networks of Innovation*, Tuomi (2002) goes even as far as stating "...if new knowledge has no impact on anyone's way of doing things—in other words, if it doesn't make any difference—it is not knowledge. Only when the way things are done change, an innovation emerges..." (Tuomi, 2002, p. 10). Consequently, innovation involves both discoveries or inventions and their successful implementation in industry.

Innovation Strategies in the Food Industry. http://dx.doi.org/10.1016/B978-0-12-803751-5.00001-5

While there are many discoveries or inventions, only a few develop into innovations. The world is full of forgotten discoveries that, if they had been implemented, might have changed the world's path of development (Inventions, 2015). It is sometimes just by chance or by the engagement of individuals or groups if discoveries are being picked up and transformed into innovations. By looking back, one might realize that a development decision in the past was not the optimal one and the realization of a competing discovery might have been better. However, due to investment activities in the selected option upon its developments over time, a reverse of developments might no longer be feasible without major losses in return. This situation is referred to as "path dependency" and has been discussed broadly in literature (Page, 2006).

This is why a regular scanning of business and sector activities regarding emerging inventions (sometimes referred to as "environmental scanning," Morrison, 1992) is of major relevance for assuring that appropriate inventions are not overlooked and that innovation support is focusing toward a "best" development direction. Environmental scanning is described as an activity dealing with the "...careful monitoring of an organization's internal and external environments for detecting early signs of opportunities and threats that may influence its current and future plans..." (Businessdictionary, 2015).

It is obvious that environmental scanning is supported by networks where enterprises communicate with each other, with experts or with customers and consumers. This has been acknowledged by policy and business services that support networking especially among SMEs through the support of network initiatives and the establishment of meeting platforms of various kinds such as the European Enterprise Network (http://een.ec.europa.eu/), The latest was set up with the explicit mission of helping SMEs to innovate.

Many of the challenges faced by the food sector such as the delivery of sufficient food are not challenges as such. They become challenges because of responsibilities placed by society on the food sector. Food is a basic human need that cannot be left to decisions by industry alone. This is where the responsibility of policy comes in, and the justification for policy engagement in facilitating initiatives toward improvements in innovation support. Specifically, they include the support of networks that could facilitate the emergence of innovations in the food sector for SMEs.

However, while networks may provide an environment that could support initiatives (and especially initiatives by SMEs) toward devising and realizing inventions, the actual drivers for engaging in such initiatives are linked to the pressures and challenges the sector and its actors are facing. The drivers in combination with a fitting environment may lead to inventions that could in principle support the sector in coping with the challenges. However, the further development of inventions toward usability in routine operations as a basis for potentially reaching innovation status is based on the availability of enabling technologies or concepts. Inventions that could not build on suitable enabling technologies or concepts would remain in an invention status and could not contribute to an innovative development of the sector.

The following chapters will discuss the suitability of network support, outline some priority drivers and enablers for innovations in the food sector, and discuss some examples where the combination of drivers and enablers has opened the way for emerging inventions with innovation potential.

1.2 THE NETWORK ENVIRONMENT FOR INNOVATION SUPPORT

1.2.1 NETWORK FOCUS

As supported by studies and literature (Greve et al., 2015), networks have a crucial role in innovation support. This support could build on the creation of inventions or supporting its utilization and implementation. A comprehensive study on the innovation ability of networks that involved 32 network cases from nine European countries (Deiters and Schiefer, 2012) allows a first analysis on the relationship between the organizational and managerial concepts of networks and their sustainability and innovation support. It provides the basis for the following discussions.

The selection of case studies is based on an initial specification of structural indicators and an identification of networks that could assure a broad variation in network alternatives. The initial indicators include age, source of finance, spatial orientation (local/global), positioning in value chain, focus of innovation, network driver, and scope (see Table 1.1).

The table provides an impression on the network variety included in the study. Most networks were already some years in existence. This is due to the fact that the evolvement of networks requires some time for reaching a network status with some organizational routines. However, a few newcomers allow insight into difficulties in the uptake process. Furthermore, many networks are quite mixed in their characteristics, eg, they might have a vertical but also a horizontal focus or combine formal and informal elements. They are with some exceptions predominantly domestic in reach. This fits SME network support as networks need to build on trust and personal interaction that is at least in initial phases best served through networks with members within easy reach. It is striking that there are almost no networks that are solely based on private funding. The majority are either based on public–private partnerships or on public funding only.

The networks could be categorized according to their focus activities and their relationship with innovation generation as networks

1. creating an atmosphere for innovation through socialization activities
2. developing an innovation ability through training activities that support learning
3. promoting the identification of innovation opportunities through the support of knowledge exchange
4. pushing toward innovation implementation through activities that facilitate coordination and management of joint business activities

The ordering represents in principle an increase in dedication and commitment but also a decrease in an atmosphere that supports intuition. However, the ordering represents as well a network development path starting with a socialization period creating ideas and trust development, followed by a learning period and trusted knowledge exchange until a dedicated business-oriented innovation phase can emerge. So a network best suited for supporting innovation from the beginning to the end would be one that includes all four of them.

Participating in networks with socialization activities requires very limited dedication but provides, in turn, opportunities for exchange and the development of relationships ("networking") that may provide benefits much beyond costs. As socialization is an ongoing process, such networks do not follow a classical life cycle approach with a natural ending phase.

Training initiatives acquaint participants with the current situation in issues of interest. This is a service toward learning that, if asked for, could involve a positive cost–benefit ratio depending on the price asked for the services. A network, building on to provide a teaching program with continuously changing subjects of interest, is a learning platform. In a sector with newly emerging challenges and changing business environments such networks could remain attractive indefinitely if training services remain attractive and priced below perceived benefits.

Networks with a clear focus on the implementation of business-related activities involve the highest level of dedication. The development of arrangements and the dedication of resources involve investments by participating stakeholders (research, enterprises, and public bodies) that are usually based on

Table 1.1 Summary of Network Characteristics (Network IDs in Appendix)

		Denmark			Sweden			France			Italy			Netherlands			Ireland			Belgium					Germany			Hungary			Summary
		I	II	III	I	II	III	I	II	III	I	II	III	I	II	III	I	II	III	I	II	III	IV	V	I	II	III	I	II	III	
Age	old (>2y)	1		1	1	1	1	1	1	1	1	1	1	1	1		1	1	1	1	1	1	1	1	1		1		1	1	**25**
	new (<2y)		1													1										1		1			**4**
Finance	public		1	1	1		1	1	1	1				1	1																**9**
	private	1						1	1	1						1			1						1		1				**8**
	PPP					1					1	1	1				1	1		1	1	1	1		1	1		1	1	1	**15**
Orientation	local (dom.)	1	1	1	1	1	1	1	1	1	1	1	1		1	1	1	1		1	1	1	1	1	1	1	1	1	1	1	**27**
	global (int.)												1	1			1										1			1	**5**
Target	horizontal	1	1	1	1	1	1			1	1	1	1					1	1	1	1	1	1		1	1	1	1	1	1	**22**
	vertical		1			1		1	x		1	1	1	1	1	1	1									1		1	1	1	**14**
Focus Innov.	product	1	1	1	1	1	1		1								1	1		1		1	1	1		1	1	1	1	1	**18**
	process	1	1			1			1	x	1	1	1	1	1	1	1	1		1		1	1	1		1	1			1	**19**
	organiz.							1		x	1	1	1	1			1	1			1				1	1					**10**
	market	1	1		1	1				1							1	1	1							1					**9**
Driver	industry	1				1							1				1			1			1	1		1	1	1		1	**11**
	research			1			1						1				1			1				1		1			1	1	**9**
	focal comp.												1																1	1	**3**
	public body, group		1	1	1		1	1	1		1	1	1	1	1		1	1		1		1	1				1			1	**18**
	smes					1		1		1	1	1	1			1		1	1		1					1		1	1	1	**14**
Formal	formal	1	1	1	1	1	1	1	1	1	1	1	1	1	1	1	1	1	1	1	1	1	1	1	1	1	1		1		**27**
	informal							1	1	1	1	1	1													1		1		1	**9**
Scope	food	1	1	1	1	1	1	1		1	1	1	1				1		1	1			1	1	1	1	1	1	1	1	**22**
	other		1		1					1									1			1	1						1	1	**8**

Deiters, J., Schiefer, G., 2012. Network learning and innovation in SME formal networks, International Journal on Food System Dynamics 3, 201–213.

a clear cost–benefit evaluation by stakeholders, which keeps the networks active. Such networks may deal with innovations in products, processes, resource use, or services. Close cooperation with suppliers and customers may lead to a network cooperation commonly referred to as a "food chain." Cooperation in sharing resources such as transportation units, warehouses, or marketing services are usually developed as a business-to-business approach and not as a network approach with a clear focus on organizational or managerial innovation.

1.2.2 INCEPTION AND DYNAMICS

In the study by Deiters and Schiefer (2012) it was realized that most networks evolve out of real or perceived pressures that ask for action such as access, eg, to new technology, problems in staffing, deficiencies in consumer trust, market protection, risk in developing new markets, or emerging competition. The pressures help to overcome the investment barriers involved in creating a network.

As the creation of networks is a resource-consuming effort, network approaches are, in principle, long-term initiatives. However, differences in network attractiveness over time for different stakeholders from research, business, or public institutions may create critical conflicts in emerging networks that, if not overcome, may jeopardize their long-term sustainability.

It is easy to see that in emerging networks where innovations are still dependent on research activity the immediate benefits are not with business but with research. The latter gains better access to industry and gets engaged in training and research initiatives from the beginning. Business benefits need time to develop as they depend on project outcomes and the subsequent transformation of project outcomes into business environments.

The delay in network benefits that business members may reap from research activities is the basis for initial public support of network development, which is more common than an exception. Networks, especially those dedicated for supporting innovation and knowledge exchange, seem to depend on public support at least during the initial phase. This is especially relevant for SMEs, which are dependent on public support during the initial research phase but also for covering the transformation investments that allowed them to pick up research results for utilization in their business environment.

In contrast to large companies that can easily link up with research and commit internal resources for development activities, SMEs need network support for communication among themselves as well as with research and their institutional environment. They usually do not have the resources for the initiation and management of suitable networks. In addition, they depend on support in network creation and management. This is where public institutions come in as they have interest in promoting development and competitiveness of the sector as well as in rural development and the improvement of the labor market.

Furthermore, beyond the initial phase geared toward the creation of inventions, research results need to enter a transformation phase, which is crucial for moving inventions toward solutions that fit business needs and could provide the basis for broad adoption and sector innovation. Usually research does not engage in such transformation phases while inventions might still need considerable investments by industry for reaching operational value. As a consequence, network cooperation between research and business actors involves a phase where neither research nor business actors are actually benefitting from cooperating in innovation networks. Case study experiences show that when public support is limited to the period during which inventions are being developed and presented, network activities—especially those involving SMEs—might dissolve, keeping enterprises and especially SMEs from reaping the benefits of network engagements.

Whatever focus, networks usually dissolve or loose engagement of members if objectives have not been reached within an "acceptable" period of time as defined by their membership or, alternatively, after they have reached their objectives (eg, developed a new product). This is supported by the study of Deiters and Schiefer (2012) where most networks have been active only for a limited period of time after which they dissolve or lose their development dynamics. Exceptions are networks with a built-in reinvention concept. This is the case for networks based on an umbrella organization. They build on networks within networks. Two typical types can be considered as two opposite ends of a range of organizational alternatives.

One type builds on a well-defined structure of regional networks, and the other type is open for the establishment of inside networks that develop in response to needs and dissolve if needs have been served. Both types of networks represent a network flexibility that supports its sustainability beyond a focus presently of interest.

An umbrella network with a well-defined regional structure is represented by the network of the "Maschinenringe" (network for joint use of machinery; www.emr.maschinenringe.com). In analyzing its development over time it can be shown that while the umbrella organization remained stable throughout the years, there were dynamic changes within the cluster of regional networks. Activities within the cluster networks increased and decreased depending on needs, interests, or innovation opportunities. With the established structure an increase of activities in a region that was inactive before did not require much of an investment. This facilitated the uptake of innovation opportunities and kept the network as a whole on a sustainable development path.

An umbrella network with a "floating" internal network configuration is represented by the Swedish Skane Food Innovation Network (www.skane.com/en/livsmedelsakademin). It allows the creation of changing internal "networks on the go" according to changing needs and interests. This is an attractive opportunity from a theoretical point of view. However, it requires a higher level of organizational investment and organizational experience by those who invent and organize such networks.

1.2.3 NETWORK PERFORMANCE AND LIMITATIONS

Discussions on the performance of a network require a clear and quantifiable definition of objectives. An objective "innovation" is not of much value if it does not relate to positive impacts with stakeholders. One might formulate an artificial objective as a reference and evaluate networks according to the reference. However, as stakeholders will follow their own interests, the evaluation would have to link up with these objectives. For a business-oriented network involving research, industry, and public bodies, a "typical" scenario of objectives as supported by case studies (Deiters and Schiefer, 2012) could involve the following alternatives:

1. Research is interested in the provision of research projects and contacts with industry.
2. Enterprises are interested in competitive advantage and in recruiting qualified personnel.
3. Public institutions are interested in sustainability of industry and rural development.

In general, research is gaining in initial phases, while enterprises and public bodies need to wait for innovations from research to take effect. This makes performance measurements at early stages difficult. One might look at indicators of activities (like number of meetings, participation in network activities, etc.) assuming that activities are linked to future success. While these are weak indicators with regard to the stated objectives, they give a hint about continuing interest of stakeholders and their perception of potentials in future benefits, eg, in what they still "believe."

However, as most networks discussed in the literature or analyzed in Deiters and Schiefer (2012) are still at a rather early stage of development, one cannot substantiate a clear link between indicators of activities and network performance with regard to innovation objectives.

Networks emerge with a purpose in mind. If the purpose has been reached, networks could, in principle, be dissolved. One might compare this process with life cycle development as discussed in management studies (Zamagni, 2012). However, as innovation support is a recurring need, the availability of innovation networks is as well. As a consequence, in order to remain competitive, when closing down innovation networks within the sector, new innovation networks with a newly emerging purpose in mind need to be established. As the reestablishment requires resources in coordination, organization, and public support, a circle of closing down and reestablishing networks involves a negative cost–benefit ratio.

This cannot be in the public's interest if reestablishments of networks require public funds. As in a majority of cases, public bodies are active as initiators carrying the costs of coordination and organization and, in addition, are asked to provide financial resources for the initial phase. From a comprehensive cost–benefit view building on a combined interest of all stakeholders, it is therefore of critical relevance to focus on reaching long-term positive cost–benefit effects for network cooperation through, eg, a periodic reactivation process as is the case for the umbrella organizations discussed before.

1.3 DRIVERS AND ENABLERS

1.3.1 OVERVIEW

Discoveries and inventions may arise from obvious needs or develop by chance. However, inventions by chance that are not related to obvious needs may still provide the base for innovation if they touch actual needs that just were not apparent. The more general the needs are formulated the higher the probability that inventions with innovation potential might find attention by stakeholders in industry. Based on the identification of general needs, the creation of inventions may be driven by drivers that represent characteristics that allow quantification and could be matched by specific inventions. The opportunity of quantification is a critical issue in an industry that matches its success in terms of key performance indicators (KPI; Parmenter, 2015). The realization of inventions and their uptake as innovations may be facilitated by various enablers such as information technology or new organizational concepts. As new enablers may emerge as part of development processes in technology, science, or management, new inventions might emerge and become feasible.

The relationship between drivers, enablers, inventions, and innovations is a complex one. There might be cases where a driver leads to different inventions with different consequences for relevant KPIs. An invention with lower potential for KPI return (invention A) might become an innovation as a fitting enabler allow its easy adoption by industry. A competing invention with higher potential for KPI return (invention B) might be dismissed by industry as infeasible since it could not be appropriately supported by available enablers. At a later stage of development, new enablers might emerge that would have supported the adoption of invention B dismissed before. However, investments already made for adoption of invention A may prevent industry from switching to the invention that has become feasible with the newly emerging enablers.

A famous example from nonfood industry is the development of car engines. In the initial years inventions involved electrical and combustion engines. As is common knowledge, the combustion engines were broadly adopted. With the emerging improvements in battery technology as enabling technology, electrical cars are becoming a feasible, and from an environmental perspective preferable, alternative. However, past investments into combustion engines constitute a barrier for an easy switch to the new technology (path dependency) and it might need facilitators such as policy support to realize the switch.

These discussions lead to two different and conflicting arguments that innovators might have to consider when evaluating inventions for possible adoption:

1. As the development of inventions and enablers might not always perfectly match, it is important for innovators to realize a forward-looking approach that avoids dismissing an invention with available enablers as infeasible (and deciding for a less attractive one) while one might already be able to imagine that supportive enablers will emerge in the near future. This argument is especially relevant in dynamically developing domains in science and technology.
2. Developments in dynamically changing domains in science and technology might be so dynamic that any adoption of inventions, even with a forward-looking approach, might be already outdated by the time investments in adoption have taken place. This view is especially critical in a competitive environment when an adopter might expect that a latecomer in innovation might be able to build on newer technology and gain a competitive advantage. It constitutes an innovation barrier that can only be overcome in the long run if enterprises in a sector agree on developments. Networking is a crucial success factor for solving this problem.

1.3.2 DRIVERS FOR INNOVATION

1.3.2.1 Overview

Drivers that may support sector developments toward meeting future challenges have been identified from various angles such as research, policy, and business environments. Priority drivers that are representative for different clusters of challenges include among others.

1. *assuring food security and safety* (focusing on the sector's responsibility toward consumers)
2. *providing transparency* within the ever-increasing complexity of the food system (focusing on the sector's organization, efficiency, and trust between its actors and toward consumers)
3. *coping with urbanization* of society (focusing on changes in the sector's environment)
4. *complying with society's ethical concerns* (focusing on the society's views regarding environmental and social considerations)

1.3.2.2 Food security and safety

Assuring food security and food safety in the long run is outlined as a key societal challenge in the European research framework program "Food Security, Sustainable Agriculture and Forestry, Marine, Maritime and Inland Water Research and the Bioeconomy" (EU, 2015a) and in the United Nations Food and Agriculture Organization's development goals (FAO, 2015). Its essence may be summarized as: "Providing consumers with food that is safe, affordable and of the quality and diversity they expect while respecting economic, environmental, and social needs." The first part characterizes the sector's responsibility, the second one the society's interests, which, if followed, assure the sector's sustainability.

The challenges have been captured with a focus on 2050 in the slogan "two times more with two times less," ie, producing more food with less land, waste, energy, water, and negative impact on the environment (WUR, 2011). It is obvious that such a vision cannot be met by following current production and distribution schemes. The list of keywords should make it clear that innovations are needed in sourcing, production, distribution, and sales involving the whole spectrum of food sector activities dealing with technological, organizational, and managerial aspects.

Linked with the upcoming challenges, the goals could be translated into a number of drivers for sector development such as:

1. *Efficiency and affordability*: With the increase in consumption and the limitations in resources such as land and water, food production needs to improve to assure that enough food is affordable as stated earlier.
2. *Quality and diversity*: Quality monitoring and quality assurance in agriculture as well as along the whole food chain is a prerequisite for assuring food safety and quality within the diversity of products requested by consumers.
3. *Tracking and tracing* of food along the whole food chain and toward consumers is a prerequisite for risk management and the timely elimination of food products with deficiencies in safety and quality from distribution and delivery to consumers.
4. *Economic viability*: Any development within the food sector needs to be subject to economic viability, at least in the long run. While the society as a whole might place primary interest on environmental and social needs, the food sector has to assure economic viability for its actors while realizing product price policies that match consumers' purchasing abilities and willingness.

1.3.2.3 Transparency

Transparency is the basis for trust along the chain and with consumers and one of the most complex and fuzzy issues that the sector is facing. The complexities in food products and processes are generated due to:

1. the dynamically changing open network organization of the food sector with its multitude of SMEs
2. its cultural diversity
3. its differences in expectations
4. its differences in the ability to serve transparency needs
5. its lack of a consistent appropriate institutional infrastructure that could support coordinated initiatives toward higher levels of transparency throughout the food value chain.

Transparency is not meant to know everything but to create awareness on the issues enterprises and consumers are interested in, involving information on the safety and quality of products and processes and increasingly on environmental, social, and ethical aspects (Codron et al., 2005; Van der Vorst et al., 2005; Trienekens and Zuurbier, 2008).

Transparency is closely linked with the reliability of information and the trust in it. Trust is a topic that owners of certification schemes have brought up frequently. The control of the validity of information requires resources, and the provision of guarantees that support claims on information reliability is a service that, in principle, needed to be paid for. Claims in this context represent statements on product or process characteristics (eg, on quality) that are not directly apparent to consumers upon visual inspection (Schiefer, 2002; Verbeke, 2005). However, with the scattered company infrastructure of the sector and the importance of commodity products in most food products across the sector, reaching transparency is a sector problem that requires an organized systematic cooperation scheme among actors and cannot be tackled by individual companies alone (Beulens et al., 2005).

In summary, transparency needs can be captured in two priority drivers such as:

1. *Transparency information systems*: The dependency of transparency information on the activities of all actors in the value chain requires cooperation within the sector and asks for sector-wide efforts to improve on the availability and communication of information linked to sales products, eg, the organization of sector encompassing appropriate transparency information systems.
2. *Trust*: The reliability of information and, in turn, the trust in provided information is a crucial element in all communication systems. If trust is lost, market relationships between enterprises as well as between enterprises and consumers are damaged and might be difficult or impossible to rebuild. This asks for appropriate regulatory and control systems, and as some authors claim (Fritzen, 2016), systems for control of control.

1.3.2.4 Urbanization

According to UN (2015), the world is in a process of rapid urbanization. Today already more people live in urban areas than in rural areas. By the year 2050, it is expected that two-thirds of the world population will live in urban areas. Furthermore, the concentration of the population in big cities is increasing. Today about 28 cities in the world have more than 10 million inhabitants and by the year 2030 13 more cities will join this size category. There are about 43 cities with a population between 5 and 10 million and one expects about 63 of them by 2030. If incorporating cities with a population of between 1 and 5 million, there will be about 2.2 billion of people living in big cities in 2030.

This has enormous consequences for linking agriculture and food production with retail and consumers. Especially fresh products such as vegetables, fruits, and milk products that depend on frequent deliveries to the points of sales increasingly face problems in supply, logistics, quality assurance, costs, and environmental effects.

These problem areas in assuring the provision of appropriate food supply to the urban population constitute the drivers that ask for improvements and sustainable solutions for the future. They are closely linked to drivers toward food security and safety.

1.3.2.5 Society's ethical concerns

The core of society's ethical concerns is represented by the society's expectations regarding a sector's consideration of environmental and social needs. The latest is the base for long-term acceptance of the sector's activities by society and the assurance of its production base. Climate change, animal welfare, or fair trade characterize the broad range of issues that need to be dealt with. However, it has been realized that the consideration of environmental and social concerns is closely related to food security. The avoidance of food waste, the implementation of sensible water policies, the increase in the attractiveness of farming for the coming generation (especially in developing countries with its drive toward urban living), are just a few examples where environmental and social concerns cannot just be viewed as ethical considerations but are fundamental issues in assuring long-term food security.

1.3.3 ENABLERS FOR INNOVATION

1.3.3.1 Overview

Enablers with major relevance for supporting the food sector in meeting the challenges of the future include developments in a variety of disciplines but especially in information science, information technology, natural science, engineering, organization, and management. It is especially the latter two that are frequently overlooked in innovation discussions disregarding the influence of organizational concepts in food chain and network organization (including, eg, cooperation or logistics issues) or managerial concepts dealing with, eg, leadership or the motivation of people for engagement.

It is especially the combinations of these emerging enablers from different disciplines that open new opportunities with innovation potential. This is exemplified in discussions of the "bioeconomy" (EU, 2015b) where new opportunities in information technology, natural science, and engineering meet with economic and managerial developments.

1.3.3.2 Information science and technology

Information technology (IT) provides a broad range of enabling technologies. It reaches from basic information and communication systems and networks to knowledge networks, cloud services, satellite communication, mobile communication devices (eg, smartphones), sensors, automatic control, robots, "Internet of Things," swarm intelligence systems, and much more. This broad and ever-expanding range of technologies and related services offers a similar broad range of support for potential developments with innovation potential.

As an example, a suitable IT infrastructure on which information can be collected, processed, and moved toward retail and the consumer (Verdouw et al., 2010; Wolfert et al., 2010) is a critical success factor for reaching transparency. Similarly, communication networks that build on mobile communication and network devices such as smartphones, drones, sensors, the Internet cloud, automatic control, and related elements have the potential for new cooperation systems in food production, processing, and sales but also for cooperation between all actors in the chain and between chain enterprises and consumers. Linked up with new organizational and managerial concepts, they have the potential for far-reaching changes in the way that the food sector is organized and acts.

1.3.3.3 Natural science and engineering

In natural science there are many developments with potential support for the food sector. Plant and animal genetics, the extraction of proteins from plants for artificial meat production, feeding practices that control the emission of greenhouse gases, or nanotechnology are just a few examples of developments in bioscience that have support potential for developments toward innovations.

However, natural science with potential food sector relevance reaches beyond bioscience. Examples with potential impact involve, eg, solar energy, the transformation of CO_2 into base substances for chemicals, or low-energy light provided by LED developments.

A completely different view is discussed with respect to new technologies in printing where 3D-printing is being linked to all kinds of development opportunities in- and outside the food sector; 3D printing is a technology with the potential for dramatic changes in production economy.

1.3.3.4 Management and information technology

The combination of new developments in information technology and adapted new management concepts has the potential for new organizations in logistics, operations, control, and communication between actors in the food chain. Emerging social networks and related services open new opportunities for networking within the business communities but also between the business community and customers, experts, consumers, media, policy, etc. The potential is in the combination of technology and management that allows the establishment of systems not yet possible before.

1.4 EMERGING INNOVATIONS

1.4.1 OVERVIEW

Based on enablers discussed before and linked to drivers for development, a number of emerging inventions with innovation potential have entered discussions in the food sector. The focus of this chapter is not to provide a complete list of inventions that have reached attention by the public. With the many inventions presented and tried out everywhere, such a list would be beyond the capability of this chapter. Furthermore, the list would be incomplete and even misleading as many inventions that have not yet reached public attention might, when discovered, quickly reach innovation status.

However, some examples that are meant to demonstrate the broad range of developments that are being dealt with are presented. They are primarily in an experimental stage and might not reach innovation status. However, if picked up on a broad scale they might have the potential for substantially changing the future of the food sector. They demonstrate the need to broadly and regularly scan the business environments for inventions with innovation potential and to engage in judgments on which ones might be suitable for implementation.

1.4.2 DEALING WITH THE CHALLENGE OF MEAT CONSUMPTION

The increase in meat consumption especially in emerging economies is a major challenge for food security as well as for environmental and climate concerns (Hanna et al., 2011). The production of meat requires much more resources in terms of land, water, and energy to produce calories for human consumption than food based on plant products. In addition to the inefficiency in production, the production of meat is one of the critical issues regarding its negative impacts on climate change.

The growth in population especially in emerging economies and the changes in lifestyle in these countries toward more meat-based diets will contribute to an ever-increasing problem situation. However, there are initiatives that might change the situation if the inventions reach innovation status.

1.4.2.1 Classical developments

One of the major developments deals with newly emerging opportunities in animal diets that may reduce the production of greenhouse gas emissions from animal production. This could contribute to a reduction in negative environmental effects. Ongoing efforts toward improvements in efficiency are supportive but will not offer substantial improvements.

1.4.2.2 Emerging inventions toward innovation

This is the base where radically new inventions are being offered for moving forward. As examples, meat production through laboratory meat growth is being dealt with in experimental settings (Alok, 2015). Still more radical solutions build on meat substitutes "constructed" from plant proteins from different sources. According to the available information on "burger substitutes" they are being created to look, taste, and behave like meat burgers (Rusli, 2014). It will have to be seen if these and similar inventions will initiate a real innovation drive in the food sector.

1.4.3 SERVING URBAN POPULATIONS

Serving the ever-increasing urban population of more than 2billion people in big cities but especially in the emerging mega cities of more than 10million people develops into a major challenge for food security. Traditional distribution schemes will no longer be sufficient to deal with this problem. Long distances between the areas of production and consumption take time and energy and contribute to pollution. Traffic jams and the huge quantities required for serving consumer needs contribute to the severity of the challenge.

1.4.3.1 Classical developments

There are efforts for meeting the challenge that can be considered as a continuation of past efforts but incorporating new technology. They include the development of efficient routing systems based on, eg, a continuous monitoring of traffic (eg, through satellites) and a rerouting of deliveries. A related view is realized in developments toward urban or vertical agriculture where the production area is moved into the city to reduce the delivery distance (Golden, 2013).

1.4.3.2 Emerging inventions toward innovation

These classical developments are complemented by others that are based on inventions utilizing new technologies that were not available in the past. An emerging initiative involves factory agriculture where agricultural products are being produced in-house in factory environments (or private homes) utilizing artificial light based on low-energy LED light developments (eg, Infarm, 2015). This production opportunity is further supported by LED capabilities that allow the use of different lights for managing product quality and product growth (JQR, 2015).

1.4.4 SUPPORTING REGIONAL SOURCING FOR TRANSPARENCY AND TRUST

Regional sourcing may sound like a simple approach. However, it is not. Past developments toward global supply chains realized a supply chain infrastructure where sourcing was global. This infrastructure reached down to the organization of the retail sector, where products from global or regional sources were collected at central warehouses from where they were distributed to the various retail outlets. The utilization of central warehouses reduced management complexity even in scenarios with a majority of products from regional sources.

With the development of the digital economy, management complexity is no longer a barrier for suitable organization of sourcing and delivery systems. This has wide-reaching consequences and opens the way for innovations in organization and delivery. Decentralized systems in combination with the utilization of new computer-based information and management systems may prove to be better in efficiency, customer orientation, climate effects, and sustainability than today's systems with the efficiency of the past.

1.4.5 MANAGEMENT CONCEPTS

Improvements in the efficiency and quality of cooperation between enterprises in the food value chain that could support innovation, communication, transparency, quality, the integration of SMEs, and the link with consumers are a prerequisite for dealing with the sector's challenges and development drivers. This view is supported by policies and organizations throughout the sector. From this background it is obvious that cooperation initiatives are a major part in sector developments toward sustainability.

1.4.5.1 Classical developments

The major classical development deals with networking. Tuomi (2002) argues that "...innovation emerges in a complex iterative process where communication, learning and social interaction play important roles." Furthermore, referring to Cohen and Levinthal (1989, 1990) he continues to argue that "...adoption of new innovation requires learning and development of competences..." Tuomi (2002, p. 8).

Taking these arguments together one can see that communication, social interaction, learning, and the development of competences are key factors in innovation development. This supports the view on networking as a suitable tool for innovation support. However, innovative networking developments with a horizontal and vertical view are emerging, which differ from established networks discussed before. They include cooperation in quality production with networks such as those based on the GlobalG.A.P. certification scheme, communication networks in tracking and tracing or consumer communication, production networks in organic production, or social networks for business exchange (van Aalst, 2003).

1.4.5.2 Emerging inventions toward innovation

Beyond the classical network approaches, new technologies especially in digital communication and information systems open new opportunities such as those referred to as *e-communities* and *virtual enterprises* (Camarinha-Matos et al., 2013). Virtual enterprises build on individual enterprises that might be far away from each other but that present themselves jointly as a unit toward their customers and/or suppliers. This is supported by market access through online presentations and efficient internal digital communication systems. E-communities are less rigid and are based on digital networking of whatever kind including meetings, learning, exchanges, etc. among its members.

1.4.6 CUSTOM-MADE PRODUCTS IN FOOD DELIVERIES

The individualization of consumer products and food deliveries has been an issue of interest for food producers for quite some time. In production there have been small developments especially in connection with online ordering systems, eg, the ordering of custom-made cereal mixtures (www.mixmyown.com). In general, developments were scarce. However, recent inventions might have the potential to change the situation. Of specific relevance are developments in printing technology and especially the so-called 3D printing technology that allows the printing of products on demand. It is a technology in its infant stage but with potential for reaching innovation status as it meets a certain need (Lipson and Kurman, 2013).

First applications in food have experimental character but 3D printing might find its way into large-scale food processing as well as into food retail with a smaller-scale dedicated food production application.

1.4.7 OPEN INNOVATION FOR COMMUNICATION SUPPORT

Classical innovation is based on enterprise initiatives and may be supported by networking with other business partners. New technologies in IT and social media allow new forms of interaction between enterprises and expert groups or between enterprises, customers, and consumers. Interaction between enterprises and customers for support in innovation development has been discussed under the term *open innovation* (Martinez, 2013). Open innovation to be successful requires not just an appropriate technological base but also an organizational and managerial concept for interaction, the coordination of analysis, and the integration of supportive information into the company's innovation strategy. There are concepts in use that have experimental status but the development of convincing concepts with innovation potential is still a challenge for inventors in the management domain.

1.5 CONCLUSIONS

The food sector faces challenges that cannot be dealt with by the sector's present state in organization, technology, cooperation, and communication. While the challenges involve a variety of different needs, the magnitude of challenges is sometimes captured by the slogan "twice with half" where the foreseeable future requires a doubling of food production while production resources will decline. The sector can only meet the challenges if it is not only open for innovation but supports the creation of inventions or discoveries and their transformation into broadly adopted innovations.

For activating the sector toward innovation, support of the creativity inherent in SMEs as the majority of enterprises in the sector is of crucial relevance. Networking and the establishment of network organizations for supporting communication between enterprises is considered a core initiative for activating SME involvement in innovation development. However, the focus activities of networks may differ in terms of content and involvement of participants. In developing network engagement, networks may follow an intensification path that starts with social interaction and gradually develops toward more intensive engagements.

In dealing with the challenges the sector is pushed forward by drivers related to requirements from markets, consumers, or society. These drivers are presently matched by a variety of enablers that could facilitate and support sector initiatives in meeting the challenges ahead. Such sector initiatives are already evolving as inventions and discoveries. However, they may still need to be picked up, evaluated, adopted, and transformed into innovative developments. The magnitude of challenges and the radicality of emerging inventions may initiate a major reorganization in the sector's organization, processing, and management.

It is therefore of critical importance for the sector to evaluate the proposed inventions, to select the appropriate enablers, and to agree on coordinated moves and investments toward innovative changes with appropriate effects.

REFERENCES

Alok, J., 2015. Synthetic Meat: How the World's Costliest Burger Made It on to the Plate. http://www.theguardian.com/science/2013/aug/05/synthetic-meat-burger-stem-cells (30.08.15.).

Beulens, A.J.M., Broens, D.-F., Folstar, P., Hofstede, G.J., 2005. Food safety and transparency in food chains and networks – relationships and challenges. Food Control 16, 481–486.

Businessdirectory, 2015. http://www.businessdictionary.com/definition/environmental-scanning.html#ixzz3k0kQzr4Z (27.08.15.).

Camarinha-Matos, L.M., Afsarmanesh, H., Rabelo, R., 2013. E-Business and Virtual Enterprises: Managing Business-to-Business Cooperation. Springer, New York.

CIAA, 2007. European Technology Platform on Food for Life: Strategic Research Agenda 2007–2020. CIAA, Brussels.

Codron, J.-M., Grunert, K., Giraud-Heraud, E., Soler, L.-G., Regmi, A., 2005. Retail sector responses to changing consumer preferences – the European experience. In: Regmi, A., Gehlhar, M. (Eds.), New Directions in Global Food Markets. Electronic Report From the Economic Research Service. United States Department of Agriculture, USA.

Cohen, W., Levinthal, D.A., 1989. Innovation and learning: the two faces of R&D. The Economic Journal 99, 569–596.

Cohen, W., Levinthal, D.A., 1990. Absorptive capacity: a new perspective on learning and innovation. Administration Science Quarterly 35, 128–152.

Deiters, J., Schiefer, G., 2012. Network learning and innovation in SME formal networks. International Journal on Food System Dynamics 3, 201–213.

EU, 2011. Innovation Union, Innovation Union Competitiveness Report. EU DG Research and Innovation, Brussels.

EU, 2015a. Food Security, Sustainable Agriculture and Forestry, Marine, Maritime and Inland Water Research and the Bioeconomy. http://ec.europa.eu/programmes/horizon2020/en/h2020-section/food-security-sustainable-agriculture-and-forestry-marine-maritime-and-inland-water (30.08. 15.).

EU, 2015b. What Is Bioeconomy? http://ec.europa.eu/research/bioeconomy/policy/bioeconomy_en.htm (30.08.15.).

FAO, 2015. Millenium Development Goals. http://www.fao.org/post-2015-mdg/mdg/en/ (30.08.15.).

Fritz, M., Schiefer, G., 2008. Food chain management for sustainable food system development: a European research agenda. Agribusiness 24, 440–452.

Fritzen, S., 2016. Relevance of control in the food sector. (PhD dissertation), Bonn, forthcoming, (in German).

Golden, S., 2013. Urban Agriculture Impacts: Social, Health and Economic – an Annotated Bibliography. Agricultural Sustainability Institute, University of California, Davis.

Greve, H., Rowley, T., Shipilov, A., 2015. Network Advantage – How to Unlock Value From Your Alliances and Partnerships. Wiley.

Hanna, L., Tuomisto, H.L., de Mattos, M.J.T., 2011. Environmental impacts of cultured meat production. Environmental Science & Technology 45, 6117–6123.

Infarm, 2015. www.infarm.de.

Inventions, 2015. https://www.quora.com/What-are-the-most-significant-lost-inventions; http://www.toptenz.net/top-10-lost-technologies.php (30.08.15.).

JQR, 2015. http://jqrmag.com/en/technology/takumi-eng/a-new-approach-to-food-production-planned-vegetable-production-a-reality-plant-factory/ (30.08.15.).

Lipson, H., Kurman, M., 2013. Fabricated: The New World of 3D Printing. John Wiley & Sons.

Martinez, M.G., 2013. Open Innovation in the Food and Beverage Industry. Elsevier, Amsterdam.

Morrison, J.L., 1992. Environmental scanning. In: Whitely, M.A., Porter, J.D., Fenske, R.H. (Eds.), A Primer for New Institutional Researchers. The Association for Institutional Research, Tallahassee, Florida, pp. 86–99.

Page, S.E., 2006. Path dependence. Quarterly Journal of Political Science 1, 87–115.

Parmenter, D., 2015. Key Performance Indicators: Developing, Implementing, and Using Winning KPIs. Wiley.

Rusli, E.M., 2014. The secret of these new veggie burgers: plant blood. The Wall Street Journalhttp://www.wsj.com/articles/the-secret-of-these-new-veggie-burgers-plant-blood-1412725267 (30.08.15.).

Schiefer, G., 2002. Environmental control for process improvement and process efficiency in supply chain management – the case of the meat chain. International Journal of Production Economics 78, 197–206.

Trienekens, J.H., Zuurbier, P., 2008. Quality and safety standards in the food industry, developments and challenges. International Journal of Production Economics 113, 107–122.

Tuomi, I., 2002. Networks of Innovation. Oxford University Press, New York.

UN, 2015. The World Population Prospects: 2015 Revision. UN Department of Economic and Social Affairs, New York.

van Aalst, H.F., 2003. Networking in Society, Organisations and Education. OECD, Paris.

Van der Vorst, J.G.A.J., Beulens, A.J.M., van Beek, P., 2005. Innovations in logistics and ICT in food supply chain networks. In: Jongen, W.M.F., Meulenberg, M.T.G. (Eds.), Innovations in Agri-food Systems. Wageningen Academic Publishers, Wageningen, pp. 245–292.

Verbeke, W., 2005. Agriculture and the food industry in the information age. European Review of Agricultural Economics 32, 347–368.
Verdouw, C.N., Beulens, A.J.M., Trienekens, J.H., Wolfert, J., 2010. Process modelling in demand-driven supply chains: a reference model for the fruit industry. Computers and Electronics in Agriculture 73, 174–187.
Wolfert, J., Verdouw, C.N., Verloop, C.M., Beulens, A.J.M., 2010. Organizing information integration in agri-food - a method based on a service-oriented architecture and living lab approach. Computers and Electronics in Agriculture 70, 389–405.
WUR, 2011. http://www.wageningenur.nl/en/Research-Results/Themes/theme-food-production.htm (30.08.15.).
Zamagni, A., 2012. Life cycle sustainability assessment. The International Journal of Life Cycle Assessment 17, 373–376.

APPENDIX: CASE STUDY NETWORK IDENTIFICATION (DEITERS AND SCHIEFER, 2012)

Country	Identification
Belgium	*Wagralim* (www.wagralim.be)
	Reseau-Club I (www.ichec-pme.be)
	Innovatech (www.innovatech.be)
	Flanders Food (www.flandersfood.com)
	VLAZ (www.vlaz.ugent.be)
Denmark	*Food Club* (www.foedevareklubben.dk)
	Spis Nord (www.spisnord.dk)
France	*3D Network* (www.generations3d.com)
	Agency of Innov. (www.aripicardie.org)
	BioBourgogne (www.biobourgogne.fr)
Germany	*Cluster Food NRW* (www.exzellenz.nrw.de)
	Maschinenringe (www.maschinenring.de)
	GlobalG.A.P. (www.globalgap.org)
Sweden	*Green Tech Park (GTP)* (www.gtps.se)
	Skane Food Innovation Network (www.livsmedelsakademin.se)
Hungary	*MOSZI (Soda Producers) (I)*
	Pharmagora Cluster (II)
	Fish product and tech innovation (III)
Ireland	*Food for Health* (www.fhi.ie)
	Plato Ireland (www.plato.ie)
	Cais (www.irishcheese.ie)
Italy	*Parmigiano Reggiano Cheese* (www.parmigiano-reggiano.it)
	Prosciutto di Parma (www.prosciuttodiparma.com)
	Fruit sector networks (www.apoconerpo.com)
Netherlands	*Cluster A (I), green biotech*
	Cluster B (II), agrifood (www.foodvalley.nl)
	Cluster C (III), high-tech (www.dspvalley.com)

CHAPTER

2 OPEN INNOVATION AND INCORPORATION BETWEEN ACADEMIA AND FOOD INDUSTRY

B. Bigliardi, F. Galati

University of Parma, Parma, Italy

2.1 INTRODUCTION

Innovation in the food industry combines technological innovation with social and cultural innovation (Earle, 1997). The result of a food innovative process can be a novelty (eg, a new flavored chocolate bar), an improvement (eg, the incremental improvement of instant soup), or a fundamental change (eg, the development of functional foods). It occurs throughout the entire food system (including production, primary and secondary processing, and manufacturing and distribution) and can be focused in one specific area of food technology. In any case, its impact is on the food system as a whole. In a similar way, innovation that occurs in an industry different from the food industry may determine a change in the food industry. For instance, innovation of gas, electric, and then microwave heating determined changes in kitchens, cooking methods and ultimately changed the food industry forever (Earle, 1997) (Fig. 2.1). Different studies have highlighted that innovative new food products are more successful than traditional ones based only on incremental innovations, such as line extensions that generally generate only short-term, low-margin benefits (Knox et al., 2001; Van Trijp and Meulenberg, 1996). Food companies innovate for several reasons: to improve the quality, the range of goods and services as well as the flexibility for producing goods or services; to increase market share or enter a new market; to increase the capacity for producing goods or services; to replace outdated products; to reduce labor costs; and to improve health and safety (Giannoulidis, 2013).

The increasing importance of innovation in the food industry is mainly due to the recent changes that the sector has faced, both in the nature of food demand and in the supply chain organization, together with a more and more competitive environment in which the food companies have to operate. The food industry has always been regarded as a mature and slow-growing sector with low research intensity, and quite conservative in terms of type of innovations introduced to the market (Christensen et al., 1996; Martinez and Briz, 2000). While in the past it was traditionally focused on the minimization of production costs, thus devoting little attention to customer needs (Lienhardt, 2004), the food industry has recently changed from a supply-based approach to a demand-based approach, the so-called "chain reversal," in which the consumers tell producers what they want to eat (Bigliardi and Galati, 2013a,b; Boland, 2008). Moreover, factors such as globalization, the need to ensure food health, and consumers' demand for convenience, variety, and quality of food products, have all led the industry

Innovation Strategies in the Food Industry. http://dx.doi.org/10.1016/B978-0-12-803751-5.00002-7

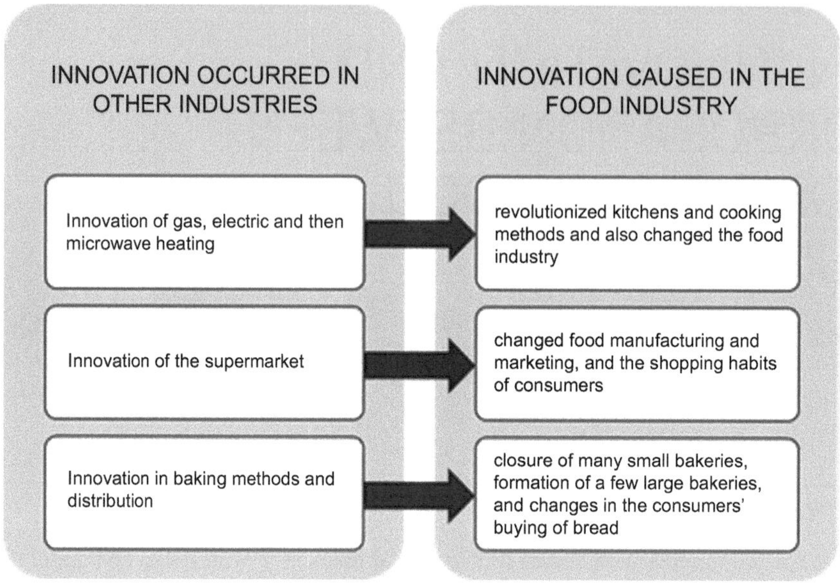

FIGURE 2.1

The impact of innovation in other industries on the food industry.

to pay more attention to creating products that meet consumers' demands (Aguilera, 2006; Bigliardi and Galati, 2013a,b; Costa et al., 2007; Omta and Folstar, 2005). In addition, the pressure to ensure food safety and be compliant with regulations forced the establishment of close relationships with external actors in order to access the knowledge needed (Bigliardi and Galati, 2013b). In such a scenario, also the continuous advancements in other sectors as well as in scientific fields like biotechnology, nanotechnology, and preservation technologies represent important opportunities for food applications (Omta and Folstar, 2005). Thus, innovation plays a central role in the food industry (Omta and Folstar, 2005; Traill and Meulenberg, 2002), while both scholars (Capitanio et al., 2010; Menrad, 2004; Rama and von Tunzelmann, 2008) and the sectoral confederation (eg, CIAA, 2008) agree in stating that innovation plays a key role in enhancing their competitiveness and satisfying fast-changing consumer expectations.

These changes have led food companies to introduce more sophisticated marketing techniques in order to gather a better understanding of the always new and differentiated consumer needs (De Jong et al., 2006), and to develop a new radical kind of product, thus testing and implementing innovative technological solutions and new business models. Even if the innovation processes in the majority of food companies are still based on internal innovation efforts, there is a growing number of food companies developing their new products by adopting factors that reside outside their boundaries, given their difficulties in leveraging only on internal factors in order to innovate (Bigliardi and Galati, 2013a; Sarkar and Costa, 2008).

All these considerations provide considerable evidence that a new paradigm, defined as "open innovation" (OI) (Chesbrough, 2003), is needed for addressing the new challenges. The formalization of the OI paradigm was provided by Chesbrough in 2003, meaning the use of purposive inflows and outflows of knowledge to accelerate internal innovation and to expand the markets for external use of innovation, respectively. OI implies the systematic exploration of a wide range of internal and external sources for innovation opportunities (West and Gallagher, 2006). Consequently, the effective implementation of the OI paradigm requires the proper management of knowledge. Such knowledge could reside inside or outside the organizational boundaries; if relevant knowledge is found outside the company, managers need to recognize, identify, capture, and manage such knowledge, choosing an appropriate integration mechanism (Bercovitz and Feldman, 2007; Wallin and Von Krogh, 2010). The OI paradigm can be adopted by firms in three different ways (Enkel et al., 2009). Firstly, they can engage in inbound OI activities, enriching their own skills and knowledge by integrating external actors into the internal innovation process. Examples of actors include suppliers, customers, or universities. Secondly, they can carry out outbound OI activities, such as technology licensing, thus earning profits by bringing ideas, patents, and other forms of intellectual property rights to the market. Finally, firms can adopt a coupled process of OI by combining the two herein reported logics (ie, inbound and outbound), thus involving complementary partners in the cocreation of innovative products/processes.

The aim of this chapter is to provide an overview on OI as implemented in the food industry and on the role of academia in such a context. With this purpose in mind, the chapter is structured as follows. First, the extant literature on this matter was reviewed by highlighting the peculiarities that distinguish the food industry in terms of OI. Next, the most recent contributions concerning cases of food companies that opened up their innovation process are analyzed. Then, the role of academia is explored prior to discussing the main challenges and paradigm shifts, as well as identifying implications for the future of OI in the food sector.

2.2 OI IN THE FOOD INDUSTRY

For decades, food companies innovated following a "closed innovation" perspective. Such a perspective refers to a situation in which all innovation is developed internally and kept inside the company. In other words, no knowledge flows go into or out of the company. More recently, empirical evidence has shown that food companies have revised their innovation strategy by moving toward an OI model. As defined by Chesbrough in his pioneering work, "Open Innovation is the use of purposive inflows and outflows of knowledge to accelerate internal innovation, and expand the markets for external use of innovation, respectively" (Chesbrough, 2003). By adopting this paradigm, food companies moved from the closed mindset "If we create the most and best ideas in the industry, we will win" to the more open one "If we make the best use of both internal and external ideas, we will win." Thus, a company only has to do part of the work itself; other firms do the rest. OI is becoming increasingly important in the food sector (Sarkar and Costa, 2008), and a move from a closed to an open approach to innovation is emerging (Fig. 2.2).

There are several reasons why food firms may rely on external knowledge to support their own innovation process (Acosta et al., 2013; Kijek, 2014). First, the characteristics of food firms themselves cause that the food industry is thought to draw not only on R&D but also on interactions with other actors. Second, the knowledge a food company may need to acquire originates in multiple scientific and technological industries, such as pharmaceuticals, chemicals and agriculture, machinery, tools, and electrical product sectors, as well as emergent scientific fields, such as nanotechnology or biotechnology. Third, the progressive pathway to new models of innovation in which the relationships between

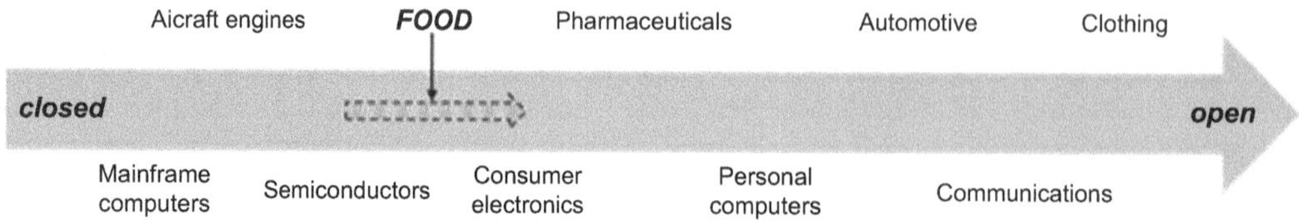

FIGURE 2.2

Open innovation in different industries.

various institutions (such as universities and laboratories) are emphasized is strongly related to the OI processes. Thus, the purpose of OI may range from merely access to external sources of knowledge to actively taking part in the creation of interorganizational knowledge and skills. In order to access knowledge, innovative food companies may establish more or less formal agreements with other actors of the supply chain, as well as with other external actors (Bigliardi and Galati, 2013a). There are different forms that companies can use in adopting the OI paradigm (Williamson and De Meyer, 2012): collaboration, networks, and innovation "ecosystems" in which participants retain their knowledge and collaborate informally. OI also offers a variety of ways in which a food company may develop and take to market an idea: in-licensing, out-licensing, cross-licensing, joint R&D agreements, corporate venture capital, joint ventures, and inorganic growth through acquisition. All of them depend on clear and predictable intellectual property (IP) arrangements. More recently, other activities (ie, incubation, spin-offs or spin-ins, crowdsourcing) have been proposed (Pénin et al., 2011).

Examples of OI implementation mainly arise from high-tech industries, such as the open-source software (Henkel, 2006; Remneland-Wikhamn et al., 2011), information and communication technology (ICT) (Bigliardi et al., 2012), biotech (Chiaroni et al., 2009), or pharmaceutical (Chiaroni et al., 2011). Even if relatively little research exists on how food companies implement OI in practice, examples of companies that adopted the OI paradigm are available. Some of these are schematically reported in Table 2.1.

As highlighted in Table 2.1, many external actors may be involved in the food innovative process. They can be grouped into four main categories:

1. individuals, such as experts, thought leaders, and consultants
2. the academic world, such as universities and research centers
3. supply chain partners, such as suppliers, consumers, retailers, and distributors
4. companies belonging to other industries

Generally, food companies prefer to engage in partnerships with actors belonging to their own sector, and thus they are characterized by similar knowledge, skills, and competences. However, companies belonging to high-tech industries are also involved in the food chain; many of the innovations introduced in new food products have been developed outside the food industry (Maula et al., 2006), for instance in the biotechnology or nanotechnology industries (Juriaanse, 2006; Maula et al., 2006). Food companies are thus demonstrating a greater openness to external knowledge, new organizational models and principles, with a view of accelerating innovation. However, OI is still often contrasted with a closed innovation model, based on the development of innovations within the company's boundaries. Given the increasing number of players and relationships, an important challenge of supply chain management is to manage the whole chain in an integrated way (Bigliardi et al., 2010).

Table 2.1 Evidence of Open Innovation Implementation in the Food Industry

Food Company	Open Innovation Implementation				References
	Why	How	With Whom	Results	
International Flavors & Fragrances	to obtain information about its customers' tastes	outsourcing part of its new product design to customers by developing an innovative toolkit	customers	valuable information on consumer tastes, costly marketing research avoids and trial-and-error cycles reduction	Thomke and von Hippel (2002)
Procter & Gamble	to develop the market for new products	adopting the Connect and Develop strategy	external research institutions, customers, suppliers, individuals, and competitors	time to market reduction for a new line through the insourcing of a technology for printing edible images and applying it to a new type of Pringles potato chips	Huston and Sakkab (2006)
Calgene	to develop a new genetically modified tomato for the fresh market (radical innovation)	establishing a wide network of relationships	actors of the food supply chain, in particular, seed companies, farmers, packers, and consumers	gaining of the complementary assets and know-how required for the success of its innovation process	Vanhaverbeke and Cloodt (2006)
Houston Farms	to provide the highest quality food products	co-innovation	suppliers and retailers	cost reduction and value added	Bonney et al. (2007)
Cargill	to expand the range of products	adopting a Web-based application to manage box of ideas	consumers	significant increasing of the number of products	Awazu et al. (2009)
Nestlé	to boost its innovative capacity in the medical nutrition space	adopting the Sharing is Winning strategy	universities and research institutes, start-ups and individual inventors, a selected number of key strategic suppliers and consumers	acceleration obtained by codevelopment of a sustainable innovation process, collaborations with motivated, talented, and highly skilled experts	Traitler and Saguy (2009)
Italian multinational food company	to answer to consumers' requirements, mainly in terms of health and safety	adopting the Food–Machinery framework	mainly suppliers, to a lesser extent universities and research centers	competitiveness increasing, possibility of increasing the yearly number of new products developed	Bigliardi et al. (2010)
A Dutch arable farm	in response to the growing changes in the competitive context and to enter in the market with very high-quality foods, with the final aim to acquire technologies externally	establishing a wide network of collaboration and adopting the Living-Lab OI model	suppliers, consumers, ICT companies, service providers and universities	exchange of theoretical and practical knowledge	Wolfert et al. (2010)

Continued

Table 2.1 Evidence of Open Innovation Implementation in the Food Industry—cont'd

Food Company	Open Innovation Implementation				References
	Why	How	With Whom	Results	
One of the largest manufacturers of pasta in Europe	in response to the growing changes in the competitive context and to enter in the market with very high-quality foods, with the final aim to acquire technologies externally	establishing a wide network of collaboration	suppliers, consumers, companies belonging to other industries, and universities	better quality of raw materials, acquisition of expertise in functional foods, and process biosensors	Bigliardi et al. (2011)
Mars	to increase the efficiency and effectiveness of the innovation process	adopting the Want, Find, Get, Manage framework	universities, research institutes, supplier networks, scouting networks	acceleration of the innovation process, improvement of the innovation quality, increase of innovation capacity, and reduction of innovation risk	Garcia (2011)
A consumer goods company specializing in sports nutrition	to obtain the necessary complementary expertise to compete	entailing collaborative strategies (partnership)	pharmaceutical company	commercialization of a new whey protein that improves utilization of glucose by muscle cells	Bröring (2013)
Lindt	to renew existing products or develop new ones	opening only certain aspects of the innovative process to collaboration	preferred suppliers (flavors development), ingredient suppliers (recipes development), panel of selected expert consumers (experimentation phase), graphic design agencies and consumer focus groups (packaging), restricted number of suppliers of machinery (production stage)	integration of outside information gathered by its marketing unit from customer focus groups, fairs, and vendors into the process	Lazzarotti and Manzini (2013)
Heinz	to create value	establishing a network of collaborations with a clear and explicit definition of the aim and scope of each partnership to ensure agreement with the company's strategy	universities and research centers, companies operating in different industries (idea generation and experimentation phases); suppliers, technical and scientific services (engineering, manufacturing and commercialization phases)	enhancement of product range	Lazzarotti and Manzini (2013)

Riso Scotti	to complete the innovation process	establishing a network of collaborations with different partners	ingredient suppliers, machinery suppliers, packaging suppliers, regulatory consultants, retailers	enhancement of product range	Lazzarotti and Manzini (2013)
MCTC-UK	to remain competitive	adopting the Value Cocreation OI framework	consumers, suppliers	escape from the commoditization trap and create consumer value	Martinez (2014)
Starbucks	to reduce paper cup wastage	crowdsourcing and cocreation	consumers	a huge number of innovative ideas and lower time to market	Arcese et al. (2015)
Backs to the Roots	searching for product diversification and development	crowdfunding	consumers	product customization, better fit to customer needs, and validation of the market demands	Arcese et al. (2015)
Unilever	to make the world a better place	strategic alliances, crowdsourcing	consumers and supply chain partners	finding of new technologies and innovations, better understanding of customer needs	Arcese et al. (2015)
Nestlé	to improve the sustainability performance of Nespresso	strategic alliances	technical experts, business partners, and other key stakeholders	higher suitability performances	Arcese et al. (2015)
Coca Cola	to understand consumer recycling behaviors at home	collaboration, strategic alliances, crowdsourcing	consumers and universities	help improve recycling rates at home	Arcese et al. (2015)
Kraft	to develop new ideas, to enhance and speed up the innovation process	cocreation and strategic alliances	companies operating in other industries	upcycling and managing packaging waste, speed up innovation and company growth	Arcese et al. (2015)
Molinos Rio de la Plata	to pursue high standard of product quality and process efficiency	crowdsourcing	consumers	healthier food products for better social sustainability	Arcese et al. (2015)
Tate and Lyle	to meet consumers' health needs	strategic alliances	suppliers, companies operating in other industries	healthier food products for better social sustainability	Arcese et al. (2015)
Spanish yogurt manufacturer	to design new food products as required by the consumers and to ensure that these products will reach the consumer in time and in adequate quantity	direct interaction with consumers and lead users (crowdsourcing) by adopting the Consumer-centric OI framework	consumers, retailers and distributors, selected suppliers	a better primary packaging, a better and more versatile product, sales improvements	Tsimiklis et al. (2015)

2.3 MODELS OF OI IMPLEMENTATION

Several models of OI implementation have been proposed by authors describing how food companies adopt such a paradigm. Among these are:

- the Connect and Develop model (Huston and Sakkab, 2006),
- the Sharing is Winning model (Traitler and Saguy, 2009),
- the Food-Machinery framework (Bigliardi et al., 2010),
- the Living-lab OI model (Wolfert et al., 2010),
- the Want, Find, Get, Manage model (Garcia, 2011),
- the Selective Sharing OI model (Lazzarotti and Manzini, 2013),
- the Value Cocreation OI model (Martinez, 2014), and
- the Consumer-centric OI model (Tsimiklis et al., 2015).

A description of each model is provided below, together with a final schematic representation of the same.

2.3.1 THE CONNECT AND DEVELOP MODEL

The Connect and Develop model, derived from the work of Huston and Sakkab (2006), shows how the use of technology may support the movement toward OI. Specifically, the authors proposed the Procter & Gamble case to describe how the company reshaped the way to manage the innovation process by means of new technologies, which were referred to with the term "innovative technologies." They created a technology brief that defined the problem to be solved and circulated it throughout global networks (research institutions, customers, suppliers, individuals, and even competitors) to discover if anyone in the world had a ready-made solution. Moreover, the creation of a Website (www.pgconnectdevelop.com) helped the company to communicate with the resources outside; anyone who was interested or had the solution could propose their ideas and get assessed by a specialized team. In some cases, suppliers' researchers would come to work in the company's labs, and in others, company's researchers would work in theirs. Thus, the model may be seen as an example of cocreation, a type of collaboration that goes well beyond typical joint development.

2.3.2 THE SHARING IS WINNING MODEL

The Sharing is Winning (SiW) model was proposed by Traitler and Saguy (2009) in the food industry where they show an application in Nestlé. This model is grounded on cocreation and involves three main typical codevelopers, both upstream and downstream partners: (1) universities, academia, research institutes, and medical centers; (2) start-ups and individual inventors; and (3) a selected number of key strategic suppliers and consumers.

SiW is based on three main pillars: the value creation along the value chain, the building of goodwill, and the establishment of trust and winning respect. It represents a paradigm shift toward accelerating codevelopment of sustainable innovation, with alignment of the entire value chain with consumer-centric innovations. The effective adoption of the SiW model led to several benefits, like the acceleration obtained by codevelopment of a sustainable innovation process or collaborations with motivated, talented, and highly skilled experts, etc. Moreover, SiW is a powerful and sustainable model because it allows to keep low the risk of making financial commitments too early in the innovation project.

This model extends the definition of OI to sustainable and enhanced processes of co-innovation (Traitler et al., 2011), and specifically it should be utilized as a platform offering co-innovation opportunities for all participants in the entire value chain. Its execution and implementation is prompted by the recognition that the partners of collaboration, (eg, universities, academia, start-ups, biotech companies, and large industrial suppliers in the case of Nestlé), are important sources of codevelopment and partnerships. In particular, among these partners academia has been shown to play a very significant role in most breakthroughs.

2.3.3 THE FOOD-MACHINERY FRAMEWORK

The Food-Machinery model was introduced by Bigliardi et al. (2010) and is one of the most adopted in the food supply chain. The specificity of the model refers to the innovative role of suppliers (a food–machinery company in their work) within the OI process, and thus it focuses on the relationships established between customer (that is, the food company) and its supplier. The model proposes the reciprocal interaction between the different actors of the food supply chain, identifying the main OI practices adopted by each of them. Moreover, it shows that in addition to suppliers, food companies are also used to collaborate (to a lesser extent) with universities and research centers. A prerequisite of this model is the fact that the food company and its suppliers are characterized by the same values: added values to shareholders, consumer focus, quality, and innovation.

2.3.4 THE LIVING-LAB OI MODEL

This model, proposed by Wolfert et al. (2010), represents a user-centric OI approach for sensing, prototyping, validating, and refining complex solutions suitable in multiple and evolving real-life contexts. In the Living-lab model, innovation from start to finish is embedded in the real-life context of users and all organizations involved in a network that are collaborating from the start of innovation. Specifically, several actors are engaged, from suppliers and processors to ICT companies and universities, cooperating with each other in interactive learning networks in which tasks for innovation are gradually implemented. The peculiarities of this iterative, design-oriented OI approach is that incremental solutions are reused, and that the role of the end user shifts from research object to a proactive position. In other words, user communities are cocreators of product and service innovations. In such a model, ICT is a key enabler.

2.3.5 THE WANT, FIND, GET, MANAGE MODEL

The Want, Find, Get, Manage (WFGM) model, originally introduced by Slowinski (2004) and applied by Garcia (2011) to the Mars case, determines how and when external knowledge is required and used in the innovation process. It consists of four steps: Want, Find, Get, Manage. The first (Want), refers to the necessity for the firm to understand the knowledge to be accessed. The second stage (Find) consists of selecting the right partner, who possesses the right knowledge. The third step (Get) consists of acquiring the knowledge identified in the Want step from the actors selected in the Find phase. Finally, the step Manage aims to coordinate and integrate the partners' resources to meet their own goals, ensuring that partners understand who does what and how to exchange the necessary information. Specifically, the process starts with the identification of the resource needs and determining which ones should/could be developed internally (in this case no OI relationship is established) and which assets have to be sought externally. As a rule, if the Want is a part of company's core areas planning to activate in the long term, then it should decide to develop the knowledge required internally. Conversely, if the WANT refers to a short-term activity, the company may decide not to build in-house expertise but rather to acquire it externally. To find the suitable partners, a Make/Buy/Partner decision-making protocol (Slowinski and Sagal, 2010) may be adopted to benchmark internal developments against external alternatives in order to ensure the selection of the best option. Once the suitable partners have been identified, a 20-item criterion is used to select the right partner(s) based on trust, expertise, uniqueness, and existing relationship. Finally, the selected partner(s) and their resources have to be coordinated and integrated in order to meet the strategic objectives of the company. As stated by Garcia (2011), its wider application could help reduce the high failure rate of new food products through the new knowledge management methods: knowledge and learning acquired in previous OI projects are easily transferred and applied to new external collaborations, avoiding typical delays and failures in OI projects.

2.3.6 THE VALUE COCREATION MODEL

The Value Cocreation model was proposed by Martinez (2014). The assumption on which the model is grounded is that creation of value is key to any company's survival. As a consequence, the voice of the consumer is central. Thus, companies adopting such a framework have to identify products that answer different consumers' needs and eventually decide to phase out those failing to contribute to the value creation. The model is an example of evolution from a product-driven innovation process toward a consumer-driven one. It requires working collaboratively with a network of partners to create value to the end consumer, thus the main challenge for companies is to find innovation partners willing to cocreate value throughout the innovation process. To select the collaborating partners, a Make/Buy/Partner decision-making protocol (Slowinski and Sagal, 2010) should be adopted, as in the WFGM model. Then, the solution has to be screened with consumers, recognized as a source of competence, which consequently act as a coproducer.

2.3.7 THE SELECTIVE SHARING OI APPROACH

The Selective Sharing OI approach was introduced by Lazzarotti and Manzini (2013) as a graded approach to OI. The strategy at the basis of the model is that only certain aspects of the innovative process are open to collaboration. A company adopts this model to develop some innovation aspects entirely internally and to avoid risks of losing control over critical information (eg, new product concepts in the case of Lindt). In addition, it does integrate outside information from external actors such as consumer focus groups, fairs, and vendors into the process. Similarly, a specific phase or process may suppose some degree of cooperation. To cite the case of Lindt, the company has developed an elaborate collaboration system with suppliers for the development of flavors. At the basis of the model, the development of effective knowledge management systems resides. The approach supposes a clear and explicit definition of the aim and scope of each partnership to ensure agreement with the company's strategy. Moreover, it may also be adopted by companies that are reluctant to use the OI process (eg, Riso Scotti) but recognize that some collaborations are necessary to complete the innovation process.

2.3.8 THE CONSUMER-CENTRIC OI MODEL

The Consumer-centric OI model was proposed by Tsimiklis et al. (2015) as a new ICT-based framework. The model incorporates the end consumer in the innovation process; his inputs, priorities, and needs are used to drive the innovation process, from the ideation to the development and commercialization. ICT plays a key role: social Websites, Intranets, and blogs are used to ask consumers to express their necessities, as well as to ask selected suppliers, retailers, distributors, and firms' employees for new ideas. Its application represents a clear example of crowdsourcing.

Fig. 2.3 shows the main characteristics and peculiarities of these models. In order to depict the models, a general food (open) innovation process has been considered, composed of four general phases: idea generation, research, development, and commercialization.

2.4 THE ROLE OF THE UNIVERSITY

Universities and academic research centers play an emerging role in almost all of the just-discussed OI models. Surprisingly, universities are typically rated quite low as knowledge sources and potential partners by firms, notwithstanding their higher impact in terms of innovative outcomes with respect to other actors of the food supply chain, like customers, suppliers, and consumers, rated as primary sources of innovation (Howells et al., 2012). For instance, in the Food-Machinery framework (Bigliardi et al., 2010), the actors of collaboration are mainly suppliers, while collaborations with universities are seldom.

However, in the food context typical examples of university-food industry (U-FI) collaboration exist. To cite only some, Nestlé established a wide network of collaborations with a very large number of important universities and research centers around the world (Traitler and Saguy, 2011). Similarly, Procter & Gamble engaged in alliances with the state universities of Ohio, signed an agreement with the state universities of Michigan, and finally built relationships with Durham University in the United Kingdom, Fraunhofer in Germany, and CSIR in India; Unilever has collaboration agreements with the University of Cambridge, allowing the integration of its R&D resources with those of the university (Dodgson et al., 2006). Although academia is only one partner, it is a very important contributor for numerous significant innovation processes and therefore deserves to be specifically highlighted.

Traditionally, the U-FI collaboration refers to the transfer of knowledge and technology, minor university inventions, and generated IP such as patenting and licensing, and to short-term consultation with a relatively low relational involvement (Perkmann and Walsh, 2007). In an OI environment, this concept refers to the relations fostering research-based collaborations, creating and managing interorganizational relationships, and promoting the cocreation of value. Technology advancement, creativity, and a growing emphasis on innovation have played a major role in the evolution of the food industry (Kim-Soon et al., 2014). In such a context, collaboration with a university represents an alternative for food companies to access sources of skills, expertise, talents required, as well as funds and facilities for their innovation process. From the university perspective, the increasing pressure to raise funding for their research from industry (Etzkowitz et al., 2000) together with the fact that the university is expected not only to produce new knowledge but also to contribute to the development of technological innovations (Laredo, 2007) makes collaboration with the food industry an essential element.

Generally, two main types of collaboration U-FI may be identified: research partnerships and research services (Perkmann, 2007). As far as research partnerships are concerned, they are designed to generate outputs, highly relevant from an academic standpoint and useful for academic publications.

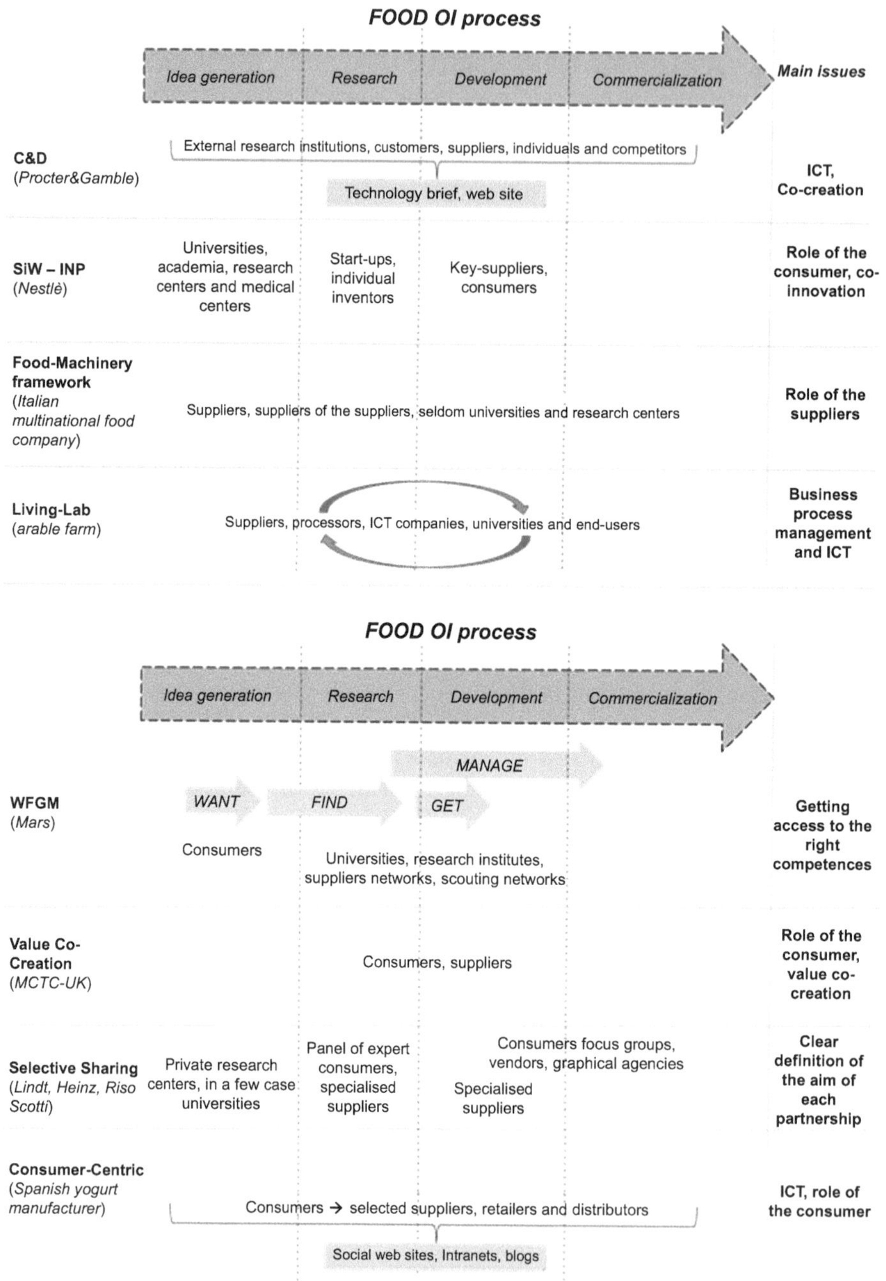

FIGURE 2.3

A schematic representation of the OI adoption models described.

Examples of this form of collaboration include sponsored research and U-FI research centers. As far as research services are concerned, they are provided by academic researchers under the direction of industrial clients and are generally less exploitable for academic publications (eg, contract research).

The traditional function of academia within the economic system has always been to provide teaching and conduct research. Recently, it has changed significantly in response to the changes in the food innovation process and in the food industry itself. In other words, the university has changed from what has been perceived as an ivory tower in the past (Wells et al., 2009) to that as a "catalytic agent of change" to drive the competition in the global economy. This new role of the university in combination with the need of the food industry to sustain its business growth is thus creating a more dynamic approach by stakeholders in U-FI collaboration, making such collaborations fundamental for technological innovation and economic development.

Unfortunately, as happens in almost all kind of collaborations, most food companies face different barriers in collaborating with academia. A first and immediate reason is that probably they are not ready to share their knowledge and IPs with external actors (Traitler, 2009; Traitler and Saguy, 2009; Traitler et al., 2011). Analysis of the extant literature on this matter highlights that the main barriers refer to academia's characteristics and to the substantial differences between the academic and the industrial values and norms (Saguy, 2011). Factors like organization and culture are the most obvious constraints negatively influencing the U-FI collaboration (Melese et al., 2009). Indeed, academia is still considered too insulated from "real-world problems" with respect to the food industry, and it is also often accused of working in isolation (Saguy, 2011). The goals of a researcher often contrast with those of the food industry (Saguy, 2011). Researchers' activities mainly focus on basic research—teaching knowledge and making discoveries that increase the depth of scientific understanding of specific phenomena are the main objectives. Conversely, industry's activities focus on applied research—industry values the benefits from transforming knowledge into practical innovation (Melese et al., 2009).

In addition, the value chain represents a basic difference between academia and the food industry; while the food industry is driven by the market and aims mainly at profit maximization, market share increasing, and consumer acceptance (Bigliardi and Galati, 2013a), academia is driven by the famous "publish or perish" problem and its main goals are basic science and student education (Saguy and Sirotinskaya, 2014). Likewise, governments and private equity have their responsibilities in limiting the U-FI collaboration. In fact, internal university funds are in most cases insufficient to establish the desired partnerships with the industry, but both governments and private equity often do not help in this regard (Saguy and Sirotinskaya, 2014).

Finally, a lack of effective IP-management rules is recognized as a hampering factor (Deschamps et al., 2013), as both food firms and academia seek full IP rights. With the progress in OI research, the metaphorical model of two separate identities, for which the only way of communication is through a technology transfer office retaining the IP (Bigliardi et al., 2015), is no longer feasible or possible. Both partners need to cocreate a shared IP strategy and to continuously interact in each stage (Saguy and Sirotinskaya, 2014).

Nevertheless, the U-FI collaboration may provide both academia and food companies several advantages. As stressed by Kim-Soon et al. (2014), collaboration between university and industry can yield synergy. Besides producing competent human capital, universities do industry-based research which will yield more customized study programs. Vice versa, industry can provide funds and help to enhance and develop curriculum by certifying a study program (Junaini et al., 2008). As a whole, this synergy model will yield mutual benefits for both universities and industry.

Moreover, from a food company's standpoint, collaboration with academia could have a direct impact on firms' innovation capacity, productivity, and profitability. Specifically, it could lead to the acceleration of the innovation process obtained by codevelopment activities, the risk of sharing innovation ideas, and the possibility to engage in collaborations with motivated, talented, and highly skilled experts (Saguy, 2011). In addition, the low absorptive capacity shown by numerous food firms (particularly by the small-sized ones) has been stressed as one of the most effective ways of building absorptive capacity (Saguy and Sirotinskaya, 2014). Conversely, from an academic standpoint, such a collaboration is often considered too applied in nature by some scholars, especially those belonging to top leading universities, and the perceived benefits of these collaborations are underestimated or marginalized. Thus, apparently, industry and academia are two separated entities (Saguy, 2011; Saguy and Sirotinskaya, 2014).

In order to effectively manage these collaborations and to obtain successful and long-term relationships, a continuous dialogue between the innovation partners, technology road maps, and the integration of value chains are required. Moreover, as stressed by Traitler et al. (2011), the fundamental prerequisite for the successful sharing of knowledge, information, and technologies is that "the solution seeker has to let the potential solution provider know its precise needs and requirements from the beginning of the collaboration." Thus, academia should have available, open, and clear sets of information regarding the food company. A successful U-FI collaboration has to be based on the identification of common goals and strategic plans that fit mutual needs. To make this possible, two main keywords have to be kept in mind from both academia and industry: "mutual vision" and "trust."

Taking these pros and cons into consideration, to let the U-FI collaboration work well under the OI lens, a structural change is needed (Muscio and Nardone, 2012). Specifically, two main significant changes and paradigms shifts are necessary.

The first steps of the (open) innovation process (that is, "ideation" and "research" in Fig. 2.3) often rely on fundamental research, which is carried out mainly in academia and will continue to define its main goal (Saguy, 2011). The following steps, namely "development" and "commercialization," are long and with significant obstacles. Markham et al. (2010) stressed that a void (also called the "Valley of Death"), exists between basic research and commercialization of a new food product. This concept refers to the resource gap between R&D labs and commercialization within academia and industrial organizations, due to the lack of capabilities to transform ideas into commercial products (Saguy, 2011).

Recalling the simple representation of the OI process previously proposed, Fig. 2.4 shows the actual situation (on the left) and the future situation after the paradigm shift (on the right). Quoting Saguy (2011), we called this paradigm shift "breaking down the wall between academia and industry." Fig. 2.4 depicts the OI innovation process as divided into two macro stages: ideation and research, and development and commercialization. In the first macro stage, basic research provides what is known as the "front end" of innovation and represents the goal of academia. The second macro stage includes all the activities carried out by the food company in order to transform basic research outcomes into potentially marketable food products throughout their commercialization. Between these stages, a wall exists in the actual situation, while a bridge would be required to successfully implement a U-FI collaboration. A dedicated transfer unit that provides specialized support services, partner searches, the management of IP and business development, the technology transfer office (TTO) may help in developing the bridge required (Bigliardi et al., 2015). Another important implication would be to design more effective policies with a final purpose of bridging the gap between academia and industry, particularly with a reference to SMEs (Muscio and Nardone, 2012). A proactive collaboration requires a proactive role of both academia and industry in each step of the innovation. To do that, academia needs to recognize that its role does not end with

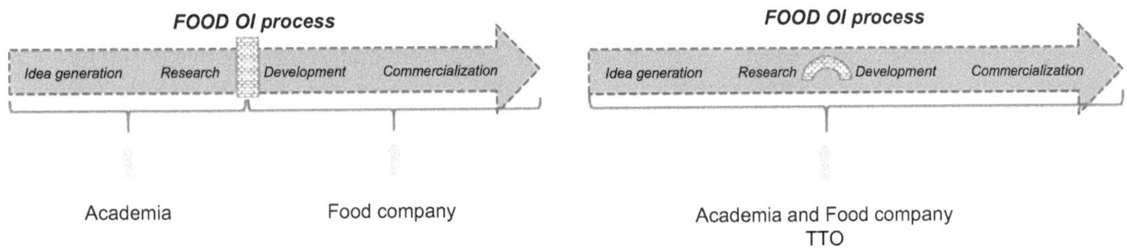

FIGURE 2.4

The first paradigm shift: breaking down the wall between academia and industry.

conducting and excelling in basic and fundamental research. Conversely, it has to learn the industry's needs and drives the invention over the research stage (Markham et al., 2010).

The second required paradigm shift refers to IP. Usually, OI innovation cannot exist without a clear IP rights definition (Bigliardi and Galati, 2013a). In the past, most companies owned all IPs, which very often remained noncommercialized (Alexy et al., 2009). This nonutilization highlights the opportunity of sharing IP rights (Saguy, 2011). The latest opportunity is even more relevant in an OI environment. In such a context, U-FI partnerships play a key role. Specifically, different IP-management strategies are available and required (Hunter and Stephens, 2010). The TTO may act as a broker between academia and the food industry in two different ways: first, by providing expertise, and second, by managing the commercialization process related to patenting and licensing (Ambos et al., 2008; Bigliardi et al., 2015). However, TTOs have emerged with a poor or inadequate consideration of issues related to the industry in particular (Bigliardi et al., 2015) and with a scarcity of resources to effectively operate (Tyler, 2009).

Thus, considering IP rights, a proactive role of both academia and industry is required. As for the role of academia, it should provide leadership by the personal example of its researchers. As far as the industry role is concerned, it has to take a more proactive role in guaranteeing adequate funding. Social responsibility is required from both sides.

To conclude this overview on the role of academia in the OI process, the following recommendations (that may serve academics) are proposed in order to establish a profitable collaboration with the industry:

1. Define a new role for industry (eg, Saguy, 2011): this result may be reached for example by supervising joint theses carried out even partially within the industry, opening the door to industrial internships, fellowships, advanced education, and so on. In addition, industries may become proactive in teaching graduate courses, mentoring research, serving on university committees and boards, and contributing to the strategic thinking of the universities, etc. In such ways industrial involvement will be improved.
2. Define a new role for the academics (eg, Saguy, 2011): as stressed for the industry, the presence of academics in the food industry should also be enhanced. By devoting part of their working time in the food industry, academics would be able to eliminate all the barriers, become a "member" of the industrial team and thus play a proactive role. This new role of academics will in turn allow students to carry out their theses or research activities in the industrial setting, where their professor is actively involved, too.

3. Enhance the status and role of applied research, thus producing research valuable to the food industry (eg, Saguy, 2011; Tyler, 2009): in such a way, teaching quality may be improved by focusing on topics for the food industry, and students' ability to interact directly with the food industry may be attractive.
4. Improve relationships with private companies, and in general intensify interactions with actors belonging to the "real world" (eg, Dosi et al., 2006; Muscio and Nardone, 2012): this could be more effective if associated with an increase in their research quality and the possibility to publish in prestigious referred journals.
5. Develop more effective TTOs (eg, Bigliardi et al., 2015; Muscio and Nardone, 2012): a TTO should be considered as the boundary of the U-FI relationship. However, as previously stressed, many of them are focused more on the university side, omitting numerous industrial opportunities. Thus, more emphasis and increasing investments in terms of resources are needed in this direction.

All these recommendations, if adopted, will enable academia to enhance collaboration with industry, which in turn will enhance the university's reputation (Tyler, 2009).

2.5 AGENDA FOR FUTURE RESEARCH

The analysis of the extant literature leads to the identification of several challenges and needs for the successful implementation of the OI paradigm in the food industry. It is widely recognized that most food companies face a series of challenges when implementing OI (Bigliardi and Galati, 2013a; Saguy, 2011; Sarkar and Costa, 2008).

The first challenge refers to the collaborations that a food company may establish. For instance, which are the right partners and how a food company may meet their (often) contrasting needs? The OI paradigm calls for exploring outside the boundaries of the organization, thus a company needs to determine which are the best actors with whom to collaborate, as well as how to respond to their needs and requirements. In doing so, it has to consider that an OI approach usually involves more than one partner depending on the knowledge, technologies, and information required in the different stage of the innovation process. Thus, a balance among numerous and often contrasting needs of the different partners has to be reached.

Second, the "management challenge" answers the question: Which management strategies have to adopt a food company in an OI environment? Indeed, the food company has to identify a supporting management process. Thereafter, a change in management is required to address the new OI mindset in order to fully exploit collaboration and networking, and to eliminate the barriers of collaboration. Management strategies may refer to IP strategy, knowledge, and value chain management.

A third important challenge considers the tools (in particular the ICT ones) that a food company may adopt in order to implement an OI approach. As previously highlighted, numerous models of OI implementation consider ICT as a key enabler (eg, Connect & Develop, Living-lab, Consumer-centric). Traditional techniques for OI have been implemented, but there is the need for identifying new and advanced ICT tools to be applied and to exploit all the advantages of an open innovation approach. Consequently, the "tools challenge" arises: How may a food company use ICT tools to support OI?

Finally, yet importantly, the last challenge refers to sustainability, which is a matter that is acquiring significant relevance in the food context. Adopting a sustainable approach to innovation could lead to a reduction of environmental impacts, healthier and safer food for an increasing population, as well as a reduction of costs and time to market. Consequently, the adoption of an "open sustainability innovation"

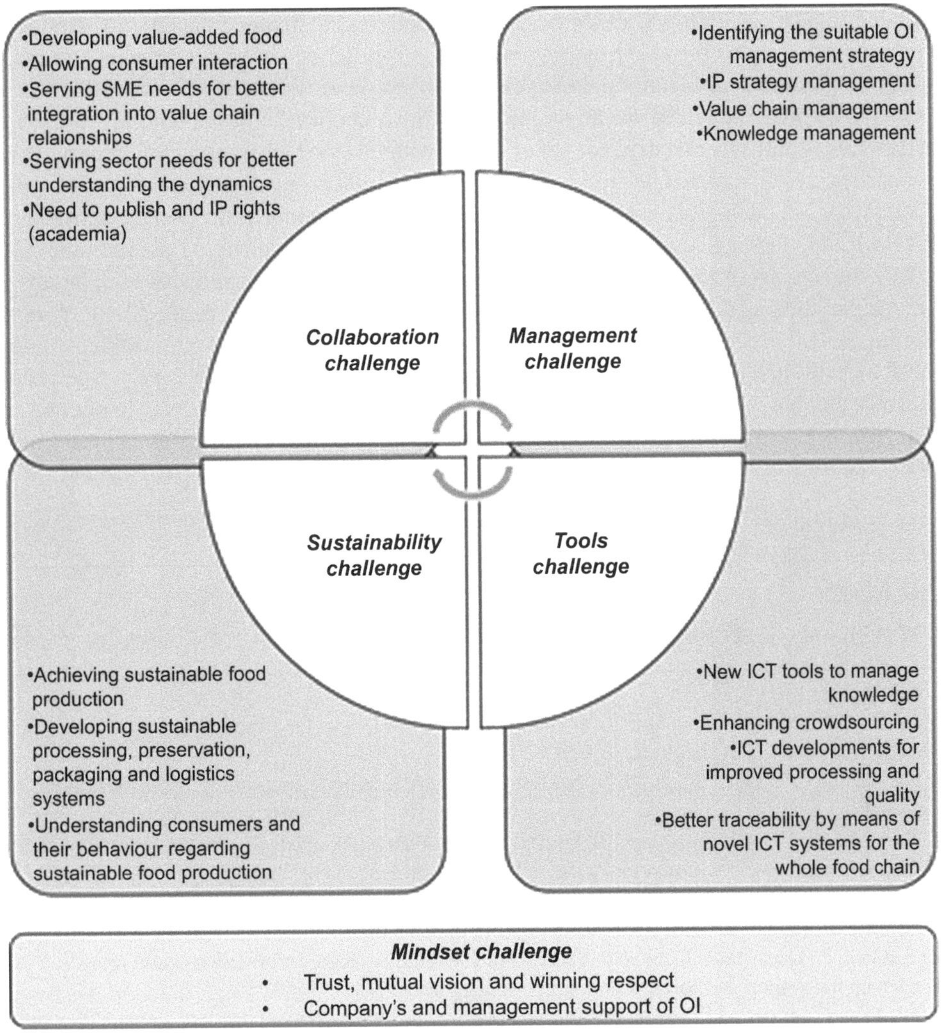

FIGURE 2.5

The main challenges and needs to the right implementation of OI in the food industry.

approach could be a source of strategic advantage for food companies. Thus, the "sustainability challenge" is generated: How may a food company adopt a sustainable OI model?

In addition to the previous challenges, the preliminary challenge, and probably the most obvious one, is the one labeled as "mindset challenge": How can a food company and its partners (mainly academia) adopt OI, winning the hearts and minds of their personnel?

Fig. 2.5 provides a schematic representation of these challenges and respective needs.

The overview provided in the previous sections can serve as a useful reference point for guiding further research. As suggested by Vauterin (2012), a better understanding of boundary roles between academia and industry will help to decrease the perceived market demand uncertainty. Besides, more in-depth research is needed for understanding how the collaboration between academia and industry is realized. On the other hand, according to Saguy and Sirotinskaya (2014), a win–win OI partnership is more common between large and medium-sized food firms and academia, in contrast to the situation with SMEs. This is probably due to a more open and structured R&D environment in large and medium-sized companies, even if a more openness is emerging among SMEs (eg, Saguy and Sirotinskaya, 2014). Other factors (eg, availability of resources and funding) may play an important role, too. Although SMEs find the search and scanning costs for implementing OI too high and challenging to overcome (Gassman et al., 2010), the scarcity of resources allows SMEs to benefit more from the effective implementation of the OI paradigm compared to large and medium-sized food firms (Howells et al., 2012). Nevertheless, more studies are still needed to investigate how to reduce the gap between academia and the food industry, as well as how such a collaboration can be developed and coordinated effectively. Specifically, the definition of a suitable collaboration model to be adopted in the food industry may help both researchers and managers in constructing a successful relationship. Indeed, building bridges between university researchers and businesses is critical, but OI models like SiW or WFGM could pave the way.

REFERENCES

Acosta, M., Coronado, D., Ferrándiz, E., 2013. Trends in the acquisition of external knowledge for innovation in the food industry. In: Martinez, M.G. (Ed.), Open Innovation in the Food and Beverage Industry. Elsevier, Cambridge, UK, pp. 3–24.

Aguilera, J.M., 2006. Seligman Lecture 2005: food product engineering: building the right structures. Journal of the Science of Food and Agriculture 86, 1147–1155.

Alexy, O., Criscuolo, P., Salter, A., 2009. Does IP strategy have to cripple open innovation? MIT Sloan Management Review 51, 71–77.

Ambos, T.C., Makela, K., Birkinshaw, J., D'Este, P., 2008. When does university research get commercialized? Creating ambidexterity in research institutions. Journal of Management Studies 45, 1424–1447.

Arcese, G., Flammini, S., Luchetti, M.C., Martucci, O., 2015. Evidence and experience of open sustainability innovation practices in the food sector. Sustainability 7, 8067–8090.

Awazu, Y., Baloh, P., Desouza, K.C., Wecht, C.H., Kim, J.Y., Jha, S., 2009. Information-communication technologies open up innovation. Research Technology Management 52 (1), 51–58.

Bercovitz, J.E.L., Feldman, M.P., 2007. Fishing upstream: firm innovation strategy and university research alliances. Research Policy 36 (7), 930–948.

Bigliardi, B., Bottani, E., Galati, F., 2010. Open innovation and supply chain management in food machinery supply chain: a case study. International Journal of Engineering, Science and Technology 2 (6), 244–255.

Bigliardi, B., Dormio, A.I., Galati, F., 2012. The adoption of open innovation within the telecommunication industry. European Journal of Innovation Management 15 (1), 27–54.

Bigliardi, B., Galati, F., 2013a. Models of adoption of open innovation within the food industry. Trends in Food Science & Technology 30 (1), 16–26.

Bigliardi, B., Galati, F., 2013b. Innovation trends in the food industry: the case of functional foods. Trends in Food Science & Technology 31 (2), 118–129.

Bigliardi, B., Galati, F., Marolla, G., Verbano, C., 2015. Factors affecting technology transfer offices' performance in the Italian food context. Technology Analysis & Strategic Management 27 (4), 361–384.

Bigliardi, B., Galati, F., Petroni, G., 2011. Collaborative modes of R&D: the new challenges for personnel management. International Journal of Business, Management and Social Sciences 2 (3), 66–74.

Boland, M.J., 2008. Innovation in the foods industry: personalisation and mass customisation. In: Marcure, J., Moughan, P., Bruhn, C. (Eds.), Innovation, Management Policy and Practice, 10, pp. 1–132.

Bonney, L., Clark, R., Collins, R., Fearne, A., 2007. From serendipity to sustainable competitive advantage: insights from Houston's Farm and their journey of co-innovation. Supply Chain Management: An International Journal 12 (6), 395–399.

Bröring, S., 2013. The role of open innovation in the industry convergence between foods and pharmaceuticals. In: Martinez, M.G. (Ed.), Open Innovation in the Food and Beverage Industry. Woodhead Publishing, Oxford.

Capitanio, F., Coppola, A., Pascucci, S., 2010. Product and process innovation in the Italian food industry. Agribusiness: An International Journal 26, 503–518.

Chesbrough, H., 2003. Open Innovation: The New Imperative for Creating and Profiting from Technology. HBS Press, Boston, MA.

Chiaroni, D., Chiesa, V., Frattini, F., 2009. Investigating the adoption of open innovation in the bio-pharmaceutical industry: a framework and an empirical analysis. European Journal of Innovation Management 12 (3), 285–305.

Chiaroni, D., Chiesa, V., Frattini, F., 2011. The Open Innovation Journey: how firms dynamically implement the emerging innovation management paradigm. Technovation 31 (1), 34–43.

Christensen, J.L., Rama, R., Von Tunzelmann, N., 1996. Study on Innovation in the European Food Products and Beverages Industry. European Innovation Monitoring System. EIMS publication, 35. European Commission, Directorate General XIII, Luxembourg.

CIAA, 2008. Review of Key Competitiveness Indicators for the European Food Industry. CIAA, Brussels.

Costa, A.I.A., Schoolmeester, D., Dekker, M., Jongen, W.M.F., 2007. To cook or not to cook: a means-end study of the motivations behind meal choice. Food Quality and Preference 18, 77–88.

Dodgson, M., Gann, D., Salter, A., 2006. The role of technology in the shift towards open innovation: the case of Procter & Gamble. R&D Management 36 (3), 333–346.

Dosi, G., Llerena, P., Sylos Labini, M., 2006. The relationships between science, technologies and their industrial exploitation: an illustration through the myths and realities of the so-called 'European Paradox'. Research Policy 35, 1450–1464.

Deschamps, I., Macedo, M.G., Eve-Levesque, C., 2013. University-SME collaboration and open innovation: intellectual-property management tools and the roles of intermediaries. Technology Innovation Management Review 33–41 March.

Earle, M.D., 1997. Innovation in the food industry. Trends in Food Science & Technology 8 (5), 166–175.

Enkel, E., Gassmann, O., Chesbrough, H., 2009. Open R&D and open innovation: exploring the phenomenon. R&D Management 39, 311–316.

Etzkowitz, H., Webster, A., Gebhardt, C., Terra, B.R.C., 2000. The future of the university and the university of the future: evolution of ivory tower to entrepreneurial paradigm. Research Policy 29 (2), 313–330.

Garcia, M., 2011. Future R&D strategies in food and drinks: evolution from orthodox approach to open innovation. In: IFAMA Annual World Symposium, 20–21 June 2011 Frankfurt, Germany.

Gassman, O., Enkel, E., Chesbrough, H., 2010. The future of open innovation. R&D Management 40, 213–221.

Giannoulidis, N., 2013. Trends and Innovation Needs in the European Food and Drink Industry. http://www.innofoodsee.eu/downloads/trends_and_innovation.pdf (retrieved June 2015).

Henkel, J., 2006. Selective revealing in open innovation processes: the case of embedded Linux. Research Policy 35 (7), 953–969.

Howells, J., Ramlogan, R., Cheng, S.L., 2012. Innovation and university collaboration: paradox and complexity within the knowledge economy. Cambridge Journal of Economics 36 (3), 703–721.

Hunter, J., Stephens, S., 2010. Is open innovation the way forward for big pharma? Nature Reviews Drug Discovery 9, 87–88.

Huston, L., Sakkab, N., 2006. Connect and develop: inside Procter and Gamble's new model for innovation. Harvard Business Review 84, 58–66.

De Jong, J., van Kleef, E., Frewer, L.J., Renn, O., 2006. Perceptions of risk, benefit and trust associated with consumer food choice. In: Frewer, L.J., van Trijp, H.C.M. (Eds.), Understanding Consumers of Food Products. Woodhead Publishing, Cambridge, England, pp. 125–150.

Junaini, S.N., Fadzir, S.F.S., Sidi, J., Khiri, M.J.A., Othman, R.M., April 2008. Harnessing university-industry collaboration in Malaysia through industrial training. In: 3rd International Conference on Information and Communication Technologies: From Theory to Applications, 2008. ICTTA 2008, pp. 1–5.

Juriaanse, A.C., 2006. Challenges ahead for food science. International Journal of Dairy Technology 59, 55–57.

Kijek, T., 2014. Open Innovation Practices in Food Industry: Challenges and Opportunities, Intensive Programme "Sustainability and Innovation in Rural Development". http://www.asu.lt/ev/en/61925 (retrieved June 2015).

Kim-Soon, N., Anwar, J., Razzaly, W., Rahman, A., 2014. Drivers of change of university and Industry collaboration from the perspective of the food manufacturers: a foresight case. In: Proceeding of the 23rd International Business Information Management Association (IBIMA) Conference, pp. 13–14.

Knox, B., Parr, H., Bunting, B., 2001. Model of 'best practice' for the food industry. Proceedings of the British Nutrition Society 60.

Laredo, P., 2007. Revisiting the third mission of universities: toward a renewed categorization of university activities? Higher Education Policy 20 (4), 441–456.

Lienhardt, J., 2004. The Food Industry in Europe, Statistics in Focus: Industry, Trade and Services, 39/2004. Eurostat, European Communities, Luxembourg.

Lazzarotti, V., Manzini, R., 2013. The tension between traditional innovation strategies and openness: Lindt's controlled open innovation approach. In: Martinez, M.G. (Ed.), Open Innovation in the Food and Beverage Industry. Elsevier.

Markham, S.K., Ward, S.J., Aiman-Smith, L., Kingon, A.I., 2010. The valley of death as context for role theory in product innovation. Journal of Product Innovation Management 27 (3), 402–417.

Martinez, M.G., 2014. Co-creation of value by open innovation: unlocking new sources of competitive advantage. Agribusiness 30 (2), 132–147.

Martinez, M.G., Briz, J., 2000. Innovation in the Spanish food & drink industry. International Food and Agribusiness Management Review 3 (2), 155–176.

Maula, M., Keil, T., Salmenkaita, J.P., 2006. Open innovation in systemic innovation contexts. In: Chesbrough, H.W., Vanhaverbeke, W., West, J. (Eds.), Open Innovation: Researching a New Paradigm. Oxford University Press, Oxford, pp. 241–257.

Menrad, K., 2004. Innovation in the food industry in Germany. Research Policy 33 (6–7), 845–878.

Melese, T., Lin, S.M., Chang, J.L., Cohen, N.H., 2009. Open innovation networks between academia and industry: an imperative for breakthrough therapies. Nature Medicine 15, 502–507.

Muscio, A., Nardone, G., 2012. The determinants of university–industry collaboration in food science in Italy. Food Policy 37 (6), 710–718.

Omta, S.W.F., Folstar, P., 2005. Integration of innovation in the corporate strategy of agri-food companies. In: Jongen, W.M.H., Meulenberg, M.T.G. (Eds.), Innovation in Agri-Food Systems. Wageningen Academic Publishers, Wageningen, pp. 223–246.

Pénin, J., Hussler, C., Burger Helmchen, T., 2011. New shapes and new stakes: a portrait of open innovation as a promising phenomenon. Journal of Innovation Economics 7, 11–29.

Perkmann, M., Walsh, K., 2007. University–industry relationships and open innovation: towards a research agenda. International Journal of Management Reviews 9 (4), 259–280.

Rama, R., von Tunzelmann, N., 2008. Empirical studies of innovation in the food and beverage industry. In: Rama, R. (Ed.), Handbook of Innovation in the Food and Drink Industry. Haworth Press, New York/London.

Remneland-Wikhamn, B., Ljungberg, J.A.N., Bergquist, M., Kuschel, J., 2011. Open innovation, generativity and the supplier as peer: the case of iPhone and Android. International Journal of Innovation Management 15 (01), 205–230.

Saguy, I.S., 2011. Paradigm shifts in academia and the food industry required to meet innovation challenges. Trends in Food Science & Technology 22, 467–475.

Saguy, I.S., Sirotinskaya, V., 2014. Challenges in exploiting open innovation's full potential in the food industry with a focus on small and medium enterprises (SMEs). Trends in Food Science & Technology 38 (2), 136–148.

Sarkar, S., Costa, A.I.A., 2008. Dynamics of open innovation in the food industry. Trends in Food Science and Technology 19, 574–580.

Slowinski, G., 2004. Reinventing Corporate Growth. Alliance Management Press, Gladstone, NJ.

Slowinski, G., Sagal, M.W., 2010. Good practices in open innovation. Research Technology Management 38–45.

Thomke, S., von Hippel, E., 2002. Customers as innovators: a new way to create value. Harvard Business Review 80, 74–80.

Traill, W.B., Meulenberg, M.T.G., 2002. Innovation in the food industry. Agribusiness 18 (1), 1–21.

Traitler, H., Saguy, I.S., 2009. Creating successful innovation partnerships. Food Technology 63 (3), 22–35.

Traitler, H., Watzke, H.J., Saguy, I.S., 2011. Reinventing R&D in an open innovation ecosystem. Journal of Food Science 76 (2), 62–68.

Tsimiklis, P., Ceschin, F., Green, S., Qin, S.F., Song, J., Baurley, S., Makatsoris, C., 2015. A consumer-centric open innovation framework for food and packaging manufacturing. International Journal of Knowledge and Systems Science 6 (3), 52–69.

Tyler, J.E.I., 2009. Advancing university innovation: more must be expected more must be done. Minnesota Journal of Law, Science & Technology 10, 143–212.

Van Trijp, H.C.M., Meulenberg, M.T.G., 1996. Marketing and consumer behaviour with respect to foods. In: Meiselman, H.L., McFie, H.J.H. (Eds.), Food Choice, Acceptance and Consumption. Blackie Academic & Professional, London.

Vanhaverbeke, W., Cloodt, M., 2006. Open innovation in value networks. In: Chesbrough, H.W., Vanhaverbeke, W., West, J. (Eds.), Open Innovation: Researching a New Paradigm. Oxford University Press, Oxford, pp. 258–281.

Vauterin, J.J., 2012. The Demand for Global Student Talent: Capitalizing on the Value of University-Industry Collaboration. Lappeenranta University of Technology.

Wallin, M.W., Von Krogh, G., 2010. Organizing for open innovation: focus on the integration of knowledge. Organizational Dynamics 39 (2), 145–154.

Wells, P., Bristow, G., Nieuwenhuis, P., Christensen, T.B., 2009. The role of academia in regional sustainability initiative: Wales. Journal of Cleaner Production 17, 1116–1122.

West, J., Gallagher, S., 2006. Challenges of open innovation: the paradox of firm investment in open-source software. R&D Management 36 (3), 319–331.

Williamson, P., De Meyer, J., 2012. A ecosystem advantage: how to successfully harness the power of partners. California Management Review 55, 24–46.

Wolfert, J., Verdouw, C.N., Verloop, C.M., Beulens, A.J.M., 2010. Organizing information integration in agri-food. A method based on a service-oriented architecture and living lab approach. Computers and Electronics in Agriculture 70 (2), 389–405.

CHAPTER

OPEN INNOVATION OPPORTUNITIES FOCUSING ON FOOD SMEs

3

I.S. Saguy, V. Sirotinskaya
The Hebrew University of Jerusalem, Rehovot, Israel

3.1 INTRODUCTION

This chapter cites largely from a previous publication that focused on open innovation (OI) and the food industry. The main focus here is on small and medium enterprise (SME) applications, OI utilization, and examples (Saguy and Sirotinskaya, 2014). OI, as also highlighted by Bigliardi et al. (2015), was originally defined by Chesbrough (2003): "valuable ideas can come from inside or outside the company and go to market from inside or outside the company as well," to be expanded later as "the use of purposive inflows and outflows of knowledge to accelerate internal innovation, and expand the markets for external use of innovation, respectively" (Chesbrough et al., 2006). Today, OI is essential to surviving and gaining a competitive advantage in most business environments, where companies must use both external and internal ideas, open channels for knowledge access, employ external technology and solutions, and purchase or license inventions (Saguy and Sirotinskaya, 2014; Traitler and Saguy, 2009; Traitler et al., 2011).

We believe that today OI has expanded beyond its previous definition. For instance, the European Union (EU) has a new paradigm, Open Innovation 2.0 (OI2), based on a "Quadruple Helix Model" in which government, industry, academia, and civil participants work together to cocreate the future and drive structural changes far beyond the scope of what any single organization or person could do alone. This model also encompasses user-oriented innovation models to take full advantage of the cross-fertilization of ideas, leading to experimentation and prototyping in a real-world setting (http://ec.europa.eu/digital-agenda/en/growth-jobs/open-innovation). The Quadruple Helix Model (previously termed "fourth helix"), originally suggested by Saguy (2013), states that for OI to succeed, especially in SMEs, all participants of the OI ecosystem (university, industry, government, and private sector) need to take a proactive role. The private sector (eg, banks, private and venture capital, other private and government funds, angels) has a paramount role in making OI a reality for SMEs, justifying its inclusion (Saguy et al., 2013; Saguy and Sirotinskaya, 2014).

It is also worth noting that OI practices were initiated in software, electronics, telecom, biotech, and pharma industries before eventually spreading to others, including the food industry (Gassmann et al., 2010). Today's paradigm consists of diverse sets of practices encompassing different dimensions, such as open business model, OI intellectual property, strategy, collaboration, crowdsourcing, threadless cocreation (Sloane, 2011), and social responsibility (Saguy, 2011, 2013). More recently, new ideas from the software domain have emerged, such as agile operations and DevOps, which will likely make

Innovation Strategies in the Food Industry. http://dx.doi.org/10.1016/B978-0-12-803751-5.00003-9

an important contribution to OI and are covered in the section to follow that focuses on future challenges. The main topics highlighted in this chapter include:

1. review and critical assessment of recent progress in OI, its utilization and implementation in the food industry, with a focus on SMEs;
2. adaptation of OI to food industry and SMEs' needs;
3. level and degree of innovative openness, and SMEs' challenges in implementing OI;
4. roles and selection of solution brokerage houses;
5. changes in academia's roles, barriers, collaborations and mindset, and intellectual property models;
6. specific recommendations addressing the unique needs of SMEs.

3.2 SMEs AND LARGE COMPANIES

Before delving into the utilization and exploitation of OI in the food industry, we define the distinction between large companies (LCs) and SMEs. According to the EU, SMEs are defined by number of employees, and either turnover or balance sheet totals as listed in Table 3.1.

These data highlight the very limited budget and human resources with which SMEs operate, unequivocally and significantly affecting their possible spectrum of operations and their possibilities for conducting business compared to LCs. This fact calls for special consideration, impacting OI practices and utilization (described further on). Also noteworthy is the vast difference, even among SMEs themselves, in number of employees, ranging from less than 10 to up to 50 and 250 employees for micro-, small-, and medium-sized companies, respectively. To further amplify this point, the average number of employees in EU SMEs was recently estimated as 14 (McKenna, 2015). Hence, SME is a general term that requires careful assessment due to these and other dissimilarities. In this chapter, microcompanies are included as part of the SME definition.

3.3 NOVELTY STATUS OF OI IN THE FOOD INDUSTRY

The food industry faces a large number of challenges: strict safety regulations, instantaneous changes and continuous evolution in consumer needs (eg, emerging health and wellness issues and functional foods, etc.), shortened product life cycles, the time-to-market race, cluttered retail shelf space (Bellairs, 2010), and increasing difficulty in meeting the heterogeneous requirements of a growing number of chain players such as suppliers, customers, and legislators (Sarkar and Costa, 2008). OI addresses these

Table 3.1 Definition of Company Category (http://ec.europa.eu/growth/smes/business-friendly-environment/sme-definition/index_en.htm)

Company Category	Number of Employees	Turnover (€ Million)	Balance Sheet Total (€ Million)
Medium-sized	<250	≤50	≤43
Small	<50	≤10	≤10
Micro	<10	≤2	≤2

challenges and opportunities, but it is still gaining momentum in the food sector (Bellairs, 2010; Khan et al., 2013; Saguy and Sirotinskaya, 2014; Traitler et al., 2011).

To quantify OI contributions to the food industry, it is vital to assess the actual driving forces behind it. The real question is whether OI is a truly novel set of practices that the food industry has adopted by joining the overall business trend and devoting real efforts toward its implementation, or is this more of the same that has been pursued for years without the "glory" and wide acceptance of the OI label (Saguy and Sirotinskaya, 2014)? For decades, the food industry has been practicing and advocating consumer and supplier involvement in all aspects of new product development, as well as collaborations with academia and consultants. Hence, one could erroneously conclude that OI implementation was occurring long before it was first conceptualized (Chesbrough, 2003), but without the OI label. To determine whether OI offers a new paradigm or is just a fad, one should define which OI practices are indeed novel and to what extent OI has altered innovation practices and performance.

Consumer involvement was, and still is, a natural and paramount part of food industry practices long before OI's conceptualization. However, in the era of OI, consumers/customers have gained an additional, more expansive and critical role, that of cocreator. Companies in the OI era are participating in large-scale collaborative networks of actual and virtual consumer communities, where consumers share their food product experiences and evaluate product appeal, value, acceptance, and effectiveness. Moreover, they possess a set of value-creation tools, playing the role of codesigners, innovators, marketers, and branders (Romero and Molina, 2011). "Crowdsourcing" fits this description. It represents the act of a company or institution that outsources, for instance, consumer research to an undefined large network as an open call. This is often undertaken by individuals (Howe, 2006). Another possibility is utilization of "lead users" and "early adapters" (Franke et al., 2006; von Hippel, 1986, 2011), which offers valuable information for idea generation and value sharing, while promoting the diffusion of innovation. The process of value cocreation is a new way of collaborating with customers/consumers. More importantly, however, is recognizing its full potential by becoming an outstanding constellation of knowledge aggregation and product ideation and a powerful OI tool. Suppliers also play a vital role as a source of information, providing knowledge about new technologies, ingredients' functionalities, and other players and competitors (Bigliardi and Galati, 2013; Saguy and Sirotinskaya, 2014).

Early involvement of both equipment and ingredient suppliers in various new product development processes is not new, but in the OI era, this practice has been transformed into a manageable, structured, and systematic process of bringing the outside in. It mandates profound changes in organizational behavior and requires a customized managerial approach that can be implemented through some well-defined models, such as the "Sharing Is Winning" model (Saguy, 2011; Traitler and Saguy, 2009; Traitler et al., 2011), the "food-machinery framework" (Bigliardi et al., 2010, 2015), or Unilever's "Want, Find, Get, and Manage" model (Slowinski and Sagal, 2010), to count only a few. However, although the agri-food sector enjoys a spectrum of close cooperation, OI remains underutilized (Fortuin and Omta, 2009). A partial explanation is linked with intellectual property–related issues. In order to have OI become an extension of the innovation ecosystem, compelling changes have to be made in organizational behavior, whereas innovation practices should be leveraged across all organizational activities and a culture of openness should be embraced.

It is worth noting that OI flourishes within an innovation ecosystem that could be defined as a community of interacting individuals and complementary organizations that build collective value. It highlights the coevolution of suppliers, producers, and competitors within an economic community (Walsh, 2014). Hence, the typical framework of working with consumer-insight companies, suppliers, equipment manufacturing, and other projects for a fee, in most cases, does not fit the general OI scope of an innovation ecosystem. This explains why the food industry has thus far been collaborating with consumer research companies, ingredient and equipment suppliers, academia, marketing and sales experts, etc., in their pursuit of new product development, but still struggles with OI.

The previous discussion highlights the debate over whether OI is a set of novel practices or just "old wine in a new bottle" (Manceau et al., 2011). We believe that OI is novel and offers vast and substantial opportunities for the food industry—LCs and SMEs—by enabling them to seize the enormous full potential and outstanding vast possibilities to collaborate and partner, thereby becoming proactive members of the innovation ecosystem. OI is neither an "old wine" nor a "new bottle." It is indeed a "distinctive novel brew" in a new and sophisticated evolving and futurist package that the food industry, and especially its SMEs, must embrace for its survival and future success.

3.4 OI APPLICATION IN LARGE AND MULTINATIONAL FOOD INDUSTRY COMPANIES

As already noted, OI practices in LCs have been accelerated by the rise in collaborative information technologies, such as global networks and Web-based communities, enabling easy access for locating and connecting with external parties. In addition, the emergence of electronic markets has given birth to third-party market makers (Feller et al., 2012) through crowdsourcing (Sloane, 2011). These third-party intermediaries are characterized as OI solution brokerage houses (see later discussion). Despite increasingly successful implementation of OI, critical questions remain on its wide successful utilization, and on whether it mainly suits only LCs and multinational enterprises (MNEs) (Saguy and Sirotinskaya, 2014). Our clear and unequivocal answer is that OI has vast potential especially for SMEs, but its mode of operation and implementations require additional development and research. Typical examples on OI utilization by LCs and MNEs can be found in Chapter "Open Innovation and Incorporation Between Academia and Food Industry" of this book.

3.5 RADICAL OPENNESS AND DISRUPTIVE INNOVATION

The food industry's unique needs require that OI be considered part of the more general term "radical openness," providing a foundation built on four pillars (Saguy and Sirotinskaya, 2014; Tapscott, 2013):

- Collaboration—The organization's boundaries have to become porous, fluid, and open, and the conventional wisdom that a firm's talent lies inside is archaic. Exploring outside the boundaries of the organization means global scouting. Implementation of this fundamental principle dictates profound changes in most companies' inner architecture and culture and deep adjustments to the management mindset. It also means orchestrating the innovation process, creating new tools, goods, and services, and engaging with local and global partners in their pursuit of value cocreation. Cocreation is probably the most powerful component, benefiting both the participants and the entire ecosystem.
- Transparency—It amplifies the need to be overt, no longer only providing good value and great products but also having core and ethical values that sustain the entire organization and build

integrity, ethics, and trust. The latter is a most essential component in creating successful innovation partnerships (Saguy, 2011; Traitler and Saguy, 2009; Traitler et al., 2011).

- Sharing—It calls for sharing assets and even intellectual property. While the paradigm of closed innovation requires tight control and ownership of all intellectual property, OI is based on profiting from others' use of its intellectual property (Saguy and Sirotinskaya, 2014). Intellectual property is probably the most difficult barrier and calls for thinking outside the box so that some benefits can be shared. Although the Sharing Is Winning concept was coined (Traitler and Saguy, 2009) as an indispensable part of OI, it does not mean that innovation is free or that intellectual property is compromised. Despite agreement that there is no innovation without intellectual property, in OI this topic should be considered from a different perspective (see following discussion).
- Empowerment—Knowledge and intelligence are power; their dissemination leads to disaggregation and decentralization. This process gives employees full access to this power, authority, and influence and allows them to use this strength to make their own decisions. Thus, openness brings empowerment and freedom to collaborate, share, and participate. Ultimately, it reveals a real sense of interdependence and creates an awareness that will build a sustainable, transparent, and collaborative business environment.

Disruptive innovation is an emerging and strategically important practice (Christensen, 1997) that broadens and develops new markets and provides new functionalities that, in turn, may disrupt existing market linkages (Boston-Fleischhauer, 2015; Govindarajan et al., 2011; Menon, 2011; Zortea-Johnston et al., 2012). Companies utilizing OI can and probably should accelerate their innovation by partnering with a selected set of early stage, disruptive technology providers. This can not only accelerate OI to supply innovative products to existing markets but also provide new opportunities and markets. However, challenges remain on how to effectively find, collaborate, and assimilate disruptive innovation without negatively impacting the internal innovation processes, cooperation, and management. Full utilization of disruptive innovation could be of great importance to SMEs as it most definitely will provide new markets and opportunities if adequate practices are developed.

3.6 SME UTILIZATION OF OI

In contrast to the frequent and widespread utilization of OI by LCs and MNEs (as shown in the previous chapter), there are far fewer reports focusing on SMEs (Baregheh et al., 2012; Brunswicker and Vanhaverbeke, 2014; Dries et al., 2014; Hossain, 2015; Hutter et al., 2013; Krause and Schutte, 2015; Lee et al., 2010; McAdam et al., 2014; Spithoven et al., 2013; van de Vrande et al., 2009). Various factors facilitating OI implementation and enhancing success rates for SMEs have been suggested (Saguy and Sirotinskaya, 2014): using third-party intermediaries in searching for partners (Lee et al., 2010), cooperating with academia through a dedicated network (http://www.hightecheurope.com/), R&D integration and knowledge transfer, among others, for example, exploiting alternative technology applications, inventing problem-solving tools (Bianchi et al., 2010), knowledge and technology transfer, and intellectual property (Braun and Hadwiger, 2011). However, the use of OI remains a challenge for even the largest SMEs for a number of reasons: lack of resources, too small a size, limited R&D (Gassmann et al., 2010), and insufficient organizational adaptation ability (Saguy, 2011). Language barriers, time constraints, and a plethora of other real or perceived barriers have also been mentioned. Some typical, albeit scarce, examples are reported in the literature (Boesso et al., 2009; Braun and Hadwiger, 2011; Kumar et al., 2012). The infrequency of these cases amplifies SMEs' struggle with OI implementation. Academia–industry collaboration is another critically essential driver for OI implementation. However, there are numerous challenges to establishing such collaborations due to conflicting interests. For instance, while academia's main focus is on basic research driven by fundamental science and knowledge, industry is driven by maximizing profits. Other issues, such as the need to publish and intellectual property rights, are two common examples of these disparate interests. Four paradigm shifts have been suggested as an effective way of coping with these challenges (Saguy, 2011), but to date, they have not been fully implemented.

3.7 OI IMPLEMENTATION CHALLENGES IN SMEs

SMEs are the backbone of most economies. For instance, 99.1% of Europe's 274,000 food and drink companies are SMEs. They account for 49.6% of the turnover (€528 billion) and 63.3% of the employment (2.9 million jobs) in Europe's food and drink industry (2015 data: http://www.fooddrinkeurope.eu/our-actions/topic/small-and-medium-sized-enterprises-smes/figures). However, OI implementation in SMEs is still facing an uphill battle with unique challenges. SMEs and many traditional sectors are struggling with OI implementation due to their limitations and numerous real or perceived barriers, eg, a relatively low absorptive capacity (Spithoven et al., 2011), a low ability to manage perceived challenges (Larsen and Lewis, 2007; Rahman and Ramos, 2010; van de Vrande et al., 2009), a lack of adequate collaboration (Lee et al., 2010), limited financial resources for internal R&D (Brunswicker and Vanhaverbeke, 2014), inadequate human resources and competencies, absence of production facilities, modest market power, lower "status" as an innovation partner, restrictions in securing intellectual property (Deschamps et al., 2013), and a narrow business portfolio and knowledge base (Bianchi et al., 2010). In addition, many SMEs perceive the OI model as prohibitively expensive and/or complex (Saguy, 2011) and consequently give up on it even before it is seriously considered.

SMEs undoubtedly practice OI in one way or another, but their focus is mainly on short-term profitability (Vanhaverbeke et al., 2012). OI practices and the knowledge base accumulated from LCs' and MNEs' experiences are not readily transferable to the SME context. Questions of whether OI really suits SMEs, and adequate approaches for capturing its full potential, remain open. OI can fit SMEs, but a different mindset is required, ensuring implementation of a new paradigm and game-changing approaches. Recommendations include:

- Scale—a small scale offers unique opportunities (eg, agile decision-making, flexibility of entrepreneurial culture, business specialization, and niche products) (Brunswicker and Vanhaverbeke, 2014).
- Flexibility and expertise—give SMEs an advantage in building and nurturing relationships for win–win collaborations (Manceau et al., 2011).
- Uniqueness—SMEs' collaboration rules should differ from those used by LCs and MNEs, focusing on informal and personal relationships (Vanhaverbeke et al., 2012).
- Resources—shortage of resources in SMEs should provide an incentive for outsourcing technology, knowledge, and other skills beyond their own core competencies (Pullen et al., 2008).

- Growth—the exponential growth in science complexity, technology, information, and cost mandates that SMEs use partners' key assets and resources to facilitate new ways of generating and capturing value.
- Openness—this axis is a major attribute of OI and requires careful consideration to improve the company culture.

The main differences between LCs' and SMEs' innovation openness can be described by plotting partner-selection openness (abscissa) and degree of selection-topic openness (ordinate). This enables distinguishing four class quadrants (Saguy and Sirotinskaya, 2014) (Fig. 3.1):

- Class 1 (lower left)—Represents organizations (both LCs and SMEs) that use closed innovation, cooperating with preselected partners on a narrow list of topics, for example a long-term partnership with a well-known supplier or research lab.
- Class 2 (upper left)—Represents companies that are focused on collaborating with partners with preestablished relationships and some previous successful experience(s), which may lead to progressively expanding topics. For SMEs, this class is limited due to insufficient resources and/or an inability to sustain long-term collaborations.
- Class 3 (lower right)—Focuses on relevant topics for innovation that are defined and aligned with the LC's strategy, enabling wide and open searches for new or previous external partners. For most SMEs this mandates strategic commitment that is typically not feasible for many SMEs.
- Class 4 (upper right)—Applies to only a few LCs and MNEs that have been implementing OI for a long time, as required for mature collaborations with the highest level of openness. Such companies welcome any partners and are eager to address all types of innovative ideas, regardless of

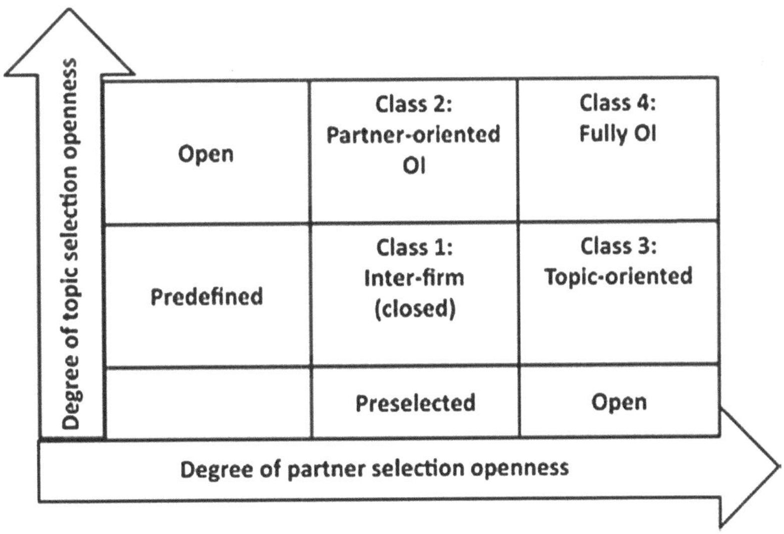

FIGURE 3.1

General types of innovation openness.

Adapted by Saguy, I.S., Sirotinskaya, V., 2014. Challenges in exploiting open innovation's full potential in the food industry with a focus on small and medium enterprises (SMEs). Trends in Food Science & Technology 38, 136–148.

their strategic priorities. Procter & Gamble is a prominent example of a Class 4 company (Lafley and Charan, 2008; Panduwawala et al., 2009; Sakkab, 2002). Class 4 is of no relevance for most SMEs, due to their various limitations (Lee et al., 2009).

Another way to characterize the degree of innovative openness relies on sourcing. It refers to how firms can use external sources of innovation (eg, customers, suppliers, academia, intellectual property experts, and network partners) (Brunswicker and Vanhaverbeke, 2014; Dahlander and Gann, 2010). In SMEs the degree of innovation openness can again be described using two axes: breadth (abscissa) and depth (ordinate). Degree of breadth represents the level of external sources or search channels that a firm relies on in its innovative activities, while the degree of depth is the extent to which it draws from various external sources (Laursen and Salter, 2004, 2006). The result is depicted as four different classes (quadrants) (Fig. 3.2):

- Class 1 (lower left)—Represents SMEs that rely on closed innovation; they can be labeled "minimal searchers" because of the limited number of external sources of innovation with which they interact and the weak and infrequent nature of these interactions.
- Class 2 (upper left)—Describes SMEs that use a limited number of external sources of innovation, but their interaction is somewhat deeper and more intensive. "Supply chain searchers" rely heavily on traditional supply chain linkages, while "technology-oriented searchers" engage in intense interactions with higher education institutions and intellectual property experts (Brunswicker and Vanhaverbeke, 2014).

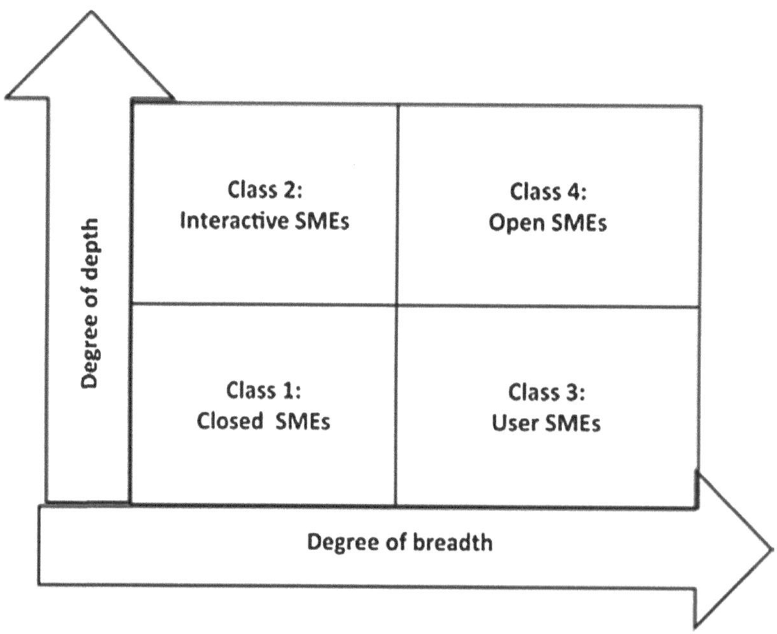

FIGURE 3.2

Classes of SMEs' external innovation openness.

Adapted by Saguy, I.S., Sirotinskaya, V., 2014. Challenges in exploiting open innovation's full potential in the food industry with a focus on small and medium enterprises (SMEs). Trends in Food Science & Technology 38, 136–148.

- Class 3 (lower right)—Describes SMEs that are open to using a large number of external sources of innovation. However, these interactions are scarce and inconsistent. Such SMEs are labeled "application-oriented searchers" that regularly interact with value-chain actors and network partners and rarely with higher education institutions or intellectual property experts (Brunswicker and Vanhaverbeke, 2014). Although collaboration may enhance innovation, for SMEs it brings with it the risk of opportunistic behavior (Zeng et al., 2010), which could lead to potential information leaks.
- Class 4 (upper right)—Represents "full-scope searchers," SMEs that have continuous interactions with numerous external sources of innovation, thus possessing an open approach to innovation (Idrissia et al., 2012) that may overcome the impediments to innovation. However, this is rarely feasible for SMEs, and even LCs are still struggling with its full utilization.

3.8 SOLUTION BROKERAGE HOUSES: ROLES AND SELECTION

The strategic decision to implement OI is a paramount first step toward making it a reality. At this point, external capabilities, knowledge, and technology should be brought into a company's innovation process. To facilitate this process, solution brokerage houses have emerged. Their main functions are to:

1. act as intelligent agents and/or "scouts" that stimulate external collaborations between businesses, scientific institutions, public and private national organizations, and international institutions that have had no prior relationships;
2. facilitate knowledge and technology transfer between solution seekers and solvers in a competent way; provide value-added services, and enable collaboration in innovation communities in a secure, controlled environment. The solution brokerage houses' typical core capabilities include formulating innovation problems, identifying where the best solution can be found, and establishing the necessary relationships. Detailed information can be found in the article by Saguy and Sirotinskaya (2014).

It is quite clear that solution brokerage houses are more attuned to the needs of LCs and MNEs, which have a greater ability to capture value from their participation. NineSigma (http://www.ninesigma.com/), InnoCentive (http://www.innocentive.com/), YourEncore (http://www.yourencore.com/), Strategic Allies (http://www.strategicallies.co.uk/), and Yet2.com (http://www.yet2.com/) are only a few of the numerous OI service providers to LCs. There are some solution brokerage houses that also offer services aimed at enhancing OI utilization by SMEs (eg, InnoCentive Showcase Challenges, Yet2.com's Ohio Technology Scouting Program). However, these typically address those SMEs that have new technical solutions or disruptive technology to commercialize, rather than those seeking special solutions for their innovation processes and products. Typical average project cost and duration are estimated at $25,000 ($15,000–$400,000) and 3 months (up to 24 months in some cases), respectively. Although solution brokerage houses should fully comply with openness and cocreation of ecosystems, direct requests for specific information to assess various typical cost structures and project durations for SMEs and to appraise their services have not been successful. This probably indicates that OI solution brokerage houses still consider their business model to be confidential and proprietary. Consequently, no specific recommendations for optimal utilization of solution brokerage houses services by SMEs are provided.

3.9 ROLES FOR ACADEMIA

Academia should play a paramount role in the utilization of OI. The four-helix concept is based on academia's role as one of the key players. However, practices and myths differ widely. The success of industry–academia collaborations depends on finding and defining common goals, negotiating plans that fit mutual needs, and obtaining financial and intellectual payoffs for both parties. Nevertheless, there are still many challenges to mutually satisfying academic and business interests. Academia has a long history of working in isolation from the food industry and has become too insulated from "real-world problems" (Saguy, 2011). The roots of this isolation are nourished by the substantial differences between the academic and industrial cultures, values and norms, which have led to a few paradoxes. Some general points: many companies do not rate universities and other higher education institutions as primary sources of innovation. Similarly, higher education institutions do not necessarily see their role as providers of scientific know-how and knowledge to the industry, as this is often considered too applied in nature (Saguy and Sirotinskaya, 2014). The goals of research for the sake of knowledge often conflict with the objectives of an industry focusing on profit and short- to very short-term research (Saguy, 2011). Academia–food industry links are characterized by low relational involvement in knowledge transfer and generation of intellectual property (Perkmann and Walsh, 2007). The size of the firm determines the span of its collaboration with academia (Howells et al., 2012). Collaboration between LCs/MNEs and academia are more common, due to a more open R&D environment and sufficient resources and funding.

There are many reasons why SMEs perform poorly in establishing industry–academia collaborations (Saguy and Sirotinskaya, 2014):

- Access barriers—SMEs find it difficult to access information and knowledge from higher education institutions because of communication difficulties and lack of a common agenda and language between the two parties (Deschamps et al., 2013). Further, the search and scanning costs of seeking collaborations are too high (Howells et al., 2012).
- Knowledge gap—A lack of adequate knowledge of intellectual property management.
- Reluctance—SMEs' reluctance to be involved in intellectual property issues.

On the other hand, in most cases, internal university funds are insufficient to develop links with industry, especially for conducting pilot-plant-scale studies. Unfortunately, governments and private equity perform poorly in this regard. To make a significant change, a new mindset is required (Saguy, 2011)—academia needs to recognize its inherent social responsibility and duty to guide SMEs in each step of their innovation process in a long-term partnership. This partnership should create value for the business, as well as provide academia with numerous benefits, such as broadening experience contributing to research, hands-on student experience, enhancing a university's reputation, and attracting more funding (Tyler, 2009).

3.10 REVISED INTELLECTUAL PROPERTY MODEL

Without any doubt, innovation cannot exist without intellectual property rights. Food SMEs' limited resources and capabilities, as well as complexities driven by consumer protection, regulations, safety, and sustainability, create gaps in technology translation. Therefore, when dealing with SMEs, diverse and adaptable intellectual property–management strategies are required (Saguy and Sirotinskaya, 2014). The burden of appropriate strategy development lies on the shoulders of both academia and government (Saguy et al., 2013). SMEs should adopt a radical change in their approach toward intellectual property and management practices, such as those recently developed in the pharmaceutical industry (Allarakhia and Walsh, 2011). A similar shift is expected in the food domain due to accelerated importance and roles of recent developments such as in functional foods, nutrigenomics, and enginomics (Saguy et al., 2013). It is evident that both academia and SMEs should adopt the Sharing is Winning model to enhance collaborative exchanges designed to assist the four-helix partners in effectively and mutually managing knowledge assets and intellectual property while considering their social responsibility duties and obligations (Saguy and Sirotinskaya, 2014).

3.11 SELECTED SME EXAMPLES

3.11.1 EU ACTIVITIES

3.11.1.1 TRAFOON project

The EU has invested significant efforts on SMEs by supporting numerous studies and networks. For instance, TRAFOON is a network (http://www.trafoon.eu/) of 30 European research institutions, technology-transfer agencies, and SME associations that covers the value chain of four groups of traditional food products based on grains, fish, vegetables, and mushrooms, and sweet fruits and olives. It aims to support the SMEs producing these products. In addition, TRAFOON interlinks researchers, knowledge-transfer agents, and SME associations in 14 European countries fostering the transfer of sustainable innovation and entrepreneurship.

The following example highlights the specific steps carried out with the olive sector (Braun and Casado Hebrard, 2015):

- Initially, the project identified and collected the most limiting concurrent SMEs' knowledge gaps in the olive sector—termed inventory of needs—based on the close communication between R&D and SMEs. This was carried out by extracting the information from 17 stakeholders from Spain and Portugal (11 olive oil SMEs, three table olive SMEs, and three olive oil and table olive SMEs).
- A multi-stakeholder workshop was set up for nine SME-representative associations and 10 TRAFOON experts. They identified the SMEs' needs/demands and matched them with the previously collected available innovations. Then, competitive production and marketing strategy was developed.
- Two training workshops were held in Spain and Portugal. These meetings addressed the SMEs' requirements identified by the inventory of needs, including training for technical innovations, generic topics on food processing, marketing, legal issues, intellectual property, certification, labeling, and shortfalls in the implementation of innovations.
- Identified and unresolved issues provided the basis for future development of a strategic research and innovation agenda for the olive sector.

The feedback and knowledge gained from this project is to be used in future development and implementation of the second round of training workshops (TWs). Furthermore, a multilingual online Information Shop (www.TRAFOON.org) has been created. The Information Shop includes information gathered by the TRAFOON network on innovations in the production and marketing of olive products and the traditional food categories. This online tool also contains databases of experts and organizations to enable potential future collaborations and SME-oriented research projects, and will include all technology/innovation knowledge transferred during the TWs and guidelines for product innovations in diverse European languages.

3.11.1.2 HighTech Europe

Another example of EU effort focusing on SMEs is HighTech Europe. This EU interdisciplinary and interactive communication and collaboration platform is facilitating direct implementation of innovations into SMEs. It is aimed at achieving long-lasting integration of European R&D activities into the high-tech food-processing sector by increasing the innovation rate, stimulating new market opportunities, and overcoming knowledge transfer barriers. HighTech Europe completed its activities in 2013, but one of its main achievements was the creation of the Food Tech Innovation Portal (Food TIP), a knowledge portal that provides access to the latest technologies, profiles of institutions and companies, testing facilities, and services in food processing. It provides a nucleus for establishing the European Institute for Food Processing that collects, maps, monitors, and links industrial needs and available R&D results. Intensive communication and cooperation among universities, applied R&D centers, and the private sector (federations, industry) permit translation of scientific results into marketable industrial applications with special benefits for SMEs (www.foodtech-portal.eu).

3.11.2 STATE SUPPORT

Another example of bridging the gap between researchers developing new technologies and SMEs comes from projects supported by the Commission for Technology and Innovation (CTI), Switzerland (https://www.kti.admin.ch/kti/en/home.html). The projects focus on private companies, in particular SMEs, to make them more competitive and improve their efficiency and performance capabilities. The CTI sponsors joint innovation projects that can fill a market need, raises SME awareness of available services and supports their initiatives, provides funding to scientists who use research findings to develop marketable products and services, encourages knowledge and technology transfer, and guides innovation leaders in the creation of start-up companies. The CTI Voucher Program, for example, is intended to improve the innovative capacities of Swiss-based SMEs by encouraging them to work more closely with eligible research institutes. This cooperation ensures that R&D projects will be managed more effectively and makes it easier for project partners to assess risk. A CTI Voucher is essentially a commitment made by the CTI to provide federal funding for an approved R&D project. All CTI Voucher payments are intended to cover the research institute's costs related to the R&D project and are paid directly to that institute. Similar programs are available in other countries, promoting collaborations between SMEs and academia and simultaneously increasing universities' relevance in conducting applied R&D.

3.11.3 ACADEMIA–SMEs

The following is a typical example from China where universities are reaching out to assist SMEs (Lee, 2015). Enactus is an international organization that brings together a diverse network of university students, academic professionals, and industry leaders who share a vision of creating a better and more sustainable world through the positive power of business. STU Enactus (launched by Shantou University in 2008) currently comprises over 80 students from seven different colleges, and is supported by faculty advisor Dennis Lee and other academic and business advisors.

Porcelain production is the primary industry in Chaozhou city (30% of its gross domestic product). It also produces a large amount of porcelain waste. Lacking an effective recycling method, this waste causes safety risks, which led to serious pollution, affecting the livelihoods of the 210,000 residents. To address this problem, STU Enactus developed a Waste Porcelain Recycling model to connect porcelain factories with ceramic powder workshops by means of a group of transportation laborers. As a result, transportation costs were reduced, productivity was increased, while distribution channels and business networks for porcelain and ceramic powders were expanded. This innovative project enabled generating additional substantial income, created 21 new local jobs, raised the social standing of sanitation workers, and improved the living environment of all Chaozhou residents.

3.11.4 LC–SMEs

This example highlights one other possibility of OI utilized to scout SMEs offering unique technological solutions (Godick, 2015). PHLburg Technologies, Inc (http://www.phlburg.com/index_en.html) is an internationally recognized technology-commercialization company representing MNEs by searching for and identifying novel technology and centers of scientific excellence that satisfy their customers' specific needs. Hence, it fits the description of an OI solution brokerage house. PHLburg is often requested to scout and identify a solution provider (ie, SMEs owing the needed technology) that is willing to enter into a collaboration with an LC. The immediate concern of many SMEs is the possible ramifications of these collaborations on their mere existence. Therefore, there is a need for a clear undesigning to avoid outcomes that may jeopardize their future.

The groups within the LCs that are typically engaged in this effort are: Business Unit—the company section with the technology need and eventual technology user; Research Team (RT)—the company's R&D unit that implements the new technology/solution; Innovation Team—a part of the RT focusing on external technology; solution brokerage house—the OI solution provider, an independent firm, specialized in technology scouting; and SME Technology Developer, intellectual property owner—inventor of the technology, identified by the solution brokerage houses to satisfy the Business Unit's needs.

A typical project flow is as follows:

1. Annual technology needs assessment—the Business Unit lists its critical need (one or more). Simultaneously, the RT assesses the need and determines if it has the expertise necessary to satisfy it. The Innovation Team makes the same assessment and determines if an SME is likely to have the technology to satisfy it.
2. The Innovation Team interviews and then retains a solution brokerage house to identify the needed technology.
3. Identification process—the solution brokerage house identifies and then qualifies technology and collaborator candidates to satisfy the Business Unit's needs.
4. Evaluation process—the Innovation Team (greater extent), Business Unit (mid-level), and RT (lesser extent), assisted by the solution brokerage houses, conduct due diligence and select the technology developer that the company will work with.

5. Intellectual property acquisition—the Innovation Team and solution brokerage houses, with assistance from the Business Unit, conclude intellectual property–acquisition negotiations and any necessary additional technology-development agreements.
6. Development process—the Business Unit and solution brokerage houses, with assistance from the Innovation Team, manage further intellectual property development with the Technology Developer.

Depending on the specific needs, technology complexity, technology uniqueness, intellectual property issues, and further development, the time to market is generally 4–30 months. Searching, identification, evaluation, and acquisition typically take 4–8 months. Development is the unknown. Experience shows that a fast-track project with simple development will take an additional 6 months. Time to market in the United States, for more challenging projects, can take 18–24 additional months. Depending on their needs and wants, SMEs can negotiate their position/acquisition, partner, supplier, and within certain restrictions, be a competitor.

This typical example describes the path taken by an EU-based food LC to obtain technology for reducing fats in their products. The solution brokerage house search for a technology solution identified developers at a US university, EU university, and a SME. The technologies were at a level that fits NASA Technology Readiness Levels #3–#8 (https://www.nasa.gov/directorates/heo/scan/engineering/technology/txt_accordion1.html). The LC requested proposals from all three entities. The US SME with experience in the industry was chosen as the commercialization partner. The SME licensed its technology to the LC. Under the terms of the exclusive (field of use) license agreement, the LC pays the SME a minimum annual royalty and additional royalties based on usage.

It should be noted that SMEs should increase their chances of being a technology partner with an LC by making themselves known to solution brokerage houses and to chief technology officers of LCs in their industry.

There are numerous other examples concerning the implementation and utilization of OI by SMEs. However, an overall assessment is that SMEs are still struggling and are faced with hurdles in adopting the principles of OI. The role of academia is therefore paramount in bridging or minimizing the gap and reaching out to SMEs.

3.12 FUTURE TRENDS, CHALLENGES, CONCLUSIONS, AND RECOMMENDATIONS

Unabated development of science and knowledge as well as increasingly intense acceleration of globalization driven by technological progress and complexity are catalyzing the frenzied internationalization of R&D, innovation, and collaboration. OI consists of vigorous networking with other companies, public research institutes, universities, and external suppliers, and sharing and accessing outside information and technology.

Additionally, the new concept that originated from agile methods also should be mentioned. Originally, agile methods were introduced in the software industry focusing on solutions evolving through collaboration between self-organizing, cross-functional teams, which promote adaptive planning, evolutionary development, early delivery, continuous improvement, as well as rapid and flexible responses to change. The basic idea of implementing agile methods focuses on people rather than processes

(Fontana et al., 2014). A recent publication highlights a number of the challenges involved in combining agile methods and OI. To fully benefit from the combination of agile and open practices to increase effectiveness of software development, a clear understanding of the inter- and intraorganizational applicability and implications of open innovation in agile systems development is needed (Conboy and Morgan, 2011; http://www.engineersjournal.ie/2014/02/13/embracing-open-innovation-in-agile-software-development/). Under the agile approach, companies are seeing increased productivity within their software-development teams, faster release of digital products and services, and improved customer experiences with continual user involvement (Bossert et al., 2015).

It is also worth noting that some distinct problems are expected to arise when two or more project teams try to combine agile and OI principles. For instance, openness is often compromised by a perceived competitive element and lack of transparency between business units. In addition, minimal documentation often reduces effective knowledge transfer while the use of short iterations, stand-up meetings, and presence of on-site customers reduce the amount of time for sharing ideas outside the team. Hence, a clear understanding of the inter- and intraorganizational applicability and implications of OI in agile systems development is required to address key challenges for research and practice (Conboy and Morgan, 2011).

DevOps is the next wave of innovation in software development and delivery and a critical enabler of agile software development. Under this product-development approach, companies seek to fully integrate their software-development functions with their information technology operations so that teams can jointly build, test, release, and maintain new digital applications more frequently and more efficiently (Bossert et al., 2015).

The agile and DevOps concepts and methods could become a high priority for the food industry, as they are excellent approaches to utilizing and building upon consumers' inputs and involvement in all new product-development stages and the need to coordinate collaboration among innovation teams and business units. For those in the food industry who have witnessed the major changes occurring from quality control to quality assurance, hazard analysis critical control points and full integration of the aforementioned concepts should be quite easy to digest. However, its implementation and full utilization in an OI environment remain to be developed and practiced.

OI's continuing progress has diffused into some areas of the food industry with far-reaching consequences that have changed the way some SMEs, LCs, and MNEs innovate. The new OI mindset has shattered many real and perceived barriers, consequently enhancing disruptive thinking and mindset, openness and local/global collaborations, identifying possible future partners, reducing development costs and time to market, and maximizing the value of their intellectual assets. However, most benefits are not shared by all, as is clearly seen with many SMEs and numerous LCs that have failed to join the new innovation ecosystem. The exponential progress of science and technology, diminishing time to market and raising the need to address increased economic pressures, mandates that the food industry, and especially its SMEs, improve their innovation processes and become proactive members of the OI ecosystem. Selected recommendations highlight some of the previously mentioned: Sharing Is Winning collaboration is a paramount aspect of OI and it should be fully adopted and practiced by academia, industry, and other key innovation members. The new innovation ecosystem calls for all parties, known as the fourth helix (industry, academia, government, and private businesses), to be proactively involved in robustly supporting innovation sustainability. New academic metrics are needed to quantify and reward important OI dimensions (eg, relevance, impact, scientific-innovation breakthroughs, applied research, rewards, partnerships, SME involvement, social responsibility, employability).

Academia also needs a revised intellectual property model to consider in addition to innovation's financial returns, the unique needs of SMEs, and its paramount role in social responsibility. It also calls for new licensing activities and strategic alliances for a proactive intellectual property strategy that will allow sharing technologies rather than accumulating intellectual properties as a defense mechanism. The new intellectual property model should also find ways to create value by licensing unused technologies or by selling off ancillary patents. Furthermore, management reform is required to address the new OI mindset, culture and strategies need to be adjusted to galvanize the impetus of change, global collaboration, networking, and breaking visible and perceived barriers. To catch up, SME managements will have to be extremely flexible, or face the consequences. Food science and technology curricula also need to be revised to incorporate the vast possibilities offered by the World Wide Web, and the continual, unique, and ever-changing needs of consumers, industries, and markets. Entrepreneurship, networking, innovation, and new product development, as well as science needs, social responsibility, and culture, should become an integral part of the courses offered.

REFERENCES

Allarakhia, M., Walsh, S., 2011. Managing knowledge assets under conditions of radical change: the case of the pharmaceutical industry. Technovation 31, 105–117.

Baregheh, A., Rowley, J., Sambrook, S., Davies, D., 2012. Food sector SMEs and innovation types. British Food Journal 114, 1640–1653.

Bellairs, J., 2010. Open innovation gaining momentum in the food industry. Cereal Foods World 55, 4–6.

Bianchi, M., Campodall'Orto, S., Frattini, F., Vercesi, P., 2010. Enabling open innovation in small-and medium-sized enterprises: how to find alternative applications for your technologies. R&D Management 40, 414–431.

Bigliardi, B., Bottani, E., Galati, F., 2010. Open innovation and supply chain management in food machinery supply chain: a case study. International Journal of Engineering, Science and Technology 2.

Bigliardi, B., Galati, F., 2013. Models of adoption of open innovation within the food industry. Trends in Food Science & Technology 30, 16–26.

Bigliardi, B., Galati, F., Marolla, G., Verbano, C., 2015. Factors affecting technology transfer offices' performance in the Italian food context. Technology Analysis & Strategic Management 27, 361–384.

Boesso, G., St Davcik, N., Favotto, F., 2009. "Health-enhancing" products in the Italian food industry: multinationals and SMEs competing on Yogurt. AgBioForum 12 (2), 232–243. http://www.agbioforum.org.

Bossert, O., Ip, C., Starikova, I., 2015. Beyond Agile: Reorganizing it for Faster.

Boston-Fleischhauer, C., 2015. Disruptive innovation: latest buzzword or new reality? Journal of Nursing Administration 45, 469–470.

Braun, S., Casado Hebrard, F.J., 2015. Private Communication.

Braun, S., Hadwiger, K., 2011. Knowledge transfer from research to industry (SMEs)–an example from the food sector. Trends in Food Science & Technology 22, S90–S96.

Brunswicker, S., Vanhaverbeke, W., 2014. Open innovation in small and medium-sized enterprises (SMEs): external knowledge sourcing strategies and internal organizational facilitators. Journal of Small Business Management 53, 1241–1263.

Chesbrough, H., Vanhaverbeke, W., West, J., 2006. Open Innovation: Researching a New Paradigm. Oxford University Press, Oxford.

Chesbrough, H.W., 2003. Open Innovation: The New Imperative for Creating and Profiting From Technology. Harvard Business Press, Cambridge.

Christensen, C., 1997. The Innovator's Dilemma: When New Technologies Cause Great Firms to Fail. Harvard Business Review Press, Boston.

Conboy, K., Morgan, L., 2011. Beyond the customer: opening the agile systems development process. Information and Software Technology 53, 535–542.

Dahlander, L., Gann, D.M., 2010. How open is innovation? Research Policy 39, 699–709.

Deschamps, I., Macedo, M.G., Eve-Levesque, C., 2013. University-SME collaboration and open innovation: intellectual-property management tools and the roles of intermediaries. Technology Innovation Management Review 3.

Dries, L., Pascucci, S., Török, Á., Tóth, J., 2014. Keeping your secrets public? Open versus closed innovation processes in the Hungarian wine sector. International Food and Agribusiness Management Review 17, 147–162.

Feller, J., Finnegan, P., Hayes, J., O'Reilly, P., 2012. 'Orchestrating'sustainable crowdsourcing: a characterisation of solver brokerages. The Journal of Strategic Information Systems 21, 216–232.

Fontana, R.M., Fontana, I.M., da Rosa Garbuio, P.A., Reinehr, S., Malucelli, A., 2014. Processes versus people: how should agile software development maturity be defined? Journal of Systems and Software 97, 140–155.

Fortuin, F.T.J.M., Omta, S.W.F., 2009. Innovation drivers and barriers in food processing. British Food Journal 111, 839–851.

Franke, N., Von Hippel, E., Schreier, M., 2006. Finding commercially attractive user innovations: a test of Lead-User theory. Journal of Product Innovation Management 23, 301–315.

Gassmann, O., Enkel, E., Chesbrough, H., 2010. The future of open innovation. R&D Management 40, 213–221.

Godick, N.B., 2015. PHLburg Technologies, Inc., President. Private Communication.

Govindarajan, V., Kopalle, P.K., Danneels, E., 2011. The effects of mainstream and emerging customer orientations on radical and disruptive innovations. Journal of Product Innovation Management 28, 121–132.

von Hippel, E., 1986. Lead users: a source of novel product concepts. Management Science 32, 791–805.

von Hippel, E.A., Ogawa, S., de Jong, J.P.J., 2011. The Age of the Consumer-Innovator.

Hossain, M., 2015. A review of literature on open innovation in small and medium-sized enterprises. Journal of Global Entrepreneurship Research 5, 1–12.

Howe, J., 2006. The rise of crowdsourcing. Wired Magazine 14, 1–4.

Howells, J., Ramlogan, R., Cheng, S.-L., 2012. Innovation and university collaboration: paradox and complexity within the knowledge economy. Cambridge Journal of Economics 36, 703–721.

Hutter, K., Hautz, J., Repke, K., Matzler, K., 2013. Open innovation in small and micro enterprises. Problems and Perspectives in Management 11, 12–22.

Idrissia, M.O., Amaraa, N., Landrya, R., 2012. SMEs' degree of openness: the case of manufacturing industries. Journal of Technology Management and Innovation 7, 186–210.

Khan, R.S., Grigor, J., Winger, R., Win, A., 2013. Functional food product development–opportunities and challenges for food manufacturers. Trends in Food Science & Technology 30, 27–37.

Krause, W., Schutte, C., 2015. A perspective on open innovation in small-and medium-sized enterprises in South Africa, and design requirements for an open innovation approach. South African Journal of Industrial Engineering 26, 163–178.

Kumar, K., Boesso, G., Favotto, F., Menini, A., 2012. Strategic orientation, innovation patterns and performances of SMEs and large companies. Journal of Small Business and Enterprise Development 19, 132–145.

Lafley, A., Charan, R., 2008. P&G's Innovation Culture. Strategy + Business Magazine, pp. 1–10.

Larsen, P., Lewis, A., 2007. How award-winning SMEs manage the barriers to innovation. Creativity and Innovation Management 16, 142–151.

Laursen, K., Salter, A., 2004. Searching high and low: what types of firms use universities as a source of innovation? Research Policy 33, 1201–1215.

Laursen, K., Salter, A., 2006. Open for innovation: the role of openness in explaining innovation performance among U.K. manufacturing firms. Strategic Management Journal 27, 131–150.

Lee, D., 2015. In: Associate Professor (Ed.), Journal of Asia Entrepreneurship and Sustainability Shantou University Business School. Private Communication, Shantou, Guangdong, China.

Lee, S., Park, G., Yoon, B., Park, J., 2010. Open innovation in SMEs—an intermediated network model. Research Policy 39, 290–300.

Lee, Y.G., Park, S.H., Song, Y.I., 2009. Which is better for a firm's financial performance: an externally oriented or inwardly oriented innovation strategy? An empirical study on Korean SMEs. Asian Journal of Technology Innovation 17, 57–73.

Manceau, D., Moatti, V., Fabbri, J., Kaltenbach, P.-F., Bagger-Hansen, L., 2011. Open Innovation: What's Behind the Buzzword? Institute for Innovation and Competitiveness i7 and Accenture.

McAdam, M., McAdam, R., Dunn, A., McCall, C., 2014. Development of small and medium-sized enterprise horizontal innovation networks: UK agri-food sector study. International Small Business Journal 32, 830–853.

McKenna, B., 2015. A strategic research and innovation agenda for very small traditional food processors. Personal Communication.

Menon, S., 2011. Linking generativity and disruptive innovation to conceptualize ICTs. Internet Research 21, 347–361.

Panduwawala, L., Venkatesh, S., Parraquez, P., Zhang, X., 2009. Connect and Develop: P&G's Big Stake in Open Innovation. University of Bath.

Perkmann, M., Walsh, K., 2007. University–industry relationships and open innovation: towards a research agenda. International Journal of Management Reviews 9, 259–280.

Pullen, A., Weerd-Nederhof, P., Groen, A., Fisscher, O., 2008. Configurations of External SME Characteristics to Explain Differences in Innovation Performance.

Rahman, H., Ramos, I., 2010. Open Innovation in SMEs: from closed boundaries to networked paradigm. Issues in Informing Science and Information Technology 7, 471–487.

Romero, D., Molina, A., 2011. Collaborative networked organisations and customer communities: value cocreation and co-innovation in the networking era. Production Planning & Control 22, 447–472.

Saguy, I.S., 2011. Paradigm shifts in academia and the food industry required to meet innovation challenges. Trends in Food Science & Technology 22, 467–475.

Saguy, I.S., 2013. Academia-Industry interaction in innovation: paradigm shifts and avenues for the future. In: Yanniotis, S., Taoukis, P., Stoforos, N.G., Karathanos, V.T. (Eds.), Advances in Food Process Engineering Research and Applications. Springer, New York, pp. 645–656.

Saguy, I.S., Singh, R.P., Johnson, T., Fryer, P.J., Sastry, S.K., 2013. Challenges facing food engineering. Journal of Food Engineering 119, 332–342.

Saguy, I.S., Sirotinskaya, V., 2014. Challenges in exploiting open innovation's full potential in the food industry with a focus on small and medium enterprises (SMEs). Trends in Food Science & Technology 38, 136–148.

Sakkab, N.Y., 2002. Connect & develop complements research & develop at P&G. Research Technology Management 45, 38–45.

Sarkar, S., Costa, A.I., 2008. Dynamics of open innovation in the food industry. Trends in Food Science & Technology 19, 574–580.

Sloane, P., 2011. A Guide to Open Innovation and Crowdsourcing: Advice From Leading Experts in the Field. Kogan Page Publishers, London.

Slowinski, G., Sagal, M.W., 2010. Good practices in open innovation. Research-Technology Management 53, 38–45.

Spithoven, A., Clarysse, B., Knockaert, M., 2011. Building absorptive capacity to organise inbound open innovation in traditional industries. Technovation 31, 10–21.

Spithoven, A., Vanhaverbeke, W., Roijakkers, N., 2013. Open innovation practices in SMEs and large enterprises. Small Business Economics 41, 537–562.

Tapscott, D., 2013. Radical openness: four unexpected principles for success. In: Ted Conferences.

Traitler, H., Saguy, I.S., 2009. Creating successful innovation partnerships. Food Technology 63, 23–35.

Traitler, H., Watzke, H.J., Saguy, I.S., 2011. Reinventing R&D in an open innovation ecosystem. Journal of Food Science 76, R62–R68.

Tyler, J.E., 2009. Advancing university innovation: more must be expected-more must be done. Minnesota Journal of Law, Science and Technology 10, 143–212.

van de Vrande, V., de Jong, J.P.J., Vanhaverbeke, W., de Rochemont, M., 2009. Open innovation in SMEs: trends, motives and management challenges. Technovation 29, 423–437.

Vanhaverbeke, W., Vermeersch, I., De Zutter, S., 2012. Open Innovation in SMEs: How Can Small Companies and Start-ups Benefit From Open Innovation Strategies?

Walsh, M., 2014. The Dictionary of Dangerous Ideas. Tomorrow Limited, Hong Kong.

Zeng, S.X., Xie, X.M., Tam, C.M., 2010. Relationship between cooperation networks and innovation performance of SMEs. Technovation 30, 181–194.

Zortea-Johnston, E., Darroch, J., Matear, S., 2012. Business orientations and innovation in small and medium sized enterprises. International Entrepreneurship and Management Journal 8, 145–164.

CHAPTER

TRANSITION TO A SUSTAINABLE AGRO-FOOD SYSTEM: THE ROLE OF INNOVATION POLICIES

4

P. Morone, L. Cottoni

Unitelma-Sapienza, University of Rome, Rome, Italy

4.1 INTRODUCTION

In this chapter, the way that innovation policies lead the system toward the adoption of a sustainable agro-food model is addressed. Firstly, the growing pressure exerted upon the agro-food sector as a consequence of the ongoing demographic and economic trends is addressed. Most notably, the world population is projected to increase by more than 2 billion people by 2050, and nearly all of this growth is forecast to take place in developing countries (FAO, 2015). At the same time, large and fast-growing economies—most notably India and China—will experience increasing wealth, with the number of people considered to be part of the "global middle class" (ie, with incomes between $6000 and $30,000 in PPP terms) projected to reach nearly 5 billion by 2030 (Kharas, 2010). This will lead to a growing demand for food and fiber to feed a growing population with a smaller rural labor force (due to the ongoing urbanization process). This trend could be complemented by a growing demand for feedstocks for a potentially huge bioenergy market (FAO, 2009).

All these changes call for a shift in the paradigm currently governing the agro-food system, allowing the introduction of a more sustainable approach, capable of providing the right answers to emerging challenges. This issue within the framework of transition theory (and, more specifically, sustainability transition theory) is addressed, allowing the identification of the change drivers within a coevolutionary framework where ecosystems, technologies, institutions, business strategies, and user practices coevolve within a multilevel (micro-meso-macro) perspective (Foxon, 2011). The chapter will then introduce some examples of innovation activities, which are effectively pushing the agro-food sector toward a sustainable transition, focusing mostly on the waste issue. The analysis begins by presenting an assessment of the major (mega) trends distressing the agro-food sector.

4.2 THE GROWING PRESSURE ON THE AGRO-FOOD SYSTEM

The growing pressure exerted by demographic and economic trends upon the agro-food system is posing new and unexpected challenges to be faced by most policy makers around the world. In order to understand how these challenges can be transformed into opportunities, the boundaries of the problem

Innovation Strategies in the Food Industry. http://dx.doi.org/10.1016/B978-0-12-803751-5.00004-0

should be clearly identified. As we proceed here, we identify three key areas of enquiry where the limits to growth[1] of the agro-food sector are identifiable: (1) the impact of population growth on food security and climate change; (2) the indirect land use change associated with the growing biofuel production; (3) the waste management issue stemming from growing global demand and production of food and manufactured goods.

4.2.1 POPULATION GROWTH, FOOD SECURITY, AND CLIMATE CHANGE

The continuous growth of the world population in the last decades (mainly due to the progress made in the medical sciences and to the "green revolution"[2] in agriculture) is putting serious pressure on the global agro-food system, with an increased global demand for food, arable land, and water resources. As of November 2015, the world population is estimated to be about 7.283 billion people,[3] and it is expected to reach 9.6 billion by 2050 (UN, 2014), exceeding 10 billion by 2100, with estimates ranging from 9.3 to 12.6 billion (Gerland et al., 2014).

At the global level, the highest growth rate of the world population was recorded in 1963 at +2.2% annually. Since then the growth rate steadily declined to the current 1.08% (CIA, 2015).[4] In spite of this generalized slowdown in the growth of the world population, some countries in Sub-Saharan Africa and Asia are still growing at a pace above 2%. Moreover, most of these countries are expected to register a significant increase in GDP per capita, with 2 billion people expected to join the so-called new "global middle class"—defined as those earning between US$6000 and US$30,000 a year on a purchasing power parity basis—by 2030. As new wealthy consumers will have higher demand for food, ceteris paribus the prices of agriculture commodities are likely to increase sharply, thus endangering food security for those who cannot afford high prices for food (Wilson and Dragusanu, 2008).

There is no clear definition of food security. However, there are at least three widely accepted definitions of it. According to the 1974 World Food Conference, food security is defined as "the availability at all times of adequate world food supplies of basic foodstuffs to sustain a steady expansion of food consumption and to offset fluctuations in production and prices." The 1996 World Food Summit defined it as the situation that exists "when all people, at all times, have physical and economic access to sufficient, safe and nutritious food to meet their dietary needs and food preferences for an active and healthy life." On the other hand, the World Health Organization[5] defines food security as the status in which "all people at all times have both physical and economic access to enough food for an active,

---

[1]As most readers know, this is the title of the challenging and path-breaking study conducted in 1972 by the Club of Rome. In their study the authors created a computing model that took into account the relations between various global development trends and produced computer simulations for alternative scenarios. Part of the modeling entailed different amounts of possibly available resources, different levels of agricultural productivity, birth control, and environmental protection. Most scenarios resulted in a steady growth of the population and of the economy until a turning point predicted to be reached around 2030. Only drastic measures for environmental protection proved to be suitable to change this trend, and only under these circumstances (ie, proactive environmental policies) did the Club of Rome manage to calculate scenarios in which both world population and wealth remain at a constant level.

[2]The Green Revolution refers to the massive spread of new technologies, occurring mostly in the middle of the 20th century, which led to marked increases in agricultural production worldwide associated with the introduction of modern or high-yielding crop varieties (Evenson and Gollin, 2003).

[3]http://www.census.gov/population/international/.

[4]CIA (2015) https://www.cia.gov/library/publications/the-world-factbook/geos/xx.html. Retrieved December 2015.

[5]http://www.foodandenvironment.com/2013/01/basic-concept-of-food-security.html.

healthy life (...); the ways in which food is produced and distributed are respectful of the natural processes of the earth and thus sustainable (...); both the consumption and production of food are governed by social values that are just and equitable as well as moral and ethical (...); the ability to acquire food is ensured (...); the food itself is nutritionally adequate and personally and culturally acceptable (...); and the food is obtained in a manner that upholds human dignity."

Building on these complementary definitions, we can observe how food security has at least four intertwined dimensions: availability, access, utilization, and stability.[6] Food *availability* is related to the supply of food, food *access* to the affordability for the households and individuals, *utilization* refers mainly to the safety and quality of the food and its preparation, and, finally, *stability* is linked to the possibility to have a stable food supply.

The concept of food security is obviously a socioeconomic one, so it can be measured through various indicators such as per capita daily caloric intake. Moreover, questions of food security (and its countermovement of food insecurity) are deeply rooted into poverty and underdevelopment issues. Currently, the international community is intensifying cooperation in order to tackle this issue (eg, Millennium Development goal no. 1), identifying the main obstacles for the achievement of food security as: water scarcity, soil degradation, poverty and scarcity of capital, pests and diseases, and ineffective public sector policies. This trend is even more severe considering that arable lands per capita are expected to decline sharply within the next decades (Bruinsma, 2009), hence, requiring a notable increase in productivity in order to satisfy the growing food demand. All of this will magnify the already significant human impact on the environment. In the particular case of the agro-food system, the negative impact on the environment can occur through contaminating pollutants, soil erosion and degradation, hazardous and/or unnecessary waste, deforestation, depletion of water resources, and overexploitation of other natural resources. This might have several consequences, also economic ones, considering that a severe degradation of the environment can, by all means, undermine economic growth in the long term (Caldwell et al., 1997).

Climate change is another major area of concern highly related to these socioeconomic issues. Indeed, often world regions that are more vulnerable to food insecurity are also highly exposed to climate change. At the same time, it is difficult to estimate the overall impact of climate change on the agro-food sector because there is a very complex network of relations between different factors that mutually affect each other, hence undermining the possibility to establish clear causal relations and making it difficult to elaborate models for the assessment of the impacts of climate change on the environment (Lohele, 2011).

Generally, climate change affects the agro-food sector both directly and indirectly. Direct effects are linked to modifications of the ecological conditions—a fact that might yield positive or negative effects. For example, in countries with lower temperatures or a humid and temperate climate, a milder climate can enhance the productivity of crops both in terms of quality and quantity. However, higher temperatures can severely damage crops in dry zones. Another way by which climate change might affect the agro-food sector is through the increased occurrence of extreme adverse climate events worldwide (including droughts, flooding, storms, etc.). These rapid changes in climate conditions can negatively affect the yield of crops. The effects of adverse events can be partially mitigated by higher investments in better and safer storage facilities or in improving the irrigation system in areas at risk of drought. However, these mitigation policies are costly, meaning that the poorest areas, which experience the

[6]http://www.fao.org/fileadmin/templates/wsfs/Summit/Docs/Final_Declaration/WSFS09_Declaration.pdf.

highest rates of food insecurity (eg, some countries in Sub-Saharan Africa and south Asia) are also likely to be more vulnerable and less able to mitigate the adverse effects of climate change.[7]

Indirect effects are related to a series of pain chains that are potentially disruptive not only for the agro-food sector but for the whole economy. The first pain chain consists of the following steps: increased extremely adverse events in food insecure regions lead to malnourishment, which in turn worsens the health of the affected populations, resulting in a substantial decrease in economic productivity and in exacerbated poverty. Another negative outcome of climate change is a more intense exploitation of the soil; in order to face a fall in crop productivity due to climate change–related disasters, farmers will need more land that was otherwise preserved (Steinbuks and Hertel, 2016). Also, an increase in winter temperatures affects the odds that many agricultural pests will survive the winter, thus attacking the crops during springtime.[8]

4.2.2 FOOD CROPS VERSUS OTHER LAND USE

The competition for land is a recent trend in the agro-food sector. Nonetheless, its overall impact (including also a theoretical increase in food prices) is not easy to estimate, also because the competition for land is strongly related to the question of crops that are purposely grown for the production of biofuels. The backbone of this competition is a phenomenon called indirect land use change. In other words, when feedstock used for the production of biofuels are food commodities, there is a risk that land originally used for the production of crops is diverted to biofuels production. This may trigger a chain reaction. For instance, if corn prices soar, also milk, beef, pork, and cheese prices will eventually soar, as corn is used to feed the livestock. This will add further pressure on the whole agro-food sector and contribute to increases in food commodities prices (Morone and Cottoni, 2016).

However, soaring prices for food commodities is not necessarily caused by the competition for land brought on by the production of biofuels, if we consider that less than 2% of the world arable land is currently used to yield crops used for this purpose (Morone and Cottoni, 2016). Other factors can contribute to fluctuations in food commodities prices including crop productivity, consumers' and producers' expectations, financial speculation, precautionary demand, price of substitute and complementary goods, adverse weather conditions, energy costs, and inappropriate public policies.

On the other hand, it is worthwhile mentioning here that soaring prices for food can actually be an opportunity of economic growth for developing countries exporting agricultural commodities, as it may eventually lead to higher incomes and to a reduction in food insecurity. However, in the case of developing countries that are already affected by climate change, food insecurity, and poverty, this can be detrimental.

The pressure on food supply added by competition for land can be reduced through the enhancement of the "next-generation biofuels." In fact, advanced biofuels hold promise of an escape from their predecessors' food-versus-fuel conundrum (Fairlay, 2011). For instance, lingo-cellulosic feedstock (eg, trees, cereal straw, the organic of some municipal solid waste, sugar cane bagasse, forest residues, etc.) can be used for ethanol production, without the need for prime croplands, hence reducing the

[7]One innovative tool established for the purpose of mitigating climate change or helping affected populations is the "green bonds" issued by the World Bank. Through this tool, the World Bank raises funds from investors, issuing fixed income bonds in order to finance projects to mitigate the effects of climate change. http://treasury.worldbank.org/cmd/htm/WorldBank-GreenBonds.html. Retrieved December 2015.

[8]http://www.fao.org/docrep/004/y3557e/y3557e11.htm.

competition with food and fiber crops (Morone and Cottoni, 2016). Other innovative feedstocks for biofuels (such as micro- and macroalgae), although still not commercially viable, are the object of intensive scientific research (Morrison et al., 2014).

However, biofuels have been heavily criticized also from the purely "scientific" point of view (Searchinger and Heimlich, 2015) mainly because of their low efficiency in converting sunlight into energy. As a matter of fact, only 0.5% of the incoming solar radiation is converted into sugar via photosynthesis and only 0.2% in ethanol, in the case of Brazilian sugar cane. Searchinger and Heimlich (2015) recommended the use of photovoltaic cells that have an estimated conversion rate of solar energy into electricity of up to 16%, far greater than the 0.2% of the biofuels. In this way the competition for land can be strongly reduced, mitigating the pressures of the other megatrends. Moreover, as opposed to biofuels production, photovoltaic technology does not need prime land with good and fertile soils and plenty of water. However, there are some serious disadvantages also in the photovoltaic technology, including intermittency, difficult storage, expensive equipment, complex and expensive waste treatment, frailty (and insurance costs), and a long-term irreversibility of the process for the land in which they are installed (up to 20 years or even longer).

Another issue that must be taken into account when considering the impact of land competition on the agro-food sector is the need to conserve land in order to offset greenhouse gas (GHG) emissions. In other words, with food demand soaring, deforestation will become harder to prevent. From this perspective, a transition toward a more sustainable agro-food production system is compulsory, not only because of the inherent finitude of the available land but also due to the need to avoid a skyrocketing food demand created by the rise of the new "global middle class" and a more intense deforestation.

A final issue to be considered here relates to fiber and timber production. With the rising global middle class these industries present a tough competitor for land to food producers. Yet, for both food and fiber producers this pressure can be minimized if an efficient system of waste management and recycling—an area of enquiry where we now turn our attention—is created.

4.2.3 MANAGING AND AVOIDING WASTE

Waste can pose risks to the agro-food sector (and the environment, in general) in various ways. For example, illegal landfills can degrade and contaminate the soil; moreover, if one-third of the food produce worldwide goes to waste, this is responsible for unnecessarily adding 3.3 billion tons of GHGs to the planet's atmosphere (UNEP, 2013). The difficulties may well become even tougher, considering population increases, economic growth, and the related rise in mass consumption. Although a huge increase in the quantity of solid waste is expected, waste management is not just a quantitative issue but rather a qualitative one, as increasingly it contains new and complex substances (eg, electronic equipment waste, chemicals, etc.) that contribute in various and unpredictable ways to the degradation of the environment. Moreover, waste is not only a major issue in the agro-food system but also a potential threat to human health and a menace to the existence of some ecosystems (contributing to about 5% of global GHG emissions[9]).

Bearing this in mind, minimization of the impact of waste is an urgent challenge for policy makers all over the world. The first steps to be undertaken in this direction are waste avoidance and minimization. At the same time, unavoidable waste should be treated in a manner that is the least harmful for human health and the environment.

[9]http://www.unep.org/greeneconomy/Portals/88/documents/ger/GER_8_Waste.pdf.

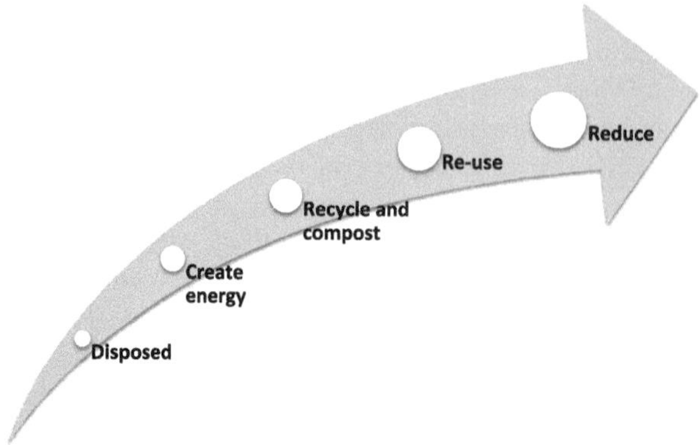

FIGURE 4.1

Waste hierarchy.

Adapted from http://www.target45plus.org.uk/waste-minimisation/waste-hierarchy/.

Recently the Environmental Protection Agency of New South Wales, Australia, approved[10] an innovative plan to pursue the highest standard of efficiency in waste treatment and minimization. The vision underlying this new policy strategy (Waste Avoidance and Resource Recovery) suggests that the primary goal is to enable local communities to improve environment and community well-being by reducing the environmental impact of waste and using resources more efficiently. Using resources efficiently and keeping materials circulating in the productive economy should, in turn, also help to create jobs and boost economic growth. This view of waste management is functional to the transition toward a circular economy, a new conceptualization of the socioeconomic system in which every resource is used with the maximum efficiency. The reverse pyramid (see Fig. 4.1) describes the waste hierarchy that is needed in order to minimize the pressure of the waste on the environment and human health.

Other solutions to improve the process of waste management can be: conservation of the resources (also avoiding excessive consumption), energy recovery from residual waste, systematized collection and segregation (in particular, of materials containing hazardous substances), reuse and recycling (avoiding the disposal of the waste in a landfill considered a destination of last resort), and improving existing processes and infrastructure for waste collection and treatment.

However, proper waste treatment or recycling can be commercially unattractive, in particular in poor countries, where innovative processes to treat waste are lacking or absent. Due to the quantity and complexity of the issues at stake, an increase in private investment would certainly help to create significant positive externalities for the population. At the same time the benefits for the investors are not guaranteed, and this can be a serious problem for the modernization and innovation in a sector where public investment is insufficient. In order to solve the problem of the scarcity of private investments, new instruments for funding innovation are being devised. The most notable examples are microfinancing and hybrid financing (combined debt and equity). The "Participatory Sustainable Waste Management Project" established in Brazil in 2006, for example, created microcredit funds from donations (Hogarth, 2009). These funds are used as working capital for financing waste transportation and

[10]http://www.epa.nsw.gov.au/resources/wastestrategy/140876-WARR-strategy-14-21.pdf.

waste-related emergency responses. The funds are also used to extend loans to waste pickers who will repay their loans after receiving payment from recycling depots (UNEP, 2011).

Other policy instruments used worldwide for the improvement of the sector include subsidies (those generating positive externalities), taxes and fees (with the aim, for instance, of discouraging behavior that might trigger negative externalities), and regulations (imposing, for instance, standards or targets for recycling; banning certain items such as plastic bags, etc.) (UNEP, 2011).

Along with these lines of intervention, another way of improving the use of the resources is the "pay as you throw" (PAYT) scheme, charging weight-based disposal fees. This solution can incentivize considerably more environmentally friendly behavior. A variation of PAYT is the "volume-based waste fee" scheme (widely used in South Korea[11]), implemented through prepaid garbage sacks in which residual waste is disposed of while recyclables are collected without paying any charges. Also in South Korea, policy makers have been planning innovative solutions including providing user cards and recycling bins with radiofrequency identification chips in order to record and weigh discarded products, billing the user monthly on the basis of the waste produced. These measures have generated significant reduction in the produced amount of waste (UNEP, 2011).

Among the different types of waste, food waste is a major problem in the agro-food sector because it prevents an optimal utilization of resources, thus exacerbating the pressures of the megatrends previously analyzed. The analysis developed so far shows how a new model is much needed to ensure both economic and environmental sustainability. A conceptual framework will highlight how a transition from an unsustainable sociotechnical regime toward a sustainable one might occur. This will also allow us to frame the role played by policy makers in the transition toward sustainability.

4.3 TRANSITION THEORY AS A CONCEPTUAL FRAMEWORK FOR SUSTAINABILITY

Sociotechnical transitions are a prominent issue in the economic and policy analysis fields. This topic is so widely studied because it allows understanding the major economic, historical, and societal changes throughout economic history. Based on a growing transition literature, we will present the approach used and identify its elements vital to a functional system analysis. To this purpose we will refer to an analytical framework developed specifically to study complex, sociotechnical systems, ie, the multilevel perspective (MLP). According to the heuristic model, transitions are the outcome of interactions between three levels: landscape, regime, and niche.

4.3.1 SOCIOTECHNICAL TRANSITION AND THE MULTILEVEL PERSPECTIVE

A technological transition is a phase in which an old technology and a new one coexist, and the new technology takes over and replaces the old one. When a new technology is first introduced in the technological array, it might not be immediately ready to compete in the mainstream market. For instance, as shown in Fig. 4.2, it took almost 60 years for the steam engine to overcome sailing, with a coexistence period lasting for more than 20 years.

[11] http://www.asiatoday.com/pressrelease/south-koreas-food-waste-solution-you-waste-you-pay. Retrieved December 2015.

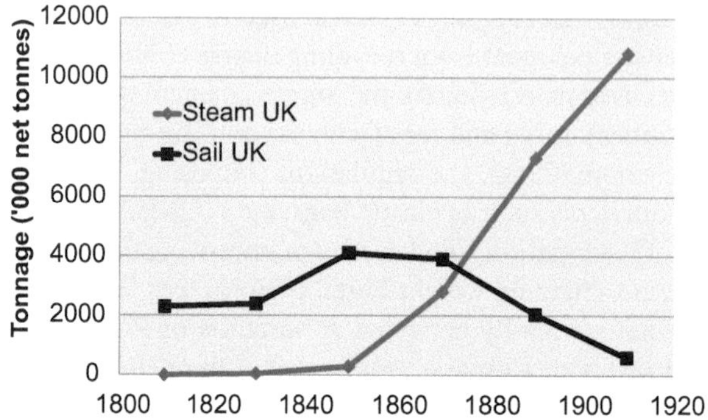

FIGURE 4.2

Tonnage of steamships and sailing ships in Britain.

Adapted from Geels, F.W., Schot, J., 2007. Typology of sociotechnical transition pathways, Research Policy 36(3), 399–417.

The MLP is a heuristic model, which covers three levels of analysis: the landscape (macrolevel), the sociotechnical regime (mesolevel), and the niche (microlevel). Technological transitions can be explained through the interaction among these three levels as the transition basically entails a shift from an incumbent sociotechnical regime to a new one, which is nurtured in the technological niche and prompted at the landscape level. Landscape and niches are *derived concepts* "because they are defined in relation to the regime, namely as practices or technologies that deviate substantially from the existing regime, and as external environment that influences interactions between niche(s) and regime" (Geels, 2011).

MLP helps understanding and analyzing complex and nonlinear phenomena such as historical and structural changes, including technological transitions, using a multidisciplinary and multidimensional perspective. Hence, MLP is not just about economics or competing technologies; it involves many other areas of investigation placing the transition process in a well-defined societal space and historical context.

In this framework, niches are the locus of innovation. A niche is like an incubation room in which new and emerging technologies have a space that protects them from competition and pressures of the selective process taking place in mainstream markets.[12] Usually rules and procedures in the niches are flexible and not formalized in order to facilitate the emergence of innovation. At the same time, the niche space is highly unstable and characterized by the coexistence of several (and often alternative) technological niches, which usually lack coordination between them and in competition among each other. However, not every niche can survive for a long time, and only few of them will get to a point where they will really challenge the incumbent sociotechnical regime. For this to happen, a technological niche should be sufficiently developed so that it reaches maturity.

Niche maturity level can be categorized into three developmental stages: embryonic, proto-niche, and fully developed (Lopolito et al., 2011). In order to assess maturity, the strategic niche management

[12]Initially, new technologies may face several problems such as high costs and low efficiency. Hence, innovation niches are an essential element of the transition, where emerging and promising technologies are nurtured, shielded, and empowered in a protected space.

approach proposes three niche mechanisms: (1) willingness (level of sharing expectations among niche actors), (2) power (based on the presence of powerful actors, namely those carrying valuable assets for niche development), and (3) knowledge (based on the density of knowledge present in the system and its flows). Maturity is reached whenever the three mechanisms coexist.

In this model transition occurs whenever a pressure at landscape level destabilizes the regime, thus creating a window of opportunity for pioneering niche technology to enter in the mainstream market. In the MLP the landscape is the least defined layer, often considered a residual concept. The landscape contains a wide and heterogeneous spectrum of factors, for instance, commodity prices, wars, environmental problems, and demographic and political issues (Geels and Schot, 2007). It is not always easy to understand whether a factor belongs to the landscape or to the regime as the literature does not provide us with clear rules to distinguish among them. As mentioned, coming from the landscape on the regime can create a window of opportunity for the technological niches. For example, high oil prices can create an opportunity to newly developed technologies in the field of nonfossil fuels.

Although possible, and sometimes desirable, destabilizing the regime is not something that can easily happen. This is mainly because the incumbent regime is a rather stable configuration, defined as a bundle of "structures made by a co-evolutionary accumulation and alignment of knowledge, investments, objects, infrastructures, values and norms that span the production-consumption divide and constitutes the prevailing mean for realizing key societal functions" (Smith et al., 2010, p. 441).

As pointed out by Geels (2002), the sociotechnical regime consists of a number of dimensions: (1) technology, (2) user practices and application, (3) the symbolic meaning of technology, (4) infrastructure, (5) policy, and (6) technoscientific knowledge. These elements are highly interlinked, making the regime a very stable reality. Regime stability is further reinforced by the so-called *lock-in effect* (Arthur, 1989), deriving from the following factors: learning by using (which accelerates technological improvements), network spillovers (the more widely a technology is used, the more applications are developed for it, and the more useful it becomes), economies of scale (which reduce unitary price), increasing informational returns (linked to learning by using, whereby the increased number of users, knowing more about the technology, makes it easier for others to learn about the technology), and development of complementary technologies (which both reinforce the position of the technology and make it more useful).

While the stability of the regime is essential for its endurance, it must also have a certain degree of flexibility in order to cope and withstand the pressure applied on it; this feature is defined as adaptive capacity. As mentioned, pressure comes both from the niches and the landscape levels, and regime actors have to use whatever is at hand in order to defend their position. The more intense or destabilizing these pressures are, the greater the amount of resources that have to be deployed to protect the regime. Regimes with a high level of adaptive capacity are more likely to survive and not be substituted for by innovations coming from the niches or destabilized by a changing landscape.

4.3.2 SUSTAINABILITY TRANSITIONS

Not every transition is a leap toward the achievement of sustainability. In the past, many transitions actually contributed to the destruction of the environment, although they led to huge economic progress. Yet, considering current environmental and socioeconomic trends, future transitions will have to address both the issue of sustainability and economic efficiency. Hence, there is a growing interest in the so-called "sustainability transitions" literature.

Sustainability transition can be defined as a long-term, multidimensional, and fundamental transformation process through which an established sociotechnical system shifts to more sustainable ways of production and/or consumption (Geels and Schot, 2010). The theoretical framework (analyzed in Section 4.3.1) is the same for both sociotechnical transitions and sustainability transitions, even if the latter can have some specific different features.

As pointed out by Smith et al. (2005), sustainability transitions can be goal oriented (and not emergent, like in the past) since they need to address specific challenges concerning issues of environmental long-term sustainability (climate change, depletion of natural resources, pollution, etc.). The main obstacle to the sustainability transitions is ensuring a commercial viability without needing the overriding support of the public sector. Sustainability can certainly bring more benefits to all; however, these benefits for the investors can be, in comparison, smaller than those generated by investing in incumbent technologies.

Goal-oriented sustainability transitions are difficult to achieve because they involve long development times, uncertainty about market demand and social advantages, scarce capital from private investors, presence of established lock-ins and path dependences in the current sociotechnical regime, and difficulty in coordinating among different institutional actors involved (government, private enterprises, researchers and scholars, other social actors, etc.). Moreover, governing transitions through public policies may not be easy as the more the sociotechnical regime is stable, the more will it resist the intervention of the public sector. Trying to impose a sustainable transition through traditional policy tools may actually have adverse effects. For instance, a cap-and-trade system in order to reduce emissions can have unintended and unforeseen consequences.[13] The regime, in fact, is influenced by a large array of factors that affect each other in unknown ways, so it is difficult for the policy maker to understand how a sustainable transition can be unlocked.

Goal-oriented sustainable transitions require extra policy efforts in order to steer the change into the desired direction. In this regard, the megatrends discussed in Section 4.2 are encouraging such efforts, determining an unprecedented pressure exerted by the landscape upon the unsustainable sociotechnical regime. As a case in point, the acknowledgment of climate change as a global problem is likely to add pressure on the incumbent regime, thus creating new opportunities for sustainable technologies under development in the niches. Indeed, the recent agreement reached at the United Nations Climate Conference in Paris in December 2015 marks an unprecedented political recognition of the risks of climate change and is likely to open a window of opportunities for new, more sustainable technologies to replace the incumbent fossil fuel–based economy.[14]

---

[13]It is clear that a binding cap on emissions will restrict the supply of all energy-intensive goods. This implies that the global price of these goods must increase, and therefore production abroad will increase, which will lead to higher emissions abroad. This effect is called "carbon leakage," and a recent study based on a large general-equilibrium model concludes that about 40% of any reduction in the production of energy-intensive goods in the European Union would be offset by higher production abroad (Veenendaal and Manders, 2008).

[14]In fact, according to John Schellnhuber, head of the Potsdam Institute for Climate Impact Research, delivering a warming of "well below" 2°C requires that global carbon dioxide emissions peak "well before 2030" and "be eliminated as soon as possible after 2050." That would represent a rate of "decarbonization" far greater than the world has yet seen. http://www.economist.com/news/international/21683990-paris-agreement-climate-change-talks. Retrieved December 2015.

4.4 TURNING CHALLENGES INTO OPPORTUNITIES: FROM WASTE TO WEALTH

Food waste and food losses are key issues to be addressed to allow the development of a more sustainable agro-food system. Although there are no official definitions of "food waste" and "food loss," the most accepted meaning of food waste is "all food produced or purchased that is discarded by humans" (Gallo, 1980), while the nature of food loss is currently being investigated (Griffin et al., 2009). Until 2000 in the European Union food waste was defined as "any food substance, raw or cooked, which is discarded, or intended or required to be discarded."[15] Since 2000, however, there is no official definition of food waste, but there are several categories of waste and each of them is subjected to a different European legislation. In 2014 the European Commission, while defining the targets and the standards in a proposal of a legislative act relating to waste management, defined food waste as food (including inedible parts) lost from the food supply chain, not including food diverted to material uses such as biobased products, animal feed, or sent for redistribution.[16]

The lack of an official definition prevents the collection of homogenous and comparable data, making the research about these phenomena quite difficult. According to a report of the Food and Agricultural Organization of the United Nations (FAO (2011)) , on a global basis food waste was estimated at 1.3 billion tons per year.[17] In the United Kingdom, the Waste Resources Action Program distinguishes three categories of food waste: avoidable (when the food is discarded even though it is perfectly edible and suitable for human consumption), possibly avoidable (eg, when the food discarded can become edible if prepared in a different way), and unavoidable (when the food discarded is not edible).

In Section 4.2, the question of waste production as a source of pressure on the socioeconomic and environmental systems was discussed. Such pressure is caused by the growing amount of municipal solid waste (a consequence of population growth and the ongoing urbanization process) and the hazardous compounds (and the GHGs) released into the environment, posing serious threats to the health of humans and other species. Food waste itself is not a threat to human health, however, it can indirectly affect it by reducing the efficiency of the agro-food system. Hence, as stated by the FAO Director-General José Graziano da Silva, "[i]n addition to the environmental imperative, there is a moral one: we simply cannot allow one-third of all the food we produce to go to waste, when 870 million people go hungry every day."[18] In this regard, food waste is primarily a socioeconomic issue rather than a mere ecological concern. Food waste and losses are severe obstacles to food security for poorly nourished and undernourished people of the world. Moreover, reducing food waste relieves the competition for land and water making these resources available for other valuable uses.

Through proper policies purposely designed, it is possible to improve the overall efficiency of the whole supply chain in the agro-food sector. An efficiently engineered supply chain can be a viable instrument to relieve food insecurity in both developed and developing countries. A more efficient supply chain can bring lower prices to consumers and lower costs for the firms. Here, the primary tool to mitigate the entity of food waste is prevention at every level: consumers, for example, can avoid

[15] http://eur-lex.europa.eu/legal-content/EN/ALL/?uri=CELEX:31975L0442. Retrieved December 2015.

[16] http://eur-lex.europa.eu/legal-content/EN/TXT/?uri=CELEX:52014PC0397. Retrieved December 2015.

[17] In the United States about 95% of food waste ends up in landfills (Environmental Protection Agency: http://www.epa.gov/recycle/reducing-wasted-food-home. Retrieved December 2015.

[18] http://www.unep.org/newscentre/Default.aspx?DocumentID=2726&ArticleID=9611. Retrieved December 2015.

discarding perfectly edible food for mere aesthetic reasons, or limit overconsumption and over storage; producers, on the other hand, can engage in a more careful life cycle analysis of products, making the supply chain more efficient using new technologies.

In fact, food waste and loss occurs throughout the whole supply chain. According to the FAO (2013), 54% of the world's food wastage occurs "upstream" during production, postharvest handling, and storage, while 46% happens "downstream," at the processing, distribution, and consumption stages. As a general trend, developing countries suffer more food losses during agricultural production (mainly due to the poor availability of appropriate and functioning storage facilities), while food waste at the retail and consumer level tends to be higher in middle- and high-income regions. In developed countries 42% of food waste is produced by households, 39% by the food industry, 14% by the food service industry, and 5% in retail and distribution (Mirabella et al., 2014). As a matter of fact, the later a food product is lost along the chain, the greater the environmental consequences are, since the environmental costs incurred during processing, transport, storage, and cooking must be added to the initial production costs.

Bearing this in mind, changes must be introduced at every level of the human food chain, including farmers and fishermen; food processers and supermarkets; local and national governments; and individual consumers. Thus, the aim is to prevent food wastage from happening in the first place and to reuse or recycle it when prevention is not possible. This process can be realized through the establishment of a "food recovery hierarchy."[19] Food recovery contributes to convert waste into wealth, thus enhancing the efficiency of the agro-food system through the valorization of the resources that otherwise would have been wasted. This vision of food waste is deeply linked to the "circular economy" and the "zero-waste economy" concepts.[20]

Higher in the food recovery hierarchy are source reduction and food rescue to feed the most food insecure. Recovery for other valued uses, such as to feed for animals or for industrial uses, are next on the list, followed by composting. Incineration and landfill are the least preferred options. Moving along this line, we shall now consider two examples of actions aiming at (1) source reduction and (2) food rescue.

4.4.1 FOOD SHARING AS A STRATEGY FOR SOURCE REDUCTION

In recent years, a number of prevention and mitigation measures have been put in place to reduce food waste. It should be noted, however, that these measures themselves are not without cost (FAO, 2014). As an alternative to more traditional policies, Belk (2007, p. 126) suggests looking at the sharing economy approach, which involves "the act and process of distributing what is ours to others for their use and/or the act and process of receiving or taking something from others for our use." Hence, a viable and possibly less costly solution could come from the application of this type of approach.

---

[19]Environmental Protection Agency: http://www.epa.gov/sites/production/files/2015-08/fd_recovery_hierarchy_lg.jpg. Retrieved December 2015.

[20]A zero-waste economy entails also recycling but actually goes beyond it by taking a systemic and holistic approach to the vast flow of resources and waste that is produced by human society. A zero-waste approach aims at maximizing recycling, minimizing waste, reducing consumption, and ensuring that products are made to be reused, repaired, or recycled back into nature or the marketplace.

Food sharing practices were first documented by anthropological studies on primitive and contemporary hunter–gatherer societies (see among others Peterson, 1993; Hunt, 2000; Ziker and Schnegg, 2005; Jaeggi and Gurven, 2013). To avoid wasting parts of the hunted animal (those that the hunter and his household would not be able to consume by themselves), the meat is shared. Although these nomad societies are based on egalitarian political organization, their practical purpose is still suitable to the nonegalitarian sedentary society we live in. That is, people share food to avoid unnecessary resource waste.

Currently, there is a renewed interest in food sharing and a wide range of successful initiatives to this aim are spread across European and American countries (see Ganglbauer et al., 2014). A public debate has been launched to discuss the way they may contribute to food waste reduction, with benefits not only for the environment but also in economic terms for household savings.

4.4.2 FOOD BANKS AS A STRATEGY FOR FOOD RESCUE

Over the last 30 years food banks—for the redistribution of donated and surplus food—have established themselves as one of the fastest-growing charitable industries in high-income countries (Riches, 2002), emerging in response to the growing twin problems of food insecurity and food waste. Food banks rely on schemes whereby voluntarily given away food (that otherwise would be lost or wasted) is redistributed to those who need it. This strategy can be applied at the production stage with crops that otherwise would go unharvested, at the manufacturing stage with production surplus, and at the distribution and market stages with food left unsold at stores and markets.

At the production stage, there are several reasons for which edible grains, fruits, and vegetables remain in the field. For instance, food might go unharvested if the price of a given crop is too low to even pay for the labor required to pick that crop and the transport costs associated with selling it. In such cases, it makes more sense for the farmer to let that food be lost. Other reasons for which crops may not be harvested include weather or pest damage, imperfections relating to shape, size, and color, etc. At the manufacturing stage, a surplus amount of food might be produced when an order placed by a retailer is reduced/canceled, while at the market stage, surplus food might be generated when a store purchases too much of a certain item that then approaches or goes past the "sell-by" or "display until" date printed by the manufacturer. For instance, fresh-cooked meals at food retail stores that are unsold at the end of a day are typically thrown away (Lipinski et al., 2013).

The main obstacles to food redistribution are related to transportation, legal, and economic factors. For instance, farmers and stores with surplus food might not be physically close enough to food banks to make the transportation of unused food an economically viable option. On the legal side, potential food donors may be concerned about the legal repercussions should the food be unsafe and its recipients suffer health consequences. As for economic questions, as already mentioned, it might be prohibitive for a farmer to harvest and sell a type of food on the market, let alone donate it. In terms of possible policy measures, there are different alternatives that can be considered. Although the transportation obstacle, for instance, can be rather difficult to address, it can be tackled by establishing additional food bank locations to shorten distances. On the other hand, in order to address legal difficulties, policy makers can pass "Good Samaritan" laws that limit the liability of donors in case redistributed food unexpectedly turns out to be somehow harmful to the consumer (Lipinski et al., 2013).

4.5 CONCLUSIONS

In this chapter, the key drivers to a sustainability transition in the agro-food sector were investigated. Firstly, the areas of pressure on the food system stemming mostly from the megatrends occurring at the outset were identified, ie, the demographic growth trends accompanied by the foreseen income and wealth increases expected over the next 35 years. These trends are most likely to generate a general increase in the demand for food and manufactured goods and, consequently, in pollution emissions levels (most notably CO_2, with great impact on climate change), intensification of land use (with the associated problems such as indirect land use change and deforestation), and increase in waste (an issue related both to the finitude of resources and natural elements and to the management and disposal of residues and waste).

All these call for a paradigm shift from the current unsustainable production and consumption model toward a new model where climate change issues, food security, and waste management are conceived unitarily and are simultaneously tackled. Building on the MLP and, most notably, on sustainability transition theory, we argued that such a change occurs when pressure exerted by landscape actors upon the regime and technological pressure exerted by innovation niches ready to replace the incumbent technological regime occur. This theoretical view stems directly from long-wave theory on technoeconomic paradigm (TEP) shifts (Freeman and Perez, 1988), in the sense that it is multidimensional and addresses structural changes. However, rather than focusing on entire economies and aggregate macroeconomic processes (which is what TEP aims at), the MLP concentrates on specific areas of transition (eg, energy, transport, agro-food systems, etc.) looking closely at the various groups, their strategies, resources, beliefs, and interactions (Geels, 2011).

These groups of actors have, therefore, a pivotal role in shaping the trajectory of the transition, stirring the change in the desired (sustainable) direction. In this sense, well-crafted innovation policy actions can accelerate the transition toward a sustainable agro-food model. As a case in point we considered two actions to be undertaken to reduce and rescue food wasted: food sharing and food banks. Both actions entail an innovative approach to food consumption, setting good examples and practices for the type of mind-set change needed to make a paradigmatic shift, turning problems into opportunities and waste into wealth.

REFERENCES

Arthur, W.B., 1989. Competing technologies, increasing returns, and lock-in by historical events. The Economic Journal 99 (394), 116–131.

Belk, R., 2007. Why not share rather than own? Annals of the American Academy of Political and Social Science 611, 126–140.

Bruinsma, J., 2009. The resource outlook to 2050: by how much do land, water and crop yields need to increase by 2050? Food and Agriculture Organization of the United Nations. Economic and Social Development Department.

Caldwell, L.K., Bartlett, R.V., 1997. Environmental Policy: Transnational Issues and National Trends. Quorum Books, Westport, CT.

Evenson, R.E., Gollin, D., 2003. Assessing the impact of the green revolution, 1960 to 2000. Science 300, 758–762. http://dx.doi.org/10.1126/science.1078710.

Fairley, P., 2011. Introduction: next generation biofuels. Nature 474, S2–S5. http://dx.doi.org/10.1038/474S02a.

FAO, 2009. The Market and Food Security Implications of the Development of BioFuel Production Background document for the FAO Committee on Commodity Problems 67th session. Rome.

FAO, 2011. Global Food Losses and Food Waste – Extent, Causes and Prevention. Rome. http://www.fao.org/docrep/014/mb060e/mb060e.pdf. Retrieved December 2015.

FAO, 2013. Food Wastage Footprint Impacts on Natural Resources. Rome. http://www.fao.org/docrep/018/i3347e/i3347e.pdf. Retrieved December 2015.

FAO, 2014. Mitigation of Food Wastage. Societal Costs and Benefits. Retrieved online at: http://www.fao.org/3/a-i3989e.pdf.

FAO, WFP and IFAD, 2015. The State of Food Insecurity in the World 2015. Rome.

Foxon, T.J., 2011. A co-evolutionary framework for analysing transition pathways to a sustainable low carbon economy. Ecological Economics 70, 2258–2267.

Freeman, C., Perez, C., 1988. Structural crisis of adjustment, business cycles and investment behaviour. In: Dosi, G., Freeman, C., Nelson, R., Silverberg, G., Soete, L. (Eds.), Technical Change and Economic Theory. Pinter, London, pp. 38–66.

Gallo, A.E., 1980. Consumer food waste in the United States. National Food Review 3, 13–16.

Ganglbauer, E., Fitzpatrick, G., Subasi, Ö., Güldenpfennig, F., 2014. Think globally, act locally: a case study of a free food sharing community and social networking. In: Proceedings of the 17th ACM Conference on Computer Supported Cooperative Work & Social Computing, February 15–19, 2014, Baltimore, Maryland, USA.

Geels, F.W., 2002. Technological transitions as evolutionary reconfiguration processes: a multi-level perspective and a case-study. Research Policy 31 (8–9), 1257–1274.

Geels, F.W., Schot, J., 2007. Typology of sociotechnical transition pathways. Research Policy 36 (3), 399–417.

Geels, F.W., 2010. Ontologies, socio-technical transitions (to sustainability), and the multi-level perspective. Research Policy 39 (4), 495–510.

Geels, F.W., 2011. The multi-level perspective on sustainability transitions: responses to seven criticisms,. Environmental Innovation and Societal Transitions 1 (1), 24–40.

Gerland, P., Raftery, A.E., Ev Ikova, H., Li, N., Gu, D., Spoorenberg, T., Alkema, L., Fosdick, B.K., Chunn, J., Lalic, N., Bay, G., Buettner, T., Heilig, G.K., Wilmoth, J., 2014. World population stabilization unlikely this century. Science (AAAS) 346 (6206), 234.

Griffin, M., Sobal, J., Lyson, T.A., 2009. An analysis of a community food waste stream. Agriculture and Human Values 26 (1–2), 67–81.

Hogarth, R., 2009. Microcapital Story: Participatory Sustainable Waste Management Project Extends Microfinance to Informal Recyclers in Brazil. Available online at: www.microcapital.org Retrieved December 2015.

Hunt, R.C., 2000. Forager food sharing economy: transfers and exchanges. Senri Ethnological Studies 53, 7–25.

Jaeggi, A., Gurven, M., 2013. Reciprocity explains food sharing in humans and other primates independent of kin selection and tolerated scrounging: a phylogenetic meta-analysis. Proceedings of the Royal Society B 280, 20131615.

Kharas, H., 2010. The emerging middle class in developing countries. OECD Development Centre, Working Paper No. 285. Research area: Global Development Outlook.

Lipinski, B., Hanson, C., Lomax, J., Kitinoja, L., Waite, R., Searchinger, T., 2013. Reducing food loss and waste. World Resources Institute, Washington.

Loehle, C., 2011. Criteria for assessing climate change impacts on ecosystems. Ecology and Evolution 1 (1), 63–72.

Lopolito, A., Morone, P., Sisto, R., 2011. Innovation niches and socio-technical transition: a case study of biorefinery production. Futures 43, 27–38.

Mirabella, N., Castellani, V., Sala, S., 2014. Current options for the valorization of food manufacturing waste: a review. Journal of Cleaner Production 65, 28–41.

Morone, P., Cottoni, L., 2016. Biofuels: technology, economics and policy issues. In: Luque, R., Clark, J., Wilson, K., Lin, C. (Eds.), Handbook of Biofuels' Production: Processes and Technologies, Second ed. Elsevier.

Morrison, G.M., Parker, N.C., Witcover, J., Fulton, L.M., Pei, Y., 2014. Comparison of supply and demand constraints on U.S. biofuel expansion. Energy Strategy Reviews 5, 42–47.

Peterson, N., 1993. Demand sharing: reciprocity and the pressure for generosity among foragers, American anthropologist. New Series 95 (4), 860–874.

Riches, G., 2002. Food banks and food security: welfare reform, human rights and social policy. Lessons from Canada? Social Policy & Administration 36 (6), 648–663.

Searchinger, T., Heimlich, R., 2015. Avoiding Bioenergy Competition for Food Crops and Land. Working Paper - Installment 9 of Creating a Sustainable Food Future. World Resources Institute, Washington, DC. Available online at: http://www.worldresourcesreport.org. Retrieved December 2015.

Smith, A., Stirling, A., Bekhout, F., 2005. The governance of sustainable socio-technical transitions. Research Policy 34, 1491–1510.

Smith, A., Voß, J.-P., Grin, J., 2010. Innovation studies and sustainability transitions: the allure of the multi-level perspective and its challenges. Research Policy 39 (4), 435–448.

Steinbuks, J., Hertel, T., 2016. Confronting the food–energy–environment trilemma: global land use in the long run. Environmental & Resource Economics 63 (3), 545–570.

UN, 2014. The World Population Situation in 2014. A Concise Report. Department of Economic and Social A airs Population Division, ST/ESA/SER.A/354.

UNEP, 2011. Waste. Investing in Energy and Resource Efficiency. Available online at: http://www.unep.org/greeneconomy/Portals/88/documents/ger/GER_8_Waste.pdf. Retrieved December 2015.

UNEP, 2013. Food Waste Harms Climate, Water, Land and Biodiversity. Available online at: http://www.unep.org/newscentre/Default.aspx?DocumentID=2726&ArticleID=9611. Retrieved December 2015.

Veenendaal, P., Manders, T., 2008. Border Tax Adjustment and the EU-ets, a Quantitative Assessment, CPB Document No. 171. Central Planning Bureau, The Hague.

Wilson, D., Dragusanu, R., 2008. The expanding middle: the exploding world middle class and falling global inequality. Goldman Sachs, New York.

Ziker, J., Schnegg, M., 2005. Food sharing at meals: kinship, reciprocity, and clustering in the Taimyr Autonomous Okrug, Northern Russia. Human Nature 16 (2), 178–210.

CHAPTER

5 INNOVATION IN TRADITIONAL FOOD PRODUCTS: DOES IT MAKE SENSE?

L. Guerrero[1], A. Claret[1], W. Verbeke[2], C. Sulmont-Rossé[3], M. Hersleth[4]

[1]IRTA, Food Research Center, Finca Camps i Armet, Monells, Spain; [2]Ghent University, Ghent, Belgium; [3]INRA, UMR1324, CNRS, UMR6265, Université de Bourgogne, Centre des Sciences du Goût et de l'Alimentation, Dijon, France; [4]Nofima AS, Ås, Norway

5.1 INTRODUCTION

One of the main economic drivers of the European Union (EU) is the food and beverage industry. The latest numbers show that 70% of raw materials of European agricultural production are mainly converted in small and medium sized enterprises (SMEs) employing 3.9 million people. In 2006, the European food industry earned a gross income of 840,000 billion euros, 60% of which were generated by SMEs (Truefood, 2009). An important part of the European food industry produces traditional foods, the commercialization of which is supported by SMEs. This means that over 70% of the total employment generated by the European food industry is located in the traditional food production. In fact, the production and sale of traditional foods constitutes a critical element of the economic boost in many European regions (European Commission, 2007).

In addition to its economic and social importance, traditional foods are a significant part of culture, identity, and European gastronomic heritage (Committee of the Regions, 1996; Ilbery and Kneafsey, 1999). In addition, they contribute to the development, diversification, and sustainability of many rural areas, protecting them from depopulation, allowing a clear product differentiation for their producers and processors (Avermaete et al., 2004), as well as providing a greater variety of food choices to final consumers. In general, traditional foods are associated with a regional identity and a particular sensory quality. A great number of them are marketed under different collective brands such as quality labels (PDO, PGI, TSG, etc.), which normally have a very good image among consumers (Guerrero, 2001).

However, traditional food producers face the challenge of improving safety, health, and convenience of their products, according to market demands, by means of different innovations that allow them to maintain and even expand their current area of influence in a highly competitive and globalized market. Indeed, European consumers ask for tasty and safe traditional foods (Cayot, 2007), but also for a greater variety, more convenient, more nutritious, and healthier options that fit better with the current needs of modern societies. However, it is important to remember that many of the technologies used in the production of traditional foods still rely on traditional production practices with low competitiveness and low efficiency (Fito and Toldra, 2006). Thus, there appear to be a number of opportunities for traditional foods that have not yet been adequately exploited.

It is widely accepted that innovation is one of the keys to success of a company. However, innovations are not enough, as long as it is imperative that the innovations are accepted by the target end users.

Innovation Strategies in the Food Industry. http://dx.doi.org/10.1016/B978-0-12-803751-5.00005-2

The introduction of a new or an innovative product on the market is not straightforward. For instance, it is estimated that providing approximately 3000 initial ideas ends up producing just one successful product (Stevens and Burley, 1997). However, acceptance or rejection of innovations and/or new technologies by consumers is the result of a complex decision process that involves evaluating the risks and benefits associated with both innovation/new technologies and existing alternatives (Henson, 1995). One of the main disadvantages that may result from implementation of innovations in traditional foods is the fact that the innovation itself could make the food lose its "traditional" nature (Caldentey and Gomez, 1997), which in turn could cause it to lose its competitive advantage and the added value that it provides to consumers. Generally speaking, consumers have a certain resistance to adopt innovations, especially when they are associated with sophisticated technologies (Dutra et al., 2007). This effect is more pronounced in the case of foods, in which cultural aspects, sociodemographics, lifestyles, or individual neophilia/neophobia (Pliner and Hobden, 1992) can be fundamental. Familiarity with food is one of the main determinants of acceptance and is, therefore, one of the most important obstacles when introducing new food products on the market (Dutra et al., 2007).

5.2 WHAT DO *TRADITIONAL* AND *INNOVATION* MEAN FOR EUROPEAN CONSUMERS?

Understanding the feelings and needs of individuals requires clear communication with them through a common language (Sokolow, 1988). Therefore, to properly explore the concept of traditional and innovation applied to traditional foods it is essential to do so from the perspective of the European consumers. For instance, it is important to know how innovations are perceived, their image, and the possibilities of theoretical success that different potentially applicable innovations may have. The analysis of these definitions is a key step to better understanding the effect that certain innovations might have on traditional foods. In addition, it allows facilitating the detection of relevant aspects affected by a particular innovation and designing a marketing strategy for the acceptance of innovations in traditional foods.

There are several definitions in the scientific literature of the concept of traditional food. According to Bertozzi (1998) a traditional food is "a representation of a group, it belongs to a defined space, and it is part of a culture that implies the cooperation of the individuals operating in that territory." Jordana (2000) defined this concept from a more sociological approach stating that "in order to be traditional, a product must be linked to a territory and it must also be part of a set of traditions, which will necessarily ensure its continuity over time." In 2006, the European Commission published the following definition of traditional when referring to foods: "Traditional means proven usage in the community market for a time period showing transmission between generations; this time period should be the one generally ascribed as one human generation, at least 25 years" (EU, 2006). More recently, the network of excellence EuroFIR (FP6) has developed a new definition of traditional food linked to product development, including aspects related to traditional ingredients, the traditional composition and type of production and traditional processing (EuroFIR, 2007; Trichopoulou et al., 2007). Within European countries the only legal definition found for traditional food is that provided by the Italian Ministry of Agriculture, under which traditional foods are "agrifood products whose methods of processing, storage and ripening are consolidated with time according to uniform and constant local use" (Ministero Agricoltura, 1999). Although all these definitions try to capture the different dimensions of the concept of traditional food, all of them lack the perspective of the final consumer of the product.

With regard to the concept of innovation, its meaning depends on the context in which it is applied, thus having a large number of different definitions. However, a common feature can be found in all of them (Fagerberg, 2004): innovation normally means the "successful introduction of something new and useful." To this initial concept other authors also added the idea of "recombination of components into new blends" in the case of foods and beverages (Moskowitz et al., 2006). To Carayannis et al. (2003) innovations are simply "the new products and services that emerge from the technology." In any case, it is important to note that in all these definitions the word *new* appears as a key aspect of the concept of innovation. Anyhow, it is remarkable that all these definitions reflect the view of technicians or experts in a particular field, whereas none of them include the perspective of the end user of these innovations.

In a study recently published by Guerrero et al. (2009a), both definitions are provided from the perspective of European consumers. So a traditional food can be defined as "a product frequently consumed or associated with specific celebrations and/or seasons, normally transmitted from one generation to another, made accurately in a specific way according to the gastronomic heritage, with little or no processing/manipulation, distinguished and known because of its sensory properties and associated with a certain local area, region or country." The concept of innovation was defined as "the addition of new or unusual ingredient; new combinations of product; different processing systems or elaboration procedures including packaging; coming from different origin or cultures; being presented and/or supplied in new ways; and always having temporary validity."

Both definitions and the feelings and needs that they may reveal are related to social and cultural aspects, which in turn may depend on the country or place of origin of the consumers. In fact, the main associations linked to each concept are strongly dependent on the country of origin (Guerrero et al., 2010a). Some of the differences between regions for the traditional concept are shown in Fig. 5.1 as an

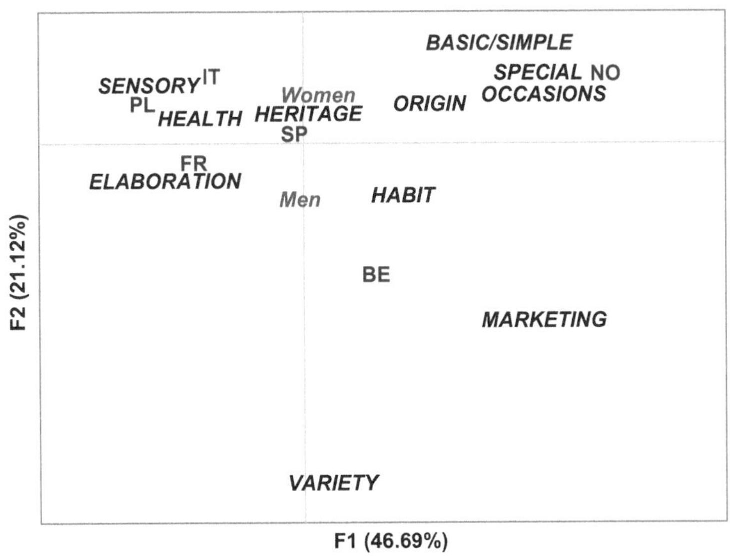

FIGURE 5.1

Main associations linked to the concept of traditional food in six different European regions: BE: Flanders (Belgium), FR: Burgundy (Dijon, France), IT: Lazio region (Italy), NO: Akershus and Østfold counties (Norway), PL: Mazovia (Warsaw, Poland), SP: Catalonia (Spain).

illustration of the cultural variability in the food domain. In this example, Sensory and Health dimensions were closer to Poland and Italy; Heritage was close to Spain; Special occasions, Basic/simple, and Origin were close to Norway; Elaboration was close to France; and Habit, Marketing, and Variety were closer to Belgium.

In the food area, culture can be regarded as one of the strongest determinants when explaining individual attitudes and behaviors (Rozin, 1990). In general, comparative studies between different cultures show important differences in everything related to food, even among relatively homogeneous countries such as those belonging to the EU (Olsen et al., 2007; Boer et al., 2006). According to Askegaard and Madsen (1998), Europe cannot be considered as a single homogeneous space with respect to food culture. Nevertheless and as stated by Guerrero et al. (2009a), both definitions can be generalized to the European population given the noticeable similarities observed between the countries involved in their study. In the case of traditional foods, some of which are locally produced, it would be reasonable to expect that the meaning of both concepts should differ considerably depending on the rural or urban character of the individuals involved. In general, urban consumers seem more willing to look for and connect with the rural aspects of food (Montanari, 1994). On the other hand, when choosing their food, rural consumers tend to focus more on civic aspects, show greater concern for everything related to the supply of products, and have a greater interest in local products (Weatherell, 2003). However, the perception of the traditional concepts and innovation does not appear to depend on the urban rural character of consumers' place of residence (Guerrero et al., 2009a).

5.3 INNOVATIONS IN TRADITIONAL FOODS

In general, the food industry has a rather slow-moving nature (Moskowitz and Hartmann, 2008) given that it is not subject to the innovation pressure that other sectors such as electronics, automobiles, or financial services are subjected to. This is particularly accentuated in traditional food products since consumers perceive traditional foods as having a strong distinctive character linked to their cultural heritage (Guerrero et al., 2009a, 2010a; Trichopoulou et al., 2007), thus being perceived as something to preserve intact for future generations. This obviously may be contradictory to the idea of innovation. This apparent incompatibility between traditions and innovations is also shown in Fig. 5.2, where both concepts were among those with the highest dissimilarity values (Euclidian distance) after performing a sorting task with European consumers (Guerrero et al., 2012).

In general, the acceptance of an innovation depends on the innovation itself as well as on the carrier product to which it is applied, especially in the food domain (Guerrero et al., 2009a). Normally if the changes are small innovations will have a greater chance of being accepted than if the changes are significant. In general, the more complex the technological processes are the more critical the consumer will be. It is worth mentioning that uncertainty may play an important role in explaining consumer propensity to adopt an innovation in food. There is ample evidence that cultures showing less uncertainty are more likely to show innovative behavior (Singh, 2006). Normally, for each food, each innovation and their possible combinations, different levels of acceptance may be found depending on the country or culture and the individual characteristics of each consumer. For example, while French and Polish consumers are not very receptive to nutritional innovations (Guerrero et al., 2009a), they are apparently well accepted by Spanish consumers (Guardia et al., 2006).

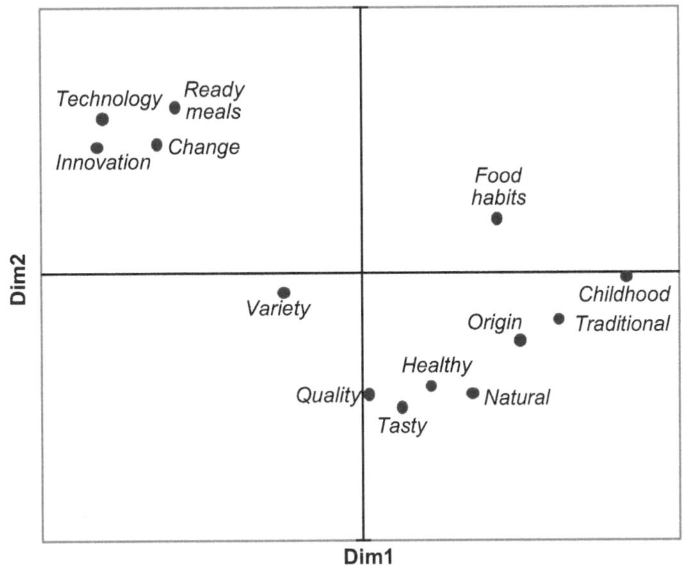

FIGURE 5.2

Similarity between key words related to the concepts *traditional* and *innovation* obtained by means of a sorting task performed with European consumers.

The resistance offered by consumers to an innovation can be explained, to some extent, by personal aversion to changes that the innovation may introduce at different levels (purchasing, eating behavior, usage patterns, norms, habits, and traditions) (Kleijnen et al., 2009). Consequently, changes introduced by any innovation in a traditional food product can be the main factor that explains its success or failure on the market. Most people act on a routine basis in food-related contexts (Aarts and Dijksterhuis, 2000), and apparently, this routine may be disturbed by the changes introduced when a product or process is innovated. In any case, what appears crucial is the need to properly inform consumers about the innovations applied. Consequently, an honest, informative, and reassuring communication could be a key to introducing different innovations in traditional foods with certain guarantees of success.

In a recent study conducted in six European countries ($N=550$ consumers) on the compatibility of certain innovations with traditional foods (Sulmont-Rossé, 2007), it was observed that, in general, innovations that increased food safety or provided significant tangible benefits were accepted, but only if they did not compromise the key features of the product. However, sensory innovations tended to be rejected (Fig. 5.3). These authors also identified three distinct segments of respondents: those seeking the authenticity of the product, those seeking healthier foods, and those that especially valued innovations oriented toward an increase in the convenience (Fig. 5.4).

More recently, a study conducted on a sample of 4828 European consumers (Guerrero et al., 2009b) showed that only the most trivial innovations were clearly rejected in traditional foods (Fig. 5.5). Specifically, only four innovations were evaluated as negatively affecting the traditional character of traditional food products. These innovations dealt with new flavors, diversification of shapes/textures, ready-to-eat dishes, and, especially, distribution of traditional food products in vending

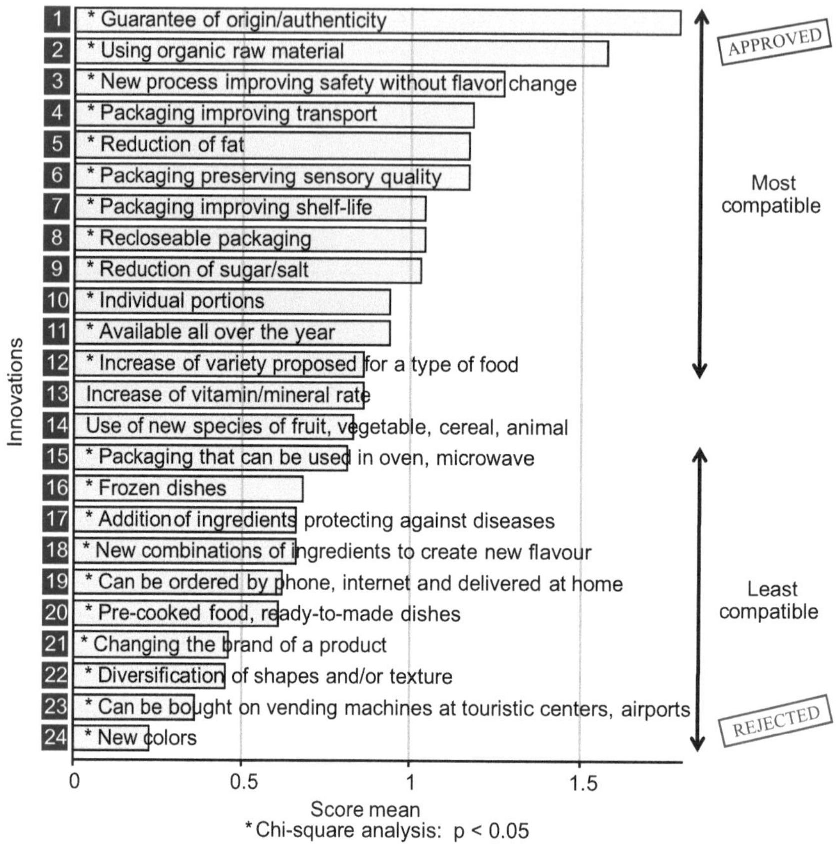

FIGURE 5.3

Compatibility between some innovations and traditional foods.

machines. The most tolerated innovation was the possibility of buying traditional food products directly from the manufacturer. A more detailed analysis of this information allowed the detection of differences between countries depending on their degree of acceptance of innovations. Consumers in Poland and Spain were the most tolerant toward the proposed innovations, while French consumers were the most reluctant to admit them. The highest disagreement between countries was detected for innovations about distribution-related issues. Regarding regional differences, two main clusters were observed, which mainly diverged in their perception of convenience-oriented innovations.

Interestingly, and despite being known for its strong traditional character, Poland was one of the countries more open to accept innovations in traditional foods in contrast with other countries like France or Belgium, which were much more reluctant to do so.

Vanhonacker et al. (2013) went one step further and projected both the damage to the traditional character and the consumer acceptance of the same innovations in a two-dimensional map (Fig. 5.6). These authors observed that a lower impact on the traditional character does not necessarily imply a higher acceptance and vice versa.

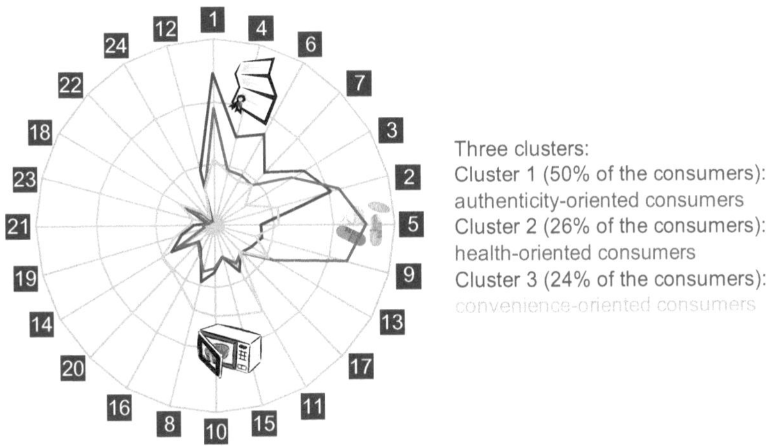

FIGURE 5.4

Clusters of consumers identified based on their acceptance/rejection of different innovations applied to traditional food products (numbers indicate the different innovations listed in Fig. 5.3).

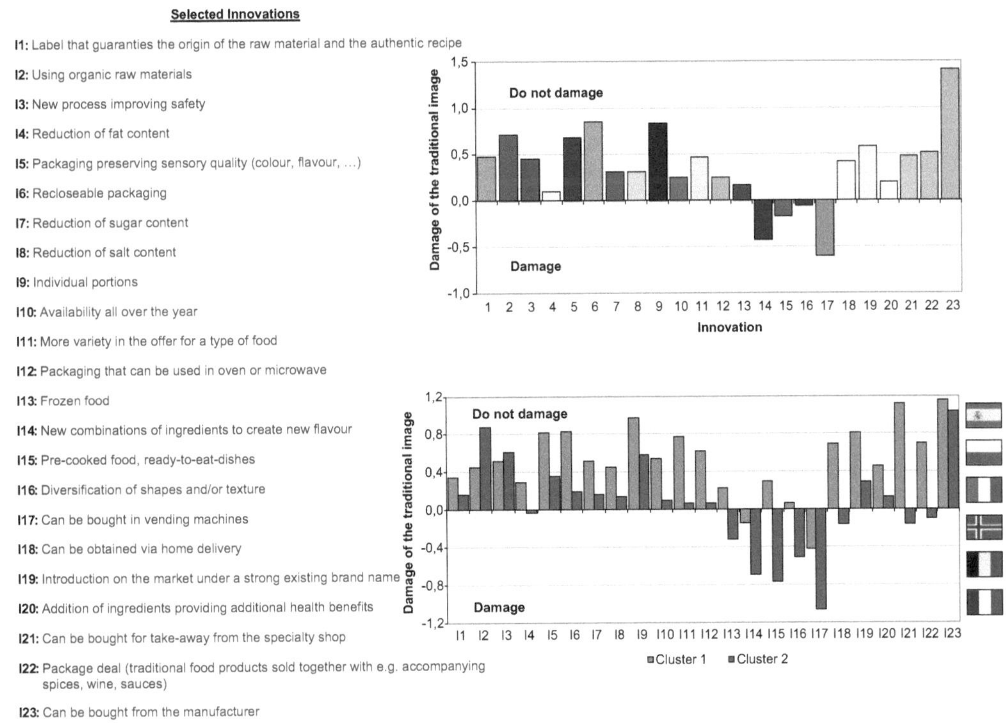

FIGURE 5.5

Damage produced by the different innovations in traditional food products.

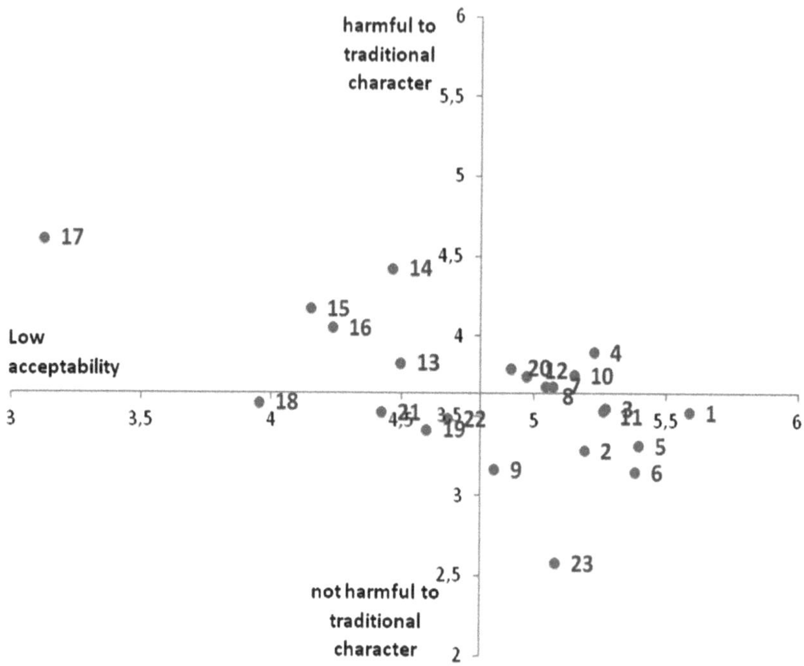

FIGURE 5.6

Consumer acceptance and perceived impact on the traditional character of the food product when applying different innovations (numbers indicate the different innovations listed in Fig. 5.5).

Nevertheless, it is important to emphasize that in all these studies the innovations assessed were valued generically, without being applied to any particular product and that there appear to exist important discrepancies between what consumers think they do, what consumers say they do, and what they actually do (Guerrero, 2010).

It is also worth mentioning that the application of specific innovations in specific foods does not necessarily have to match the results already shown. Thus, a particular innovation generically accepted may be rejected when applied to a particular product depending on the emotional linkage existing between the product involved and the consumer. This is the case in the study performed by Guerrero et al. (2010b) on two types of cheese with a marked local character in two European cities: Girona and Rome. In this study the possibility of applying two different innovations, apparently well accepted, such as the addition of omega-3 fatty acid and modified-atmosphere packaging, was evaluated in two local cheeses (Fig. 5.7). The study was conducted with consumers from Girona and Rome, and in both cases both innovations were rejected, especially the use of a modified atmosphere despite being a well-known innovation and although no apparent effect on the basic properties of the product was observed. This study showed that consumers seemed to look with suspect on product innovations for these two types of cheeses. Therefore, they were not willing to pay more for them than the standard price even though these innovations meant a product with an extended shelf life and healthier properties. The strong traditional character of the tested products might justify the rejection of these innovations.

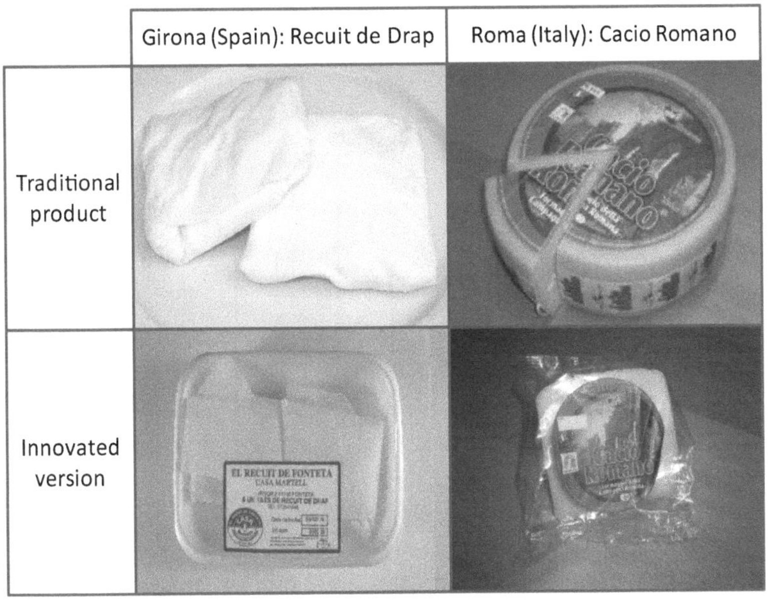

FIGURE 5.7

Samples used in the study of cheese innovation.

In another study done in Spain with dry-cured ham, two different innovations were introduced: a reduction in the salt content and the application of high hydrostatic pressure (HHP). Consumer reaction to these innovations was measured by assessing the products in three different situations: blind tasting (product without information), acceptability expectation for the different products without tasting them (informed condition), and overall acceptability for each product provided with information and after being tasted (full information condition). In addition, a conjoint analysis was also conducted in order to assess the hypothetical acceptance for each of the innovations described. Again, the same result was obtained and both innovations were rejected by Spanish consumers (Fig. 5.8 and Table 5.1, unpublished data).

In order to verify the nonhypothetical buying intention for the innovated products and the price that consumers were willing to pay for them, the previous study with innovated dry-cured hams was repeated in three Spanish cities and in France. In this case, the evaluation of the samples was performed by means of experimental auctions where consumers had the opportunity to actually buy some of the innovated products (Issanchou et al., 2010). Once more, the results showed an overall rejection by Spanish consumers to the diminished salt content, and a willingness to pay a lower price for this type of product. However, in France, hams with lower salt content were preferred and participants even were willing to pay a slightly higher price for them. On the contrary, HHP was relatively well accepted in Spain and rejected in France. These results show how the same innovation can be assessed in different ways according to the cultural and emotional involvement with the product. It seems reasonable to expect that an innovated dry-cured ham with slight sensory changes (less salty) will have less impact on an average French consumer, who is not so used to its consumption, than on an average Spanish consumer, much more familiar and involved with the product.

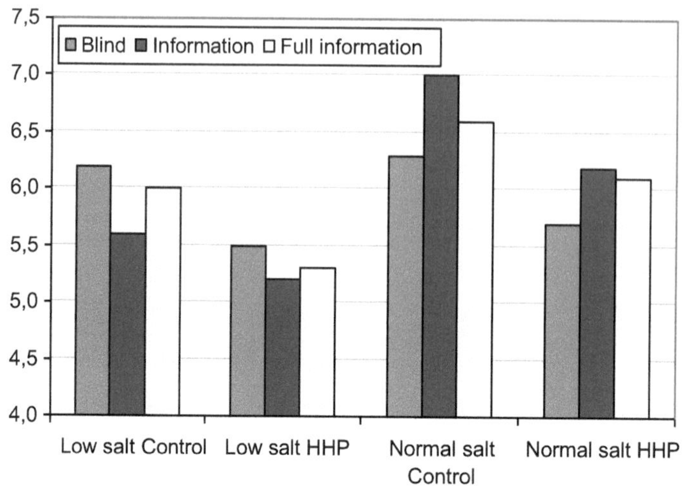

FIGURE 5.8

Consumer acceptability (blind tasting) and expectations (only product information and product information and tasting) for traditional and innovated dry-cured hams. *HHP*, high hydrostatic pressure.

Table 5.1 Hypothetical Acceptance for Two Innovations Introduced in Traditional Dry-Cured Ham: Relative Importance for Each Factor and Utilities for Each Level

		Without Tasting		After Tasting	
Factor	**Level**	**Utility**	**Importance (%)**	**Utility**	**Importance (%)**
Salt	Low	−0.83	74.2	−0.41	49.2
	Normal	+0.83		+0.41	
Treatment	Control	+0.29	25.8	+0.42	50.8
	HHP (high hydrostatic pressure)	−0.29		−0.42	

However, it should be noted than in both studies, the one dealing with cheeses and the one dealing with dry-cured hams, evident innovative segments of consumers that stated that they would accept some of the proposed innovations were obtained. These segments, in the case of Spain, represented about 25–30% of the population.

Theoretically, there are certain possibilities for introducing innovations in traditional foods without diminishing their main competitive advantage, their image, especially those related to the authenticity of the product. Generally speaking, those innovations that increase food safety and/or improve the nutritional quality and/or make it more convenient are relatively well accepted as long as they do not involve changes in the sensory quality. However, the application of innovations in specific products tends to be rejected by most consumers, most likely by the marked traditional character of the products studied and the existing emotional link between them and participants. The existence of segments of consumers with different beliefs and attitudes offers some opportunity of success to the implementation of innovations in the traditional food market. Regarding the actual purchasing behavior for innovated products, European consumers seem unwilling to pay more for the innovated version of a product even if it could be accepted.

This information could help producers of traditional foods in making decisions when applying different innovations with respect to communication strategy, product placement, and new developments.

REFERENCES

Aarts, H., Dijksterhuis, A., 2000. Habits as knowledge structures: automaticity in goal-directed behavior. Journal of Personality and Social Psychology, 78 (1), 53–63.

Askegaard, S., Madsen, T.K., 1998. The local and the global: exploring traits of homogeneity and heterogeneity in European food cultures. International Business Review 7, 549–568.

Avermaete, T., Viaene, J., Morgan, E.J., Pitts, E., Crawford, N., Mahon, D., 2004. Determinants of product and process innovation in small food manufacturing firms. Trends in Food Science and Technology 15, 474–483.

Bertozzi, L., 1998. Tipicidad alimentaria y dieta mediterránea. In: Medina, A., Medina, F., Colesanti, G. (Eds.), El color de la alimentación mediterránea. Elementos sensoriales y culturales de la nutrición. Icaria, Barcelona, pp. 15–41.

Boer, J., Helms, M., Aiking, H., 2006. Protein consumption and sustainability: diet diversity in EU-15. Ecological Economics 59, 267–274.

Caldentey, P., Gómez, A.C., 1997. Typical products, technical innovation and organizational innovations. In: Paper Presented at the Typical and Traditional Productions: Rural Effect and Agro-Industrial Problems, 52nd EAAE Seminar (June 19–21, 1997). Parma, Italia.

Caranyannis, E.G., González, E., Wetter, J., 2003. The nature and dynamics of discontinuous and disruptive innovations from a learning and knowledge managements perspective. In: Shavinina, L.V. (Ed.), The International Handbook of Innovation. Elsevier Science Ltd, Oxford.

Cayot, N., 2007. Sensory quality of traditional foods. Food Chemistry 102, 445–453.

Committee of the Regions, 1996. Promoting and Protecting Local Products: A Trumpcard for the Regions. Committee of the Regions, Brussels.

Dutra, M., Klume, L., Ferreira, G., Vieira, L., 2007. Willingness to try innovative products. The case of food products in Rio Grande do Sul, Brazil. In: Paper Presented at the 17th Annual Forum and Symposium IAMA Conference (June 2007), Parma, Italia.

EU, 2006. Council regulation (EC) no 509/2006 of 20 March 2006 on agricultural products and foodstuffs as traditional specialties guaranteed. Official Journal of the European Union L 93/1, 1–11.

EuroFIR, 2007. FOOD-CT-2005-513944, EU6th Framework Food Quality and Safety Programm. Website: http://www.eurofir.net/ (Last accessed on September 2015).

European Commission, 2007. European Research on Traditional Foods. Website: http://bookshop.europa.eu/en/european-research-on-traditional-foods-pbKI7606394/ (Last accessed on September 2015).

Fagerberg, J., 2004. Innovation: a guide to the literature. In: Fagerberg, J., Mowery, D.C., Nelson, R.R. (Eds.), The Oxford Handbook of Innovations. Oxford University Press, Oxford, UK, pp. 1–26.

Fito, P., Toldra, F., 2006. Innovations in traditional foods. EFFOST 2005 conference. Trends in Food Science and Technology 17 (9), 470.

Guàrdia, M.D., Guerrero, L., Gelabert, J., Gou, P., Arnau, J., 2006. Consumer attitude towards sodium reduction in meat products and acceptability of fermented sausages with reduced sodium content. Meat Science 73, 484–490.

Guerrero, L., 2001. Marketing PDO (products with denominations of origin) and PGI (products with geographical identities). In: Frewer, L., Risvik, E., Shifferstein (Eds.), Food, People and Society. An European Perspective of Consumers' Food Choices. Springer Verlag, Berlin, pp. 281–296.

Guerrero, L., Guardia, M.D., Xicola, J., Verbeke, W., Vanhonacker, F., Zakowska, S., Sajdakowska, M., Sulmont-Rossé, C., Issanchou, S., Contel, M., Scaveldi, L., Granli, B.S., Hersleth, M., 2009a. Consumer-driven definition of traditional food products and innovation in traditional foods. A qualitative cross-cultural study. Appetite 52 (2), 345–354.

Guerrero, L., Claret, A., Guàrdia, M.D., Verbeke, W., Vanhonacker, F., Hersleth, M., 2009b. The impact of applying different innovations on the image of traditional foods in Europe. In: 8th Pangborn Sensory Science Symposium, Florence, Italy, July 26–30.

Guerrero, L., 2010. Is consumer behaviour a qualitative affair? In: Fourth European Conference on Sensory and Consumer Research, Vitoria-Gasteiz, Spain, September 5–8.

Guerrero, L., Claret, A., Verbeke, W., Enderli, G., Zakowska-Biemans, S., Vanhonacker, F., Issanchou, S., Sajdakowska, M., Signe Granli, B., Scalvedi, L., Contel, M., Hersleth, M., 2010a. Perception of traditional food products in six European regions using free word association. Food Quality and Preference 21, 225–233.

Guerrero, L., Claret, A., Guàrdia, M.D., Contel, M., Scalvedi, M.L., 2010b. Consumer acceptance of innovated traditional cheese: an experimental study in Italy and Spain. In: Fourth European Conference on Sensory and Consumer Research, Vitoria-Gasteiz, Spain, September 5–8.

Guerrero, L., Claret, A., Verbeke, W., Vanhonacker, F., Enderli, G., Sulmont-Rossé, C., Hersleth, M., Guàrdia, M.D., 2012. Crosscultural conceptualization of the words traditional and innovation by means of sorting task and hedonic evaluation. Food Quality and Preference 25, 69–78.

Henson, S., 1995. Demand-side constraints on the introduction of new food technologies: the case of food irradiation. Food Policy 20 (2), 111–127.

Ilbery, B., Kneafsey, M., 1999. Niche markets and regional speciality food products in Europe: towards a research agenda. Environment and Planning A 31, 2207–2222.

Issanchou, S., Guerrero, L., Chaya, C., Claret, A., Godet, M., Deliza, R., 2010. Does innovation in a traditional product affect consumer's evaluation? A case study on Serrano ham with Spanish and French consumers. In: Fourth European Conference on Sensory and Consumer Research, Vitoria-Gasteiz, Spain, September 5–8.

Jordana, J., 2000. Traditional foods: challenges facing the European food industry. Food Research International 33, 147–152.

Kleijnen, M., Lee, N., Wetzels, M., 2009. An exploration of consumer resistance to innovation and its antecedents. Journal of Economic Psychology, 30, 344–357.

Ministero Agricoltura, 1999. Decreto Legislativo 30 Aprile 1998 N. 173. Decreto 785 Ministero Agricoltura 8 Settembre 1999 N. 350.

Montanari, M., 1994. The Culture of Food. Blackwell, Oxford.

Moskowitz, H., Hartmann, J., 2008. Consumer research: creating a solid base for innovative strategies. Trends in Food Science & Technology, 19, 581–589.

Moskowitz, H., Reisner, M., Itty, B., Katz, R., Krieger, B., 2006. Steps towards a consumer-driven "concept innovation machine" for food and drink. Food Quality and Preference 17, 536–551.

Olsen, S.O., Scholderer, J., Brunso, K., Verbeke, W., 2007. Exploring the relationship between convenience and fish consumption: a cross-cultural study. Appetite 49 (1), 84–91.

Pliner, P., Hobden, K., 1992. Development of a scale to measure the trait food neophobia in humans by exposure to novel foods. Appetite 19, 105–120.

Rozin, P., 1990. The importance of social factors in understanding the acquisition of food habits. In: Capaldi, E.D., Powley, T.L. (Eds.), Taste, Experience, and Feeding. American Psychological Association, Washington, DC, US, pp. 255–269.

Sokolow, H., 1988. Qualitative methods for language development. In: Applied Sensory Analysis of Foods, vol. 1. CRC Press, Boca Ratón, Florida.

Singh, S., 2006. Cultural differences in, and influences on, consumers' propensity to adopt innovations. International Marketing Review 23 (2), 173–191.

Stevens, G.A., Burley, J., 1997. 3,000 raw ideas = 1 commercial success!. Research Technology Management 40 (3), 16–27.

Sulmont-Rossé, C., Issanchou, S., Enderli, G., Verbeke, W., Vanhonacker, F., Contel, M., Scalvedi, M.L., Żakowska-Biemans, S., Sajdakowska, M., Guerrero, L., Guàrdia, M.D., Granli, B.S., Hersleth, M., 2007. Which innovations do consumers accept in traditional foods? Application of a dual sorting test. In: 7th Pangborn Sensory Science Symposium, Minneapolis, USA, August 12–16.

Trichopoulou, A., Soukara, S., Vasilopoulou, E., 2007. Traditional foods: a science and society perspective. Trends in Food Science and Technology 18, 420–427.

Truefood, 2009. Traditional United Europe Food. Website: http://www.truefood.eu/ (Last accessed on September 2015).

Vanhonacker, F., Kühne, B., Gellynck, X., Guerrero, L., Hersleth, M., Verbeke, W., 2013. Innovations in traditional foods: impact on perceived traditional character and consumer acceptance. Food Research International 54 (2), 1828–1835.

Weatherell, C., Tregear, A., Allinson, J., 2003. In search of the concerned consumer: UK public perceptions of food, farming and buying local. Journal of Rural Studies 19, 233–244.

CHAPTER

CONSUMER DRIVEN AND CONSUMER PERCEPTIBLE FOOD INNOVATION

6

H. Moskowitz[1], K. Li[1], H. Bolini[2], B. Batalvi[3]

[1]*Mind Genomics Associates, White Plains, NY, United States;* [2]*University of Campinas, San Paolo, Brazil;* [3]*SB&B Research, Toronto, ON, Canada*

6.1 INTRODUCTION

In these early days of the 21st century we hear the buzzword *innovation* again and again, in different situations, for business, for society, and even for people and personality. New is good, innovative is better, disruptive is the best. They are all easy words to say and write—new, innovative, and disruptive—but just what do these mean in the context of the food world?

There are organizations devoted to innovation, such as the International Association of Innovation Professionals, headquartered in Texas, of which author Howard Moskowitz is a board member. The membership of this organization, and indeed other organizations similar to it, comprises individuals in service businesses, individuals in manufacturing, and only occasionally, individuals in one or another way related to the world of food.

Our discussion leads naturally to the question of what then is innovation in the food industry, at least from the point of view of the consumer. Some might say that the innovation is in new ingredients. Go to any food conference where industrial suppliers exhibit and one will be inundated with new ingredients, whether these be substitutes for today's "bad boy ingredients" like sugar, fat, and salt, to quote author Michael Moss (2013). Or the innovation might be products that have new flavors to capture today's taste trends. Companies like Innova follow the product world, reporting on such trends. Other pundits and ingredient suppliers do as well. Or the innovation might be new forms of a product, whether forms designed to work in new situations where the product is consumed such as soup with containers held in one hand, or forms and packaging designed for new environments or new situations.

What then do we sense from the foregoing? What is the nature of a lot of the innovation in the food industry? The sense is that the innovative product presents the consumer with a different flavor, a different form, a different package, or a different set of ingredients to create cleaner labels that do not frighten consumers. An unhappy consequence of relative irrelevance of profound innovation in the food and beverage industry is that this innovation is greeted with a lot less celebration than the innovation in the computer industry, for example. A new snack chip is a new snack chip. A new computer chip, in contrast, could lead to fortune for the inventors, patent holders, and a leap forward leading to a new generation of machines.

Innovation Strategies in the Food Industry. http://dx.doi.org/10.1016/B978-0-12-803751-5.00006-4

6.1.1 A SHORT HISTORY OF CONSUMER RESEARCH AND HOW IT DRIVES OR DOESN'T DRIVE FOOD INNOVATION

A century ago, around the time of World War I, most food businesses were small. There were technological breakthroughs from these small businesses, usually new ingredients or new ways of preparing and storing food. Food was a staple, people had to eat, and there was little competitive pressure to sell. People would buy what was available. Fortunately for the United States, food was in abundance. A nation blessed with peace in its own territory and with rich croplands and abundant animals could afford to feed people who had the money to spend. There was competition, of course, from companies in the same space, such as biscuit companies, meat companies, and so forth, but the rule of thumb was that the demand either outweighed supply or was at least equal to supply.

In Europe the story was different. Ravaged by war, the continent suffered bouts of famine. The citizens of these war-ravaged countries grabbed the food that they could, not knowing when food would become available again. Companies arose after World War I to satisfy the consumer demand, a demand that had exploded and was far greater than the supply.

What does the foregoing mean for food? It means that the food industry found itself in a seller's market where whatever it produced had a good chance of being snatched up as long as it was reasonably edible. It meant that for all intents and purposes, once the company's product developers discovered what people liked and created the product, there was no push to expand the range of offerings. The trade, the supermarkets and small stores, had limited facings, or shelf space for individual varieties of the same product such as different flavors of tea, because in truth, there was no need to have more. What was offered by the trade was bought, generally calmly, but often with a sense of "must have just in case."

By the 1930s and certainly by the 1940s the impact of so-called consumer acceptance was already emerging. The emergence came from the impetus of the United States, specifically from the US Army Quartermaster Corps, responsible for feeding soldiers. We are not talking here of the military food rations, but rather the efforts made by the US Army to ensure that product quality, including acceptance, was up to specifications. The US Army institutes tests to measure acceptance, so that vendors from the food industry could not win choice army contracts for quality food but then supply substandard food. Such bait-and-switch tactics are common when large-scale contracts are awarded and little oversight is paid to the quality of the delivered goods. The history of military studies in food acceptance is well summarized by Meiselman and Schutz (2003).

The previous paragraphs, however, talk only about ensuring quality. There is nothing about innovation. For innovation we must move forward into the 1950s, into an era of previously unknown prosperity, when people had more money, and happily the industry was sufficiently mature to cater to people's tastes. The 1950s witnessed an evolving economy where choice was standard, rather than choice being a luxury available only to the very rich. In supermarkets and in the local grocery stores, companies were beginning to compete in earnest. There was not yet the proliferation currently seen, but there was choice. People could choose different cereals, different coffees, and different beverages, rather than being forced to scramble to purchase that which the store could make available at the time. Brands were being featured. Indeed today, 60 years later or so, many who were children in those halcyon days can recite the brand jingles from memory, often with a smile on their face when they think of how much less complicated life was, and how easily brands were able to win over loyal customers.

6.1.2 THE HISTORY: CHANGING FROM SELLING WHAT IS AVAILABLE TO ANSWERING CONSUMER DEMAND

By the 1960s the economies of the Western world had recovered to a great degree from the ravages of World War II. Companies whose production had been snatched up by the military found themselves now competing for the consumer's wallet. Modern miracles of technology made housework easier and afforded the average family an opportunity to travel in their own car. And of course, technology created the modern kitchen, the modern selection of food, the advertising of the "beautiful, average, family" sitting at meals together, the youngsters eating breakfast, and a generally smiling feeling that life was good, that all was wonderful.

This magic world produced more choice in foods. The middle class, no longer denied material goods because of the operative adage "one cannot have guns AND butter," began to demand better, tastier foods. The sedentary life was still creeping up, but in the early and mid-1950s there was enough opportunity to exercise that obesity was not yet a problem. Sugar was a problem; people who wanted beautiful figures attributed the lack of a beautiful figure to an abundance of sugar in the diet, especially in carbonated beverages.

The desire to reduce sugar was among the first of the clearly enunciated consumer demands leading to innovation. The sweet taste was a prime interest to the food world, eventuating in research spanning a century or more. Georg Cohn's monumental 1914 volume in German, *Die Organischen Geschmackstoffe*, detailed the taste qualities of many pure chemicals. Sugar substitutes had been on the market for many years, beginning with saccharin. But it was the focus on weight, on diet, on looking good—the by-products of a consumer-driven economy—that led to innovative patents on artificial sweeteners like the 10/1 ratio of cyclamate to saccharin. Saccharin and cyclamate were not alone. The innovative streak drove interest in discovering new and better-tasting sugar substitutes, such as aspartame, sucralose, and finally today's darling, stevioside.

Can these sweeteners really be called innovations? At some level they are merely substitutions of one ingredient for another, a caloric sweetener replaced by a noncaloric sweetener. The innovation was there, based on the discovery of the chemical. At the same time new products were being launched, for the most part calorie-reduced versions of regular products. Which was the innovation—the ingredient or the product incorporating the ingredient?

During this period the introduction of the microwave spurred innovation in the food industry. It is not the case that consumers demanded ever-higher performance from the microwave, or that microwave manufacturers promoted their increasing levels of performance in the same way as did and do manufacturers of consumer electronics, such as tablets, computers, players, and televisions. Consumers do not react to such technology-based claims when it comes to food and drink. Rather, the impetus of the microwave to innovation was a new way to prepare foods for consumption, safely, quickly, easily, and inexpensively. It was easier to pop a tray of food into the microwave than to prepare an oven, then heat and make a TV dinner in the oven. So the innovation on the heels of microwaves comprises innovations based on a new way to do the same thing, namely prepare good meals.

6.2 PSYCHOPHYSICAL THINKING: A MAJOR FOUNDATION FOR CONSUMER-DRIVEN INNOVATION

It is only in retrospect that we see the trees that grow from little acorns. Innovation is the same way. In the 1960s the "fad," the "method du jour" among sensory professionals in R&D was descriptive analysis—methodically, accurately, and reproducibly describing the sensory characteristics of a product (Murray et al., 2001). The activity was followed by dozens of companies, whose staff had first been instructed by the sensory professionals at consulting company Arthur D. Little, Inc. in Acorn Park, Cambridge, MA. Led by pioneers such as Stanley Cairncross, the sensory professionals at Arthur D. Little, Inc. promoted the expertise in the so-called "flavor profile" (Caul, 1957). The flavor profile would be followed in turn by more quantitative methods such as Tragon Corporation's Quantitative Descriptive Analysis (QDA) (Stone et al., 1974), and equally popular Spectrum methods (Munoz et al., 1992). Modifications of the flavor profile, to be sure, the QDA and Spectrum methods also created an "expertise" in description of sensory perceptions. But the popularity of these methods to describe products stopped the opportunity for innovation from the sensory professional, at least for a few years. The money, corporate resources on the one hand, time on the other, was devoted to ever more sophisticated methods to describe perceptions, to pages and pages of "ballots" on which every attribute was noted, quantified, and subjected to statistical tests, as well as graphed for presentation, or as would happen, simply for storage.

In the end, however, descriptive analysis would prove to be a costly blind alley for innovation in particular, and for forward looking sensory analysis in general. Descriptive analysis focused on process, the development of a trained panel, the execution of the tests, and the proper environment, without however having any vision of what to do with the data. Indeed, in the words of one vice president of R&D of a major beverage company, "I have warehouses full of descriptive data. Neither my product developers nor I have any idea how to convert these descriptions to products."

It was not the flavor profile and descriptive analysis that would lead the charge for innovation, but a wholly different point of view, psychophysical scaling, and the search for relations between physical stimuli and subjective responses. Psychophysics searches for these functional relations. In the early days, the 1940s and 1950s, it sufficed for many psychophysicists to uncover the relation between physical concentration of taste material, for example, and subjective taste intensity. There is an entire literature on these so-called psychophysical functions, linear relations in log–log coordinates that summarize how perceived subjective intensity grows with known increases in physical intensity. A summary equation describing this relation is the power function: $S = kI^n$, where S is the subjective magnitude, I is the physical magnitude, n is the exponent characteristic of the stimulus continuum, and k is multiplicative factor, usually related to the choice of numbers by the panelist (see Stevens, 1975).

By itself the psychophysical approach to studying the chemical senses did not and could not ensure innovation. What the psychophysical approach did, however, was turn the attention toward relations among variables, allowing us to avoid "opportunities" that would prove to be sterile blind alleys, namely describing the test stimulus using adjectives, or testing differences using inferential statistics. By using experiments to create and understand relations between the physical stimuli under the product developer's control and consumer responses, the psychophysical approach turned the attention from describing to engineering, from being a passive secretary recording what happened to a science of trying to find "laws," relations that ultimately could be used to create "the new."

As we write these observations in mid-2015, almost 45 years after the initial volley of shots were fired over the bow of sensory analysis, the development of psychophysics as an integral part of food innovation seems so very obvious. How could one overlook the importance of knowing the relation between physical stimuli and subject responses? Models, equations, first in one variable, then in many variables, first describing sensory impression, then describing liking and even images, would turn out to be the tools by which a great deal of innovation would develop.

6.3 APPLYING PSYCHOPHYSICAL THINKING IN THE EARLY DAYS: STUDIES OF TASTE MIXTURES

Psychophysics did not enter the world of food and food innovation fully developed as legend had Venus emerging from the head of Zeus completely formed. In the 1960s the world of psychophysics was to be found in the laboratories of psychology scattered around the world, from Harvard University to Brown University, Columbia University, and in the University of Stockholm, to name the major centers of the 1960s. The work done by psychophysicists, their search for lawful relations between physical stimuli and subjective sensory responses, was to be found primarily in the literature of experimental psychology, rarely appearing in the literature of food science, except for the descriptive analysis of a product, which found its way again and again into the food technology literature.

The insularity of psychophysics, its separation from the world of food, and its aspect of being an "ivory tower" of research ended, however, in the late 1960s, when at the US Army Natick Laboratories in Massachusetts, 17 miles west of Harvard University and the Massachusetts Institute of Technology, the late Dr Harry Jacobs was to put together a team of three young psychophysicists (at least then!), Drs Linda Bartoshuk, Herbert Meiselman, and Howard Moskowitz. Jacobs' goal was to change the way the US Army worked with food and menus, moving beyond simple food science and technology to incorporating subjective perceptions.

The key work emerging from Natick was the incorporation of sensory and hedonic (liking) ratings into studies of food, not so much simply to say that people prefer Sample A to Sample B but to see how liking changed as a product was systematically varied. The relevance, indeed criticality, of using psychophysical thinking to create a mathematical relation predicting "liking" from changes in the physical substrate seems so obvious now. Then, in the late 1960s and early 1970s, the notion was novel, rejected by many as impossible and lacking any real meaning for application.

6.4 BEYOND SIMPLE PSYCHOPHYSICS TO MIXTURE PSYCHOPHYSICS: THE JUMP TOWARD INNOVATION

The early work focused either on one variable or on the effects of one variable on another, as the variable being studied drove a sensory response. The early work was done in the spirit of the scientific method, following the belief that the way to knowledge is to isolate and study. The later worldview in psychophysics was to create models of mixtures to determine how two or more variables combine to produce a perception, either of sensory magnitude or of liking. The change in focus is subtle. The former, traditional, single-variable approach is analytic. The latter approach, studying mixtures created according to experimental design, is synthetic.

The simple psychophysics studies, relating physical stimuli to sensory or hedonic responses, present to us just the first step. Had the world of food just stopped there we would not be writing this chapter today, 50 years later. Rather, we would be parroting some of the findings, such as the rate of growth of sweetness with the concentration of sugar or saccharin, or perhaps the effect of a thickening agent such as sodium carboxymethyl cellulose on the perceived sweetness of sweetener solutions. In other words, we would have been limited to learning the effects of one variable on perception or the effect of one variable on another, as the latter drives sensory perception.

Fortunately, during the early part of the 1970s there emerged an interest in the sensory response to mixtures, and an interest in mathematical models to describe the sensory characteristics of mixtures of tastants, mixtures of odorants, and then daringly the mixtures of tastants and odorants. Rather than simply looking at the effects of one variable on another, interest subtly matured to a focus on two or more variables combined to produce a percept. The change in focus is subtle and bears repeating.

In the spirit of the new approach, driven by experimental design, a number of different studies lie at the foundation of innovation. These studies began with mixtures of sweeteners that were analyzed to predict the sweetness of the combination (Moskowitz, 1974). Such studies combining different concentrations of the sweeteners reported both on the perceived sweetness of mixtures and the acceptance of the same mixtures. The early studies on water solutions were done in the spirit of traditional psychophysics, looking for a mathematical model. There were attempts to apply economic thinking to the results, eventuating in a seminal paper in 1972, *Economic applications of sweetness scales* (Moskowitz and Wehrly, 1972), merging together dollars (economics) with sweetness and sweetener mixtures (psychophysics) to create one of the first examples of the now popular field of "behavioral economics."

The real shift from pure science to application and innovation came when the same thinking, mixtures of sweeteners, was applied to three-way mixtures of aspartame, cyclamate, and saccharin in cola (Moskowitz et al., 1979). The study was funded by Fermco Biochemics, later to be acquired by Searle, and renamed Searle Biochemics, and later to be spun off as Merisant. Fermco had a patent on a new high-potency sweetener, aspartame, and wanted to create "magic" combinations of three sweeteners with high synergy. Psychophysics would pave the way through studies of three-component, ie, ternary, mixtures. Respondents tasting cola, bottled in cooperation with RC Cola, courtesy of R&D Director Martha Jones, rated sweetness and liking. The results generated equations that predicted liking as well as sweetness. The innovation there was both methodological and substantive. The methodological innovation was using experimental design to discover new business-relevant combinations. The substantive innovation was the actual discoveries emerging about sugar sweetness.

6.5 INNOVATION THROUGH EXPERIMENTAL DESIGN, MULTIPLE PRODUCT TESTING, AND SENSORY SEGMENTATION: PICKLES, SAUCES, AND ORANGE JUICE

In the 1960s the prevailing wisdom of market researchers and sensory analysts, the two guardian groups for the sensory and consumer responses to products, dictated a number of "best practices." These best practices were that a consumer absolutely could not test more than three or four products without losing sensitivity. There was no data to support these claims, and indeed both prior research by psychophysics and subsequent consumer research by industry were to prove that these best practices were wrong, despite the strong protestations to the contrary then, in the 1970s, and even today, as of this writing in 2015.

What these erroneous best practices did, however, was slow down the potential for consumers to drive product innovation in the world of food. It was quite obvious how. Consumers were believed to be able to test only two products at most, either by rating the two products or by expressing a preference. With that limited amount of information from consumers it was difficult to construct even a simple curve relating ingredient level to liking, and it was absolutely impossible to use consumer experience with actual products to identify new taste segments.

In 1975 author Moskowitz entered the commercial world of market research, proposing that one could use consumers to test 10+ products. The testing would have to be done with consumer panelists who were recruited to participate for several hours, but once those logistical issues were solved the science itself was simple. Indeed, the approach simply mirrored the previous psychophysical experiments wherein the respondent tested many different sweetener mixtures. Instead of sweeteners, there might be different fragrances for an antiperspirant, different formulations for a cola beverage, different prototypes for a cereal, and so forth.

The studies proved successful, both in their implementation in "the field," and in the quality of the data that they generated. For example, despite the conventional wisdom that consumers could not taste more than two to three products without losing their ability to discriminate, that aforementioned research "shibboleth," the data proved again and again that the consumers could validly discriminate among products. All that needed to be done was to have a set of systematically varied levels of an ingredient, and the data showed that the expected relation occurred. Change the concentration of the chemical, holding everything else the same, present the test stimuli to respondents in random order, and the results were clear. The ratings were monotonic, despite the randomization, the many stimuli, and the many attributes on which the respondent rated the stimuli (Moskowitz et al., 1980). The validation emerged from experiments again and again, at least in those cases where the research "client" put in different levels of ingredients.

Testing many stimuli generated two other benefits, the ability to discover sensory preference segments and the ability to identify new opportunities, so-called white spaces. Both of these would prove to be valuable for food innovation (Moskowitz, 1985, 1994).

6.6 INNOVATION BY DISCOVERING AND EXPLOITING SENSORY PREFERENCE SEGMENTS

We all know that people differ from each other in what they like. There is an old Latin proverb to that effect, which rendered in English read 'of taste there is no dispute'; simple studies of taste preferences reveal these individual differences. As the concentration of sugar increases, we see different patterns. Some individuals show increasing liking with increasing sugar level or perceived sweetness, up to an optimal level, at which point any increase produces a decrease in liking. Some individuals show an optimal point at a low sweetness level, and others show an optimal point at a high sweetness level.

Such individual differences exist in water solutions, of course (Pangborn, 1970). Water solutions are model solutions, of interest to basic science, and only of slight interest to those in the food business who must create products. In the late 1970s and early 1980s, it became possible to demonstrate these individual differences in foods. The process turned out to be rather simple:

1. Select a variety of products of the same type, eg, cereals.
2. Make sure that there are at least five to six different cereals of the same type to be tested by each respondent. There might be many more cereals in the study than those tested by any one respondent, but have the respondent test as many as possible. It is in this type of study that it becomes important to recruit the respondent to participate for an extended time, eg, several hours. By paying the respondent, one can ensure that the respondent remains interested in the task and has set aside the requisite time for the evaluations.
3. Instruct the respondent to rate these different cereals on a variety of sensory attributes, encompassing appearance, aroma, taste/flavor, texture, aftertaste, and swallowing. These sensory attributes can be scaled with a simple 1–9 scale or a longer 0–100 scale, or by something more precise, such as magnitude estimation that generates a so-called ratio scale. Respondents have no problem rating the perceived intensity of a number of sensory characteristics on a simple scale. There is no need for extensive training, although it might be helpful to explain what the attribute means, just in case there is any confusion.
4. Instruct the respondents to rate these different cereals on a variety of liking scaling, such as overall liking, liking of appearance, liking of aroma, liking of taste, and liking of texture/mouthfeel. Liking is not the same as sensory intensity. Liking is one's feeling, positive or negative, about what is perceived. The same type of scale can be used as well (eg, 0=hate, 100=love).
5. Instruct the respondent to rate these different cereals on image, such as "caloric," for child versus for adult, and so forth.
6. The exercise generates a simple matrix of rating by attribute.
7. For each product tested compute the average rating on each sensory attribute, across all the respondents who tested the product. Also, compute the average liking rating across all the respondents who tested the product.
8. Fit a curve to the data from the total panel using ordinary least squares. This is shown on the left panel in Fig. 6.1. *Identify the sensory level where liking reaches its optimum.* The curve shows how overall liking covaries with sensory intensity, when all data come from averages across the respondents.
9. Now we move to the individual's liking rating versus the average sensory rating. For each respondent and for each sensory attribute, relate the liking rating assigned by that panelist to the sensory rating. The sensory values are the averages across respondents.
10. Each individual generates a separate sensory-liking curve. And in turn each person generates a sensory level where liking reaches its maximum. These appear in the right panel of Fig. 6.1.
11. Factor analyze the set of optimal sensory levels, save the factor scores, and cluster the respondents.
12. The result is a small set of clusters, representing different patterns of sensory optima. These are the sensory preference segments (Moskowitz et al., 1985).

The discovery of sensory preference segments represents an innovative step in the world of food. Consumers may all describe their ideal product, such as coffee, as being, for example, "rich and robust." One group may love bitter coffee, another may love milky coffee with low flavor, whereas yet another may love coffee with many aromatics. Yet they will all describe their ideal

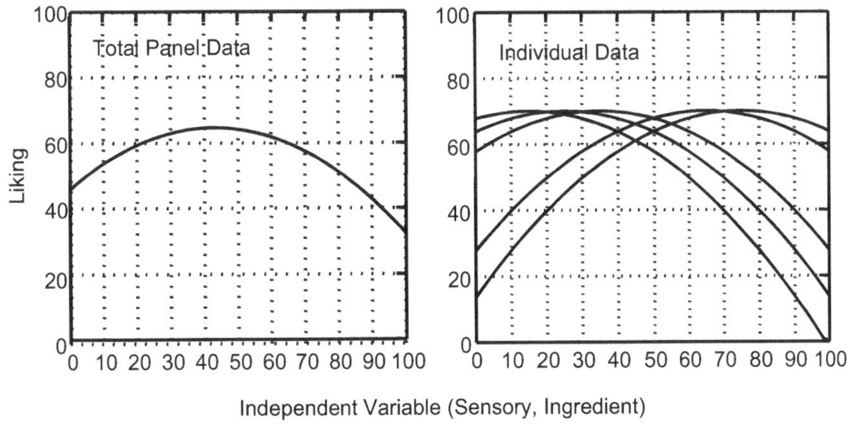

FIGURE 6.1

Representative curves showing how the typical sensory-liking relation emerges from the combination of different sensory-liking curves. The left panel shows the average curve, and the right panel shows the contributing curves. Sensory segmentation emerges from the clustering of "optimal sensory" levels for many people across different sensory attributes, when these people evaluate one type of product, eg, coffee.

coffee as "rich and robust." Through sensory preference segmentation one discovers how many "different" tongues there are in the populations, based on the pattern of preferences, and in turn the physical composition of a limited, realizable set of products to optimally satisfy and even specifically delight each segment. The sensory segmentation will reveal the real differences and allow the developer to create that new, perhaps novel, delicious product targeted for the specific sensory segment.

6.7 INNOVATION BY MODELING, REVERSE ENGINEERING AND DISCOVERING HOLES IN A PRODUCT CATEGORY

When it became obvious from experience that a respondent could test many products of the same type either in one or a few sessions held in a central-location test facility or at home, new opportunities emerged for innovation, opportunities that are being exploited now in ever-more novel ways. One of the notions was that one could "map" the different products, putting the products into a geometrical space. Mapping provides a way to visualize the relation between different stimuli, but as we will see in a moment, mapping provides a far more profound way to manipulate the data, a way that emerges from psychophysical thinking and the search for relations between variables.

Fig. 6.2 shows a schematic of 10 products mapped in a geometrical space, with the size of the circles representing the degree of liking. The approach is algorithmic, covering a range of activities starting from creating the space to finally discovering the sensory characteristics and representative "target" products. It will become clear from the exposition how the combination of psychophysical thinking, tests of many products, and statistical "modeling" have come together to create an innovative approach in the world of foods.

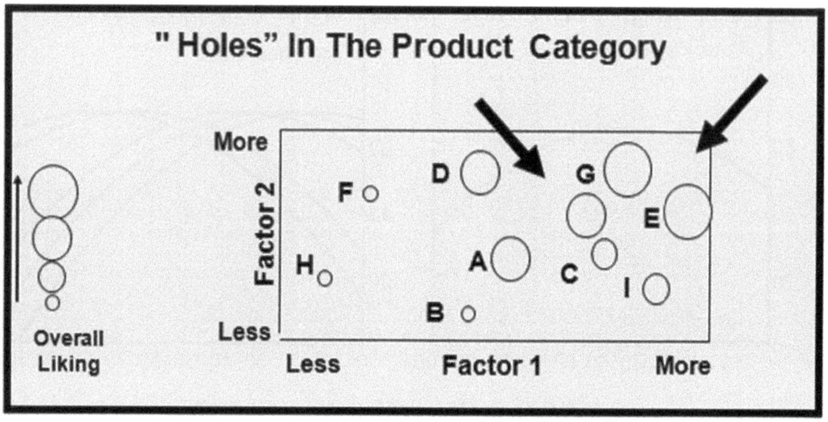

FIGURE 6.2

Two-dimensional map of 10 products in a "factor space." The factors emerge from principal components factor analysis of the sensory attributes on which the 10 products were rated. The precise location of each product corresponds to the so-called factor scores of each product on the two factors.

We follow these steps:

Part 1—Identify the sensory preference segments.

1. Lay out the data in the form of a table. The columns represent the different attributes rated by the respondents; the rows represent the different test stimuli. Keep in mind that this standard product (row) x attribute (column) is a standard, easy-to-use format, one that can be analyzed by many off-the-shelf statistical programs.
2. Using the previously discussed method for sensory segmentation, analyze the matrix to identify clusters of people with different "optimal sensory profiles." These are the sensory segments. We may end up discovering no sensory segments, that is, everyone is alike. More likely, however, we will discover two, possibly three segments, differing on the pattern of sensory attributes that they find most acceptable.
3. Keep in mind that the discovery of sensory preference segments is optional. We might well be satisfied with data from the total panel or data from key subgroups like gender, product usage, and so forth.

Part 2—Create the "basic dimensions" through principal components factor analysis.

4. Our objective in Part 2 is to locate the different products on a geometrical map, so that products "close" to each other are similar, products "far away" from each other are different.
5. We perform a principal components factor analysis on the sensory attributes only. The principal components method reduces all of the sensory attributes to a set of primary, theoretical variables. By rotating the solution using the so-called "quartimax" method, we end up with simple, clean dimensions. There are usually two to four dimensions that emerge. Fig. 6.2 shows what happens in a two-dimensional solution. In mathematics it is as easy to work in three or four dimensions as it is to work in two dimensions, but the higher dimensional "maps" are harder to visualize.

6. In Fig. 6.2 each of the 10 products has a specific pair of factor scores, locations in the two-dimensional space. These factor scores, ie, locations in the two dimensions, are not arbitrary but emerge from the principal components analysis and the quartimax rotation. Rather than locating the 10 products on a set of different sensory attributes or sensory dimensions, we locate the 10 products in a space of smaller dimensionality, where the dimensions or axes are independent of each other and then lie at 90° to each other.
7. The size of the circle in Fig. 6.2 is proportional to the level of liking. Recall that liking was not used to define the new dimensions. Only sensory attributes were used.

Part 3—Inspect the map visually for insight.

8. Often we use the map, our geometrical representation, for a visual aid to find holes, which are opportunities. It is very easy to discover holes when we work with two dimensions and just a bit harder to discover holes when we work with three dimensions. By the time we reach four dimensions we have problems if we rely on our visual inspection. Nonetheless, we can "discover" holes in four, five, or higher dimensions, but by mathematical tools, not by visual inspection. For this example, we limit our attention to two dimensions.

Part 4—Create a "product model."

9. We can apply mathematical thinking to the product and emerge with far more than merely mapping the product, although in practice the vast majority of researchers in the food industry stop with the map, happy enough to show where the products lie relative to each other. We use regression analysis to create a product model, use optimization to identify the coordinates in the two dimensional space corresponding to the "most-liked product," and introduce the notion of constrained optimization. We finish with reverse engineering, wherein we will specify a "goal profile" of the product on sensory and/or image attributes, identify the values of the coordinates (Factor 1, Factor 2) that deliver a profile estimated to be as close as possible to the goal profile, and then use the product model, ie, our equations, to estimate all the characteristics of the product.
10. We move now from creating the factors to developing the product model. The factors become our surrogate ingredients, and thus our independent variables. Of course the factors emerging from principal components factor analysis are not ingredients but rather mathematical entities, a "sort" of primaries. These factors are statistically independent of each other, independence guaranteed by the principal components factor analysis and the quartimax rotation. We might have two factors, as we see in Fig. 6.2, or perhaps three factors or even more, such as four and five factors. Each product that we tested has values on those factors, the aforementioned factor scores.
11. Through principal components factor analysis we have now created two "surrogate" ingredients for our products in Fig. 6.2, the values of which are the values of Factor 1 and the values of Factor 2, respectively. These factors were created strictly from the sensory attributes. There are other attributes as well, such as liking (overall, liking of a particular attribute such as appearance.) There are still others, image attributes. For a beverage, an example of an image attribute is "refreshing." We repeat that these liking and image attributes belong to the products but were not used to discover the two factors nor used to create the factor scores.
12. We now use ordinary least-squares regression to relate each attribute, in turn, whether sensory, liking, or image, to the factors, their squares, and cross terms.

13. For the case of two factors, 12 sensory attributes, five liking attributes, six image attributes, we would create 12+5+6 or 23 equations, one equation for each attribute. The equation is the well-known polynomial function, written as Attribute $= k_0 + k_1(\text{F1}) + k_2(\text{F2}) + k_3(\text{F1})^2 + k_4(\text{F2})^2 + k_5(\text{F1*F2})$. Estimating the 6 parameters given 10 cases, the 10 products, is easy to do. For more complex cases, say with five factors, we would have the additive constant, k_0, 5 linear terms, 5 square terms, and 10 cross terms. We might instruct the regression program to eliminate insignificant cross terms, but we would want to force in the additive constant, the 5 linear terms, and the 5 square terms, respectively. Use the product model to understand holes in the category, to optimize, and to reverse-engineer using the product model.
14. The product model, our aforementioned set of equations, relates the location in the map (values of Factor 1 and Factor 2 in our case) to the expected attribute rating. For any combination of Factor 1 and Factor 2, within the range that we have data, we can "solve" our equations. Thus we can estimate the ratings that respondents would have assigned for liking, for image attributes, and for sensory attributes, respectively.
15. Find "holes," ie, opportunities in the category (map): Holes in the category in Fig. 6.2 are empty spaces. We want to find an empty space where the boundary products achieved high liking. We can look at many different values of F1 and F2 where we "see" emptiness. The values of F1 and F2 become locations of candidate products, or more correctly possible product opportunities. For each location in this empty area we estimate the level of liking and the expected sensory and image profile. Furthermore, we can look at the different products to see which product(s) have these estimated values. These products serve as landmarks for the particular sensory attribute. It is likely that several products will serve as landmarks, each product being a landmark for a particular sensory attribute.
16. Optimize one variable subject to explicit constraints on factors and implicit constraints on sensory attributes: An optimal product in Fig. 6.2 is the location of the product expected to be most acceptable, overall, subject to specific constraints. One constraint is that the location of this optimal product must lie within the range of factor scores achieved in the actual study. We know the factor scores F1 and F1, ie, the locations, for each product. We thus know the highest and lowest values of F1, and the highest and lowest values of F2, respectively. Our optimal product must lie within those ranges and not outside the range. A second constraint is that the expected sensory profile must lie within the range tested. We know the values of the sensory attributes for each product that we tested. We therefore know the highest and lowest values for each sensory attribute. Finally, we may choose to put additional constraints onto our solution, such as ensuring that a sensory attribute lies within a tighter range than the range we tested. When all is said and done, the optimization algorithm identifies the location F1 and F2, respectively, of the product that optimizes liking (or another attribute) subject to the constraints and then prints out the expected attribute profile (liking, sensory, image). This sensory-based optimization has been discussed in detail (Moskowitz, 1998).
17. A short history of this effort, optimizing a sensory profile: In the 1980s the statistical method of "mapping," ie, representing stimuli as points in a space, became the "method du jour" in market research, based upon work in multidimensional scaling. Although the mapping of points in space, ie, products or even brand names, became popular, no one really knew what to do with the results beyond seeing where products lay relative to each other. Techniques like

mapping, starting off in market research, often migrate to sensory research. Mapping was no different. But the same question remained—what to do with the map. A review of the literature from the mid-1980s until today, mid-2015, reveals hundreds, perhaps thousands of studies, but no direction about how to use the map to design a product. The optimization method described here, and discussed 20 years ago (Moskowitz, 1994) revealed that despite the mapping, it was unclear how to make the new product. The optimization using factor scores as independent variables revealed that lack of knowledge even more clearly. One could know the "sensory profile" of the optimum point in the map, but then what? Happily, as we will see later, the mapping exercise with factors led to a mapping exercise with ingredients, and the field moved on to better applications, resulting in the development of products such as Zesty Pickle (Vlasic), Prego Spaghetti Sauce (Campbell Soup), 1892 Coffee (Maxwell House), Grovestand Orange Juice with Pulp (Tropicana), Dr. Pepper Cherry Vanilla Soda (Cadbury Schweppes), and many others.

18. Reverse Engineering: What about specifying the profile of the product to be achieved, ie, the "goal profile"? Or perhaps specifying part of the profile, and then estimating the entire profile of the product satisfying the goal? One benefit of the product model is that it can incorporate all sorts of data for a given product category. For example, for each product one can acquire the consumer ratings (sensory, liking, image, even price that one will pay), expert panel ratings (their own lexicon), and so-called "objective measures," ie, instrumental measures from test equipment, such as machine measure of texture, color, chemical composition, and so forth. One can then create the product model by applying principal components factor analysis to any set of variables, emerge with factors as we did previously, and create an expanded product model, wherein each variable, whether consumer, expert, or instrumental variable, generates its own quadratic function. Then one can specify a goal profile to be matched, with that goal profile coming from the experts, or the consumers (eg, sensory profile, image profile), or even the goal profile coming from the instrumental measures. The objective in matching is first to find the set of factors within the constraints that reproduce the goal profile "as closely" as possible. Once factor values are discovered, it is a simple task to estimate the FULL profile of that factor combination, ie, the FULL profile of the product estimated to fit the goal. We now have a way to jump across domains, predicting consumer responses from instrumental or expert responses, or even predict the full profile of attributes from a partial profile.

19. A short history of reverse engineering the product map: The mapping and reverse engineering approaches were developed in the mid-1980s as part of the effort toward combining mapping and modeling. As we noted previously, the results pointed to products but were inactionable when the goal was product design. However, Ragu, a division of Unilever, ended up using the reverse engineering for 10 years both to predict consumer acceptance from profiles of expert panels and to predict product quality of spaghetti sauces from objective physical measures. The decade of use was based upon one set of studies conducted in 1992 for the entire spaghetti sauce category. Other companies such as Borden Foods later in the 1990s used the reverse engineering approach to maintain product quality of some of their products, such as Cracker Jacks. As in the case of product optimization using the product model, experience with evaluation of disparate products in the category united through factor analysis ended up whetting the appetite of the enlightened product developer to use true experimental design, to which we now turn.

6.8 INNOVATION BY EXPERIMENTAL DESIGN COUPLED WITH SENSORY PREFERENCE SEGMENTATION

Experimental design refers to the systematic combination of variables. In the 1960s some statisticians were already recommending its use (Gordon, 1965). The time was not "ripe," however, and these voices in the wilderness were not really heard, even though they were successful in application and published in the scientific literature. Author Moskowitz also recalls an early conversation in the middle 1979s with a senior scientist, Robert Carbonell, who also mentioned that he had been using experimental design at Standard Brands since the 1950s, albeit on a sporadic basis. Moskowitz also recalls discussions with Al May of Pillsbury, again in the mid 1970's, about May's pioneering use of experimental design, and some of the reactions of management at the Pillsbury Company in Minneapolis. It appears in retrospect that the times must catch up. They eventually do for most efforts. They certainly did for innovation and experimental design.

The "magic" leading to innovation was not experimental design, per se. The notion of mixing together variables in a systematic way was recognized and often given lip service by those in corporations such as the technical director. The magic had to be something else, of business import, not just of science import. Science itself was interesting, but considered in the food industry to be secondary to sales and marketing, where the money came from, and where the stock prices and corporate success were nurtured.

The magic came in the form of new discoveries, new products emerging from experimental design, products that were considered innovative, or in groundbreaking ways to understand consumers and market to them. The best way to understand the value of experimental design is to talk about three studies, the combined effect of which was to generate billions of dollars and bring experimental design squarely into the world of food product innovation. These will be recounted as stories based upon projects done by author Moskowitz when he ran Moskowitz Jacobs, Inc. (MJI), a market research company, which sold the idea of experimental design, did the studies, and discovered the new product.

6.8.1 EXPERIENCE #1: CREATING "ZESTY" FOR VLASIC

The first experience with innovation using experimental design came from the development of what now is called Zesty, pickle number four on the scale of intensity. As in so many other cases where food innovation emerged, the case involved four individuals with foresight and courage, Dr Paul Palnitkar, Director of R&D, and then brand manager Carey Monaghan and his marketing colleagues James Dorsch and William Shaw. Carey was in charge of a number of Vlasic pickle products. As was the standard practice, when a new competitor product came into the market the brand manager would order a competitive test to see how well the product scored. In this case the product was a brand manufactured by Claussen. The study revealed that Claussen, then owned by General Foods Corporation, outperformed the corresponding Vlasic product. That discovery and the realization that in fact the pickle shelf was crowded with products that no one understood led Vlasic to commission a study with company MJI, to identify how to improve the product(s).

The foregoing is the antecedent. What happened next is part of both marketing and product development histories. Vlasic R&D, amenable to guidance by marketing and top management, created 45 different pickle brines, representing different combinations of ingredients such as salt, spice, vinegar,

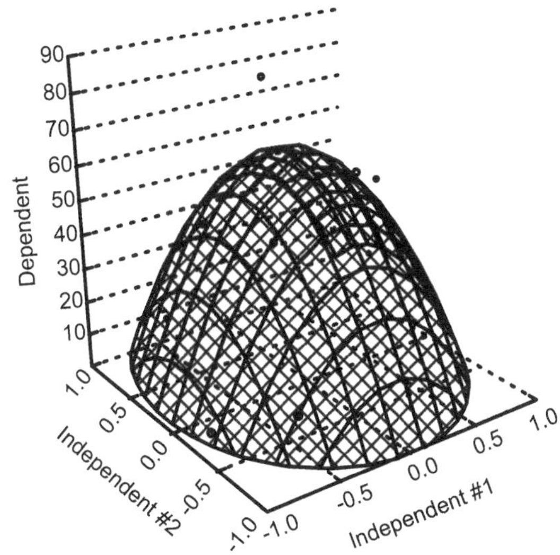

FIGURE 6.3

A response-surface profile. The optimal formulation is the combination of ingredients where liking is highest.

garlic, and so forth. These brines represented different sensory profiles. More importantly, the sensory properties of the brines, and later the pickles in the brine, could be related to the ingredients and their interactions.

The data generated a "response surface" of the form shown in Fig. 6.3. However, that was the average from the several hundred respondents. By dividing people by their preferences according to the approach pictured in Fig. 6.2, ie, the sensory level that these people liked the most, it soon became clear that there were three populations of "taste for pickles." Analysis of the surface from the different sensory segments suggested three different surfaces, each of a different shape, and with a different optimal formulation generating the highest liking.

We had discovered three groups of consumers: those who liked mild pickles, those who liked moderate-intensity flavors, and those who liked very strong, very salty, peppery, tart, almost "hot" pickles. The discovery changed the direction of product design for Vlasic, which emerged with a line of pickles varying in taste intensity. The actual intensity was communicated by the schematic picture of a thermometer, with the "temperature" shown as the "heat level" on the thermometer. The most intense pickle, Zesty, became one of the biggest sellers of pickles in history, for the simple reason that up to the early 1980s most sellers of jarred pickles for the supermarket trade were afraid of alienating customers and opted for a milder, less impactful pickle that they thought would not alienate the eater. The pickle industry had missed a major opportunity, now being recaptured.

Where then in the foregoing story is the innovation? It is in the organizing principle of sensory segmentation, that different groups of consumers have different preferences, and that these preferences are lawful and can be captured by science. And it is in the use of one fairly large experimental design to create a range of products, using that one experimental design to discover the different sensory segments and right away optimize the product design for each sensory segment.

6.8.2 EXPERIENCE #2: CREATING THREE PREGO SAUCES

Two years after the very positive marketplace performance of Zesty it was time for another major effort, this time with pasta sauce. As with Zesty, the innovative effort did not come from a programmatic institution of innovation using experimental design but rather from the bold cooperation of two remarkable people, the late Kathleen MacDonald, then brand manager at Prego, and her market research manager, Monica Wood. Monica fought the internal powers of sensory analysis who wanted to "test" and "triangulate" their way to a winning product. Monica suggested that instead of one-at-a-time testing that was suggested, or a beauty contest to select a winner, a more productive and ultimately profitable way to accomplish the objectives would be by experimental design. Both Kathleen and Monica were employees of Campbell Soup Co., which at the time owned Vlasic pickles. The experience with Vlasic helped pave the way. Distance, however, slowed down the process for 2 years. Vlasic was located in an entirely different state, Michigan, about 500 miles away from Camden, New Jersey, home of Campbell Soup. As in many corporations, there was only modest contact between Vlasic and Campbell, its mother company.

It was Monica Wood who brought the opportunity to Kathleen, with the vision that it might be possible to leapfrog the then market leader, Ragu, owned by Unilever. The effort would entail improving the current product, perhaps dramatically. What was to be discovered, however, was that there was room in the marketplace for three Prego pasta sauces, the traditional to compete with Ragu, the chunky, and the Prego spicier product. The rest is marketing history. Once again experimental design, the systematic exploration of alternative formulations with six different ingredients, not just one, was the tool by which new taste preference segments were identified through science and, at the same time, optimal products developed.

The important point for the Prego story is that it was picked up by best-selling author Malcolm Gladwell as a powerful example of human food preferences and the plurality of perfection (Gladwell, 2009). Gladwell's presentation on one of the early TED Talks in 2004 has been seen by almost 5 million people at the time of this writing, remaining a popular TED Talk (Gladwell, 2004). The TED talks, combinations of entertaining and education, as well as technology, have come to occupy a premier place in the world as a forum by which new ideas are presented by their developers or in the case of Gladwell, by their advocates.

6.8.3 EXPERIENCE #3: TROPICANA'S GROVESTAND ORANGE JUICE

Our final case history comes from the development of a (then) new orange juice with pulp, Tropicana Grovestand. The original issue was to increase consumption of orange juice by identifying different and new products. Rather than guessing, R&D Vice President Alan Boles authorized the use of experimental design, to create different prototypes varying in juice composition and pulp. Up to that time pulp was considered a waste product, to be fed to animals. The study, run in the United States as well as in France, the United Kingdom, and Germany, revealed two major segments of consumers. These segments differed in their response to pulp. There were the Low/No Pulp and the Medium/High Pulp. Table 6.1 shows the distribution of these segments in Europe. Grovestand was formulated to appeal to Segment 2, the medium/high pulp. Within 3 years the product line grew the Tropicana business from US$500 million to $1 billion. Tropicana was later bought by the Pepsi Cola Company to become part of its portfolio. The Grovestand experience suggests that optimization coupled with segmentation can drive a worldwide innovation program, as well as produce profound new, relevant knowledge about consumer responses to a product (Moskowitz and Krieger, 1998).

Table 6.1 The Distribution of Sensory Segments for Orange Juice

	Seg 1 - No or Low Pulp	Seg 2 - Medium & High Pulp
Total (across the 3 countries)	38%	62%
Country 1 = France	25%	75%
Country 2 = United Kingdom	48%	52%
Country 3 = Germany	41%	59%

6.9 INNOVATION USING EXPERIMENTAL DESIGN OF IDEAS TO CREATE NEW PRODUCTS

The previous exposition described work with products. An equally, perhaps potentially more fertile area, is innovation by the recombination of ideas. This entire field, often known by the name *conjoint measurement* (Green and Srinivasan, 1990), has been expanded by the authors to "mind genomics" (Moskowitz et al., 2005a,b). Simply put, mind genomics is the study of how we analyze and recombine ideas. The analysis and the recombination is the method. The result is innovation.

Before explicating the process, it is worthwhile recounting a short history of how this science was used to create a new product category, motorized toothbrushes, from the senior author's work with the Oral B division of the Gillette Company, in 1992. The objective was to mix and match ideas into combinations, in order to determine which combinations were most acceptable to consumers. The underlying approach created different silos or groups of individual ideas. All ideas were simple phrases describing an aspect of a product, the experience of the product, or the benefit to be gained. The benefit could be an actual physically measurable effect or an emotional benefit, a feeling. The elements were taken from two groups, tooth cleaning and mechanical cleaning and polishing by machines. The result was the synthesis of a new-to-the-world product category, the motorized toothbrush for consumer use. The full story appears in a book on consumer evaluation of personal products (Moskowitz, 1995).

Now to the specifics. Mind genomics is a new, data-driven approach to understanding the world of the everyday. The organizing principle is that knowledge is developed in a pointillist style, from the "bottom up," ie, from the intimate, profound, comprehensive understanding of specific topics, be these situations (eg, eating a breakfast), topics (eg, ethics and practice of digital piracy), products (eg, a yogurt), and so forth.

Mind genomics is best conceived as a way to understand what aspects of the topic are important to people, how people differ in the ways they respond to these aspects of the topic, and how one goes about identifying these different viewpoints of a specific topic. The applications of mind genomics range from designing the conceptual blueprint of a new product by combining elements to creating the best messages to sell the product. Beyond design and messaging come the discovery of new-to-the-world mind-set segments, people with different desires, and targeted selling by identifying specifically who in the population belongs to each mind-set.

The objectives of mind genomics are:

1. to understand reality from the "ground up," topic by topic
2. from the point of view of the person's experience

3. in a way that leads to quantification about what is important, for each person
4. so that general rules about each topic of the everyday may emerge from the empirical study
5. thus leading to the science of that topic, a science founded in human experience and perception
6. with the goal of creating an archival science, empirical, and functionally useful

The tools of mind genomics are simple, comprising:

1. the raw materials of the topic (silos or topic area, elements or specifics)
2. experimental design to create test stimuli, systematized combinations of elements
3. rating questions, which allow respondents to communicate their reactions to the test stimuli
4. the respondents, people whose "minds" are to be understood through the science
5. technology, ie, computers, to present the test stimuli, acquire the results, and do the statistics

The work products of mind genomics are:

1. a database for each topic area showing what "elements" or ideas drive interest, are valued highly, or affect one's emotions
2. a method to estimate how new-to-the-world combinations of ideas (innovations) might score
3. estimates of synergies or suppressions between ideas, guiding what to combine to achieve a breakthrough, and what to avoid combining to achieve a combination that is a "dud"
4. mind-set segments (viewpoints) for the topic area, based upon what elements are important
5. viewpoint identifying tool, to determine the mind-set segment membership of an individual; the viewpoint identifier may be personal, requiring the respondent to answer a few questions, or digital designed to assign a person to a mind-set segment based upon database information
6. the ability to sequence a person's mind on a large array of topic areas, using an array of viewpoint identifiers (personal and digital/database)

6.10 INNOVATION USING MIND-SET SEGMENTATION; TARGETED 1:1 DESIGN AND 1:1 MESSAGING

The development of mind genomics creates an opportunity for the food and beverage industry to innovate, not so much by creating products as by discovering specific messages for a person to encourage that person to eat healthfully, for example, or to eat a specific product. The big innovation here is targeted messaging, and with it the possibility of targeted product development. The order just stated now is the conventional approach, where concept precedes product development.

How does such targeted messaging work? The answer is by first doing the specific "microscience" for the product, ie, discovering for a reasonably large group of people the messages that work, that convince, and the messages that do not convince. The discovery is straightforward with the mind genomics technology, comprising experimentally designed combinations of messages for a specific food, testing on the Internet, and for each respondent deconstructing the responses into the contribution of each element. When the elements in this study comprise features of products, the developer is afforded guidance by individuals sharing the same preference patterns.

At the same time, we also obtain information about the respondent from commercial data sources, such as Experian. The data are public and can be purchased.

For each of the 300–500 respondents who participate, we now have three key pieces of information:

1. The strength of the reaction to each element, estimated using ordinary least-squares, which deconstructs the response to 48 experimentally varied vignettes, combinations of three to four messages. The deconstruction reveals the contribution of each element.
2. The mind-set of the respondent for the particular topic being studied. Typically, two to three mind-sets emerge for each topic area. Table 6.2 shows an example of a relatively recent study (2013) to identify new opportunities for pasta sauce in this increasingly competitive category. The two segments are the sensory seekers (almost 2/3 of the respondents) and package attenders (1/3 of the respondents).
3. The purchased, publicly available information about the respondent from Experian.

6.10.1 TARGETED DESIGN

When the elements of the study comprise features of the product, the researcher can either use the data from the total panel comprising all respondents or the data from each of the mind-set segments. The segments differ radically in the pattern of elements that they like the most. Consequently, the targeted design ends up with a set of products, each product fine-tuned to the preference pattern of the specific segment.

Table 6.2 Results From the 2013 Study on the Features of a Pasta Sauce That Excite Respondents, and Which They Say They Will Buy

		Tot	Seg 1	Seg 2
Base size		304	198	106
Additive constant		10	4	22
Segment 1 - focus on the sensory aspect of the product				
E2	Available in creamy texture varieties (cheese, ranch, alfredo, etc.)	**13**	**20**	−1
E4	Available in tomato-based varieties (Italian marinara, basil, garlic, pizza, etc.)	**13**	**19**	1
E5	Available in regional flavors (Southwest, Cajun, New England, Barbecue, etc.)	**12**	**19**	0
E3	Available in spicy/hot varieties (Mexican chili/salsa, Indian chutney/curry/cardamom, etc.)	9	**18**	−8
E1	Available in sweet and sour/savory varieties (Oriental/Chinese, Japanese, ginger, wasabi, soy, miso, etc.)	**11**	**17**	−2
A3	Oven-baked...tastes homemade and smells great coming out from the oven	**9**	**14**	0
A4	Combination sauce...mix wet and dry ingredients together for instant lively taste and aroma	**9**	**13**	3
C6	Good source of antioxidants...may reduce oxidative stress and the risk of chronic disease	**12**	**12**	**11**

Continued

Table 6.2 Results From the 2013 Study on the Features of a Pasta Sauce That Excite Respondents, and Which They Say They Will Buy—cont'd

		Tot	Seg 1	Seg 2
A6	Add your own extra ingredients—anything you like, whenever you want it	6	**11**	−4
F5	100% organic	9	**11**	5
C4	No added sugars…may help control diabetes	8	**11**	3
F6	Gluten-free	6	**10**	−1
A1	Cook-in sauce…complements the taste and texture of your ingredients	8	**10**	5
A2	Pour-over sauce…adds a finishing touch to create an attractive dish	8	**9**	7
F4	All natural	8	9	8
Appeals to segment 2 - eco-package oriented				
B3	Comes in a biodegradable pouch…Nothing left behind to harm the environment	9	3	**20**
B4	Aseptic (sterile) envelope keeps ingredients fresh	4	−5	**20**
B2	Plant-based bottle will disintegrate in landfill	5	1	**12**
C3	Low-sodium formula may help control high blood pressure	6	3	**10**
Lower appeal, both to segment 1 and to segment 2				
F2	Vegetarian	7	6	9
C2	Low-calorie formula may help you stick to your diet	6	6	5
D5	Refrigerate leftover preparation	6	8	2
C5	No additives/preservatives like MSG, which may cause headaches	5	8	−1
C1	Low-fat formula may help control cholesterol	3	2	5
F1	Kosher certified	3	4	2
D3	Nonrefrigerated…single-serve packets	3	6	−4
F3	Halal certified	2	4	−1
D1	Shelf stable until opened	2	4	−2
D2	Fresh chilled in store…refrigerate immediately	2	4	−3
A5	Dry sauce mix…just add to the liquid of your choice	2	6	−4
D4	Refrigerate after combining with additional ingredients…stays fresh for up to one week	2	6	−5
B6	Foil-lined paper pouch is notched for easy opening	1	−3	9
D6	Mix contents of one or more flavor packets…retains flavor for several days	1	3	−3
B1	Comes in a recyclable metal can	0	−3	6
B5	Bag in box is handy to use	0	−3	5
E6	Available in fish-based varieties (Vietnamese, Thai, Indonesian, etc.)	−1	7	−14

The Study Used the Methods of Mind Genomics to Identify the Segments.

6.10.2 THE PERSONAL VIEWPOINT IDENTIFIER

Using statistics, eg, discriminant function analysis, or technology such as machine learning, we determine which specific four to five elements from the study predict mind-set membership. This first analysis creates the questionnaire that assigns a new person to a mind-set segment, typically with correct performance 65–70% of the time. In this way, we discover the mind-set membership in a mind-set of a person in front of us, or encountered on the Internet, allowing us to give the right message or even to create a new product for people of this mind-set.

6.10.3 CREATE THE DIGITAL VIEWPOINT IDENTIFIER

Using statistics once again, go through the Experian data and identify the 15–30 pieces of information needed to properly assign a new person to a mind-set. Once this viewpoint identifier is created, we can go through very large-scale databases, scoring each person as to segment membership, and then send the person the right message.

As we see from this example, our innovation has now broken out of product design and development and into messaging. Innovation now reaches new avenues, the most important commercially today being that of communicating the right message to the right person at the right time.

6.11 INNOVATION BY CHANGING THE DEVELOPMENT PARADIGM: EMPATHY AND EXPERIMENT

Our final topic of consumer-driven innovation concerns the actual process of developing a food or beverage. For decades now product development has followed a fairly prescribed, often rigid, not particularly successful path. Those involved in innovation, working with new products, have come up with a variety of procedures to increase the odds of product success. Approaches range from the general approach, structured development paths, so-called "stage gate" and allied processes (Cooper, 1993) to actual tasks such as ethnography (Mariampolski, 2006), brainstorming (Rawlinson, 1981), or even crowdsourcing ideas (Brabham, 2008).

The continuing but rare spectacular failures, but more generally minor failures of food and beverage introductions, suggest that some other dynamic is wrong. The conventional wisdom dictates the order: develop a concept, the mental/descriptive "blueprint" of the product, and then create a product that fits the concept. Often concept development and product development are done in parallel, and only later a confirmatory test is run to see whether the combination "works," gaining consumer acceptance. The parallel path of development is clearly incorrect because one does not know whether the product and concept will match. Yet market failure continues to plague the process even when the concept is developed first and only then, afterward, is the product developed.

Is it possible that the traditional development method is wrong? Could it be that the product should be developed first, and only afterward a concept developed that both fits the product and appeals sufficiently to drive trial? This notion was presented 20 years ago at the meeting of the British Market Research Society in Birmingham, England, Spring, 1996, by Moskowitz and Gandler (1997). The response was a deafening silence. The notion of changing the order of activities in development did not make sense—or at least, not in 1996.

In 2011 authors Moskowitz and Batalvi of SB&B Research developed an alternative paradigm, empathy and experiment, based upon early discussions with Glen Backus of Topco, Inc. in Chicago, and on some early trial efforts. The guiding notion was that people do not have "concepts" guiding their enjoyment of food. Rather, people choose a food, perhaps on the basis of advertising, but also on the basis of package at the store or word of mouth from friends, or general "buzz." Those who try the product may or may not like it (Moskowitz et al., 2012).

It was the foregoing thinking that motivated the experimental design of products, ie, that product development to create a good food should precede advertising concepts. Product concepts may help the product developer, but primarily for inspiration. The role of the concept is not to guide development but rather to attract buyers. The foregoing "position" regarding how to develop products is contrary to current wisdom, which specifies a product concept, and then development toward that concept.

How then is empathy and experiment implemented in a realistic, cost-effective way in the real world, recognizing the siloed nature of professionals in the food industry, the overdependence on norms, and the almost slavish obeisance to the project calendar, to stage gate checkoff points, and so forth? In other words, can empathy and experiment provide to the food industry an agile, data-rich approach that makes the company smarter while making the company do better development? Moskowitz and Batalvi suggest a set of steps, the first being empathy, the remaining following a more rigid, stepwise, scientific process:

First, identify a product opportunity. The opportunity is identified through the empathy research phase, a proprietary technique developed by Batalvi, a researcher and as of this writing (2015), a clinician and researcher who employs psychodynamic and psychoanalytic principles in her practice and research projects. Empathy creates a unique development tool, blending psychodynamic theories with diagnostic tools adapted for consumer-listening work. Empathy uses a narrative process that is both nonlinear and nonhierarchical, a necessary and welcome counterpoint to the systematized structure that follows it.

Empathy uses language in a deeper way than do most creative approaches, making it a key first stage in creativity. When used for innovation, empathy separates the speaking "subject" from his/her speech. The resulting signifying chain, stringing together the signifier-signified, metaphor, metonymy, condensation and displacement, etc. is delinked to reveal the manifest and latent content of "desire." Resistances and limitations, generated by the symbolic, are investigated. Subsequently, the opportunity for innovation may emerge through an empathic process of semantic wordplay and free association. In some ways, empathy allows the unconscious to generate ideas that were once unfamiliar or impossible to think. Empathy may generate a random but profoundly important mix of product ideas, even from markedly different categories, in the same way that mechanized toothbrushes were invented.

After empathy has suggested the product, the task returns to the product developer, who follows these steps. Note that the steps return the process to a more rigid scientific, quantitative approach, after the empathy excursion, where the kernel of true innovation may have occurred.

1. Create ONE prototype on the bench, presenting the product developer's best guess.
2. Identify four physical variables of that prototype, where variables can appear at two levels, such as current/new, low/high, and so forth.
3. Create eight prototypes, a half replicate design. Each variable appears four times at each level. The variables are statistically independent of each other. The eight prototypes cover a wide range. The eight prototypes are not difficult to create; they may be kitchen samples. Table 6.3 shows the schematic.
4. Test the eight prototypes with 50 consumers, having the consumers rate each prototype on liking and image attributes. Other measures, eg, cost of goods, stability, and nutrition, can be input by the product developer.

Table 6.3 Product Result for Empathy and Experiment, Showing How Eight Prototypes, Systematically Varied on Four Variables, Score for the Total Panel, and for Two Sensory-Preference Segments

Prototype	Ingredient A: Density Level	Ingredient B: Chocolate	Ingredient C: Sweetener	Ingredient D: Vitamin Fortification	Like - Total	Sensory Dark	Sensory Flavor	Sensory Dense	Image Filling	Like - Seg A	Like - Seg B
1	1	1	1	1	33	47	22	65	64	34	32
2	1	1	0	0	44	27	14	48	39	32	56
3	1	0	1	0	49	15	23	49	36	29	69
4	1	0	0	1	43	24	13	51	30	27	60
5	0	1	1	0	41	19	11	33	39	25	57
6	0	1	0	1	21	17	11	37	40	27	15
7	0	0	1	1	38	31	19	31	29	29	47
8	0	0	0	0	13	27	18	28	29	10	16

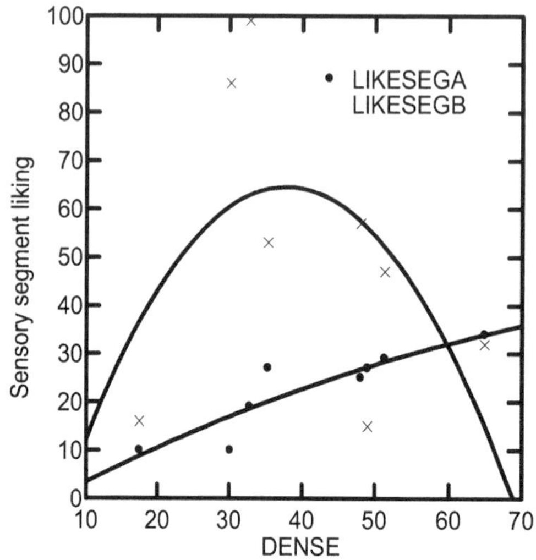

FIGURE 6.4

Sensory segments emerging from empathy and experiment. The segments (A,B) emerge with as few as 50 respondents, each evaluating eight prototypes.

5. Average the ratings to determine how the different prototypes score. When the scale for liking is an anchored 0–100 scale, the prototypes scoring 65 or higher are considered to be good (Table 6.3).
6. Using the principles of sensory preference segmentation discussed earlier, the same principles used for pickles, sauces, and orange juice, identify at least two sensory segments that "make sense," that is, the sensory segments that show different curves relating sensory attributes to liking (Fig. 6.4).
7. The greater the number of prototypes among the test set of eight products that score 65 or higher, either for the total panel or for at least one of the two emergent segments, the more likely that there are at least one or two highly acceptable products from this exercise. When only one prototype scores 65 or higher, even among the segments, it is time to return to the drawing board, and create new prototypes.
8. Only after a new and very promising product prototype has been identified do we develop a marketing concept, a set of messages that both support what the product "is," as well as drive interest in trying the product. The mind genomics approach works here. The respondent tries the product at home and later evaluates test concepts designed for advertising and promotion. The appropriate "winning elements" from the mind genomics test are those that drive both interest in purchasing the product as well as describe the prototype that won in the taste evaluation of the eight prototypes.
9. The result of the foregoing exercise is a knowledge base about the product, in terms of what physical variables drive consumer ratings of acceptance, sensory perception and image, as well as both knowledge of new-to-the-world sensory segments and a list of messages that "work," defined as fitting the product and driving purchase.

6.12 DISCUSSION: WHITHER INNOVATION IN A SLOWLY MOVING CATEGORY?

A final note is in order regarding innovation. The note invokes the opportunities in the food industry emerging out of empathy and experiment. From the senior author's experience, virtually no company he has encountered really understands how formula variables drive sensory responses, has a sense of the nature of underlying sensory preference segments, or knows the messages that "drive" consumers and potential consumers. Perhaps the greatest innovation of all might be to create such a systematic database using the principles of empathy and experiment for existing products, letting the innovations in product emerge from alternative prototypes differing in systematic ways from the current product, along with 1:1 messaging to consumers, based upon their membership in a sensory segment for the product, and the specific messages resonating among that mind genome.

There is an even grander vision, one waiting about 5–10 years to come forth. That is targeted product design and individualization of the product. Knowing the dynamics of the product itself from experimental design, and knowing how to identify an individual's mind genome using mind genomics, means that it will be possible in the not-too-distant future to create products specifically for an individual, or have the equipment to do so "on demand." More importantly, mind genomics will identify what to create for the individual and how to communicate that product to the specific individual.

ACKNOWLEDGMENT

The authors would like to thank Margie Atwater, executive editorial assistant to Dr Moskowitz, for her help in preparing this and other manuscripts.

REFERENCES

Brabham, D.C., 2008. Crowdsourcing as a model for problem solving: an introduction and cases. Convergence: The International Journal of Research into New Media Technologies 14 (1), 75–90.

Caul, J.F., 1957. The profile method of flavor analysis. Advances in Food Research 7, 1–40.

Cohn, G., 1914. Die Organischen Geschmacksstoffe. (F. Siemenroth).

Cooper, R.G., 1993. Winning at New Products: Accelerating the Process From Idea to Launch, second ed. Addison Wesley Publishing Co, Boston.

Gladwell, M., 2004. Choice, Happiness and Spaghetti Sauce. Retrieved from 18:16 www.youtube.com/watch?v=iIiAAhUeR6Y.

Gladwell, M., 2009. The ketchup conundrum. In: What the Dog Saw and Other Adventures. Little Brown & Company, New York.

Gordon, J., 1965. Evaluation of sugar-acid-sweetness relationships in orange juice by a response surface approach. Journal of Food Science 30 (5), 903–907.

Green, P.E., Srinivasan, V., 1990. Conjoint analysis in marketing: new developments with implications for research and practice. Journal of Marketing 54, 3–19.

Mariampolski, H., 2006. Ethnography for Marketers: A Guide to Total Immersion. Sage Books, Thousand Oaks, CA.

May, A. Personal Communication.

Meiselman, H.L., Schutz, H.G., 2003. History of food acceptance research in the US Army. Appetite 40 (3), 199–216.

Moskowitz, H.R., Gandler, A., 1997. Segmenting markets by communication vs. sensory preferences: the tongue has its reasons that the mind does not know. In: Proceedings of the Market Research Society. (United Kingdom), pp. 221–232.

Moskowitz, H.R., Krieger, B., 1998. International product optimization: a case history. Food Quality and Preference 9 (6), 443–454.

Moskowitz, H.R., Wehrly, T., 1972. Economic applications of sweetness scales. Journal of Food Science 37, 411–415.

Moskowitz, H.R., Wolfe, K., Beck, C., 1979. Sweetness and acceptance optimization in cola flavored beverages using combinations of artificial sweeteners—a psychophysical approach. Journal of Food Quality 2 (1), 17–26.

Moskowitz, H.R., Jacobs, B., Firtle, N., 1980. Discrimination testing and product decisions. Journal of Marketing Research 17 (1), 84–90.

Moskowitz, H.R., Jacobs, B.E., Lazar, N., 1985. Product response segmentation and the analysis of individual differences in liking. Journal of Food Quality 8 (2/3), 168–191.

Moskowitz, H.R., German, B., Saguy, I.S., 2005a. Unveiling health attitudes and creating good-for-you foods: the genomics metaphor and consumer innovative web-based technologies. Critical Reviews in Nutrition and Food Science 45 (3), 165–191.

Moskowitz, H.R., Porretta, S., Silcher, M., 2005b. Concept Research in Food Product Design and Development. Blackwell Publishing, Oxford, UK.

Moskowitz, H.R., Batalvi, B., Lieberman, L.E., 2012. Empathy and experiment: applying consumer science to whole grains as foods. In: Proceedings of the Whole Grains Summit.

Moskowitz, H.R., 1974. Sensory perception of sweetness. Cereal Science Today 19, 262–285.

Moskowitz, H.R., 1985. New Directions for Product Testing and Sensory Analysis of Foods. Food and Nutrition Press, Trumbull.

Moskowitz, H.R., 1994. Product testing 2: modeling versus mapping and their integration. Journal of Sensory Studies 9 (3), 323–336.

Moskowitz, H.R., 1995. Consumer Testing and Evaluation of Personal Care Products. Marcel Dekker, New York.

Moskowitz, H.R., 1998. Inter-relating data sets for product development: the reverse engineering approach. Food Quality and Preference 11 (1), 105–119.

Moss, M., 2013. Salt, Sugar, Fat: How the Food Giants Hooked Us. Random House.

Munoz, A., Civille, G.V., Carr, B.T., 1992. Sensory Evaluation in Quality Control. Van Nostrand Reinhold.

Murray, J.M., Delahunty, C.M., Baxter, I.A., 2001. Descriptive sensory analysis: past, present and future. Food Research International 34 (6), 461–471.

Pangborn, R.M., 1970. Individual variations in affective responses to taste stimuli. Psychonomic Science 21, 125–128.

Rawlinson, J.G., 1981. Creative Thinking and Brainstorming. Gower, Farnborough, Hants.

Stevens, S.S., 1975. Psychophysics: An Introduction to its Perceptual, Neural and Social Prospects. John Wiley, New York.

Stone, H., Sidel, J., Oliver, S., Woolsey, A., Singleton, R.C., 1974. Sensory evaluation by quantitative descriptive analysis. Food Technology 28 (11), 24–34.

CHAPTER

IMPLEMENTATION OF EMERGING TECHNOLOGIES

7

F.J. Barba[1], V. Orlien[1], M.J. Mota[2], R.P. Lopes[2], S.A. Pereira[2], J.A. Saraiva[2]
[1]University of Copenhagen, Frederiksberg, Denmark; [2]University of Aveiro, Aveiro, Portugal

7.1 INTRODUCTION

The food industry has largely invested in processing facilities relying mostly on conventional thermal technologies, which show well-established reliability and efficacy (Mújica-Paz et al., 2011). However, these thermal processing approaches usually cause a negative impact on important food quality attributes (eg, destruction of important nutrients, development of off-flavors, and color changes). In recent years, the consumer is increasingly demanding fresh-like food products, with high organoleptic and nutritional quality. Consequently, novel technologies for food processing have been developed attempting to complement or replace the conventional thermal techniques. These novel pasteurization methods include high-pressure processing (HPP), pulsed electric fields (PEF), ohmic heating, microwave, radiofrequency, and ultrasound.

Nowadays one of the most well-established nonthermal processing techniques is HPP, which uses pressures in the range of 100–1000 MPa at a large temperature spanning from freezing temperature to just below boiling temperature. The main operating principles behind this technology are the Le Chatelier and the isostatic one (Ramaswamy et al., 1999). The first states that high pressure affects any phenomenon where volume changes are involved, favoring reactions that cause a decrease in volume, while reactions involving an increase of volume will be inhibited (Cheftel, 1995; Ledward, 1995). According to the principle of isostatic processing, the product is compressed by uniform pressure from every direction and then returns to its original shape when pressure is released (Olsson, 1995). As a result, HPP treatments are independent of product size and geometry, and the effect is uniform and instantaneous, being mass/time independent (Alemán et al., 1996; Knorr, 1994, 1995; Zimmerman and Bergman, 1993), offering a great advantage to HPP over the conventional thermal technologies.

Commercial applications of HPP use 300–700 MPa for less than 3–5 min at room, or preferably refrigerated, temperature for an enhanced microbial inactivation and quality retention (Torres and Velazquez, 2005). Pressure has a limited effect on covalent bonds but substantially affects noncovalent bonds (hydrogen, ionic, and hydrophobic interactions). Consequently, compounds with high molecular weight are sensitive to HPP, while low-molecular-weight components (eg, color and flavor compounds) are usually not affected (Carlez et al., 1994). Since HPP allows a high retention of food-quality attributes, namely sensorial and nutritional properties (Barba et al., 2012; Oey et al., 2008a,b), food products treated with HPP usually have fresh-like characteristics (Chawla et al., 2011). However, HPP also has disadvantages, such as the resistance for inactivation of some food enzymes, leading to possible enzymatic and oxidative degradation of food components during storage and distribution (Thakur and Nelson, 1998).

Innovation Strategies in the Food Industry. http://dx.doi.org/10.1016/B978-0-12-803751-5.00007-6

In order to ensure financial feasibility and environmental sustainability, HPP treatments must be kept short. This can be obtained using the concept of hurdle technology, with pressure-assisted thermal sterilization (PATS) being one of the most common procedures (Mújica-Paz et al., 2011). PATS involves the combined application of high pressure (600 MPa or greater) and high temperature (60–90°C). Compared to conventional techniques, the adiabatic compression caused by pressurization allows an almost instantaneous temperature increase, and the subsequent decompression also results in rapid cooling, thus shortening the process times. Additional advantages are the lower maximum temperatures and a more uniform radial temperature distribution, reducing the local temperature over processing. Therefore, in addition to acceleration of spore inactivation, PATS also shows a higher quality and nutrient retention in selected products compared to the conventional thermal processing methods (Barbosa-Cánovas and Juliano, 2008; De Vleeschouwer et al., 2010; Hoogland et al., 2001; Knoerzer et al., 2007; Matser et al., 2004; Ramirez et al., 2009; Raso and Barbosa-Cánovas, 2003; Sevenich et al., 2013; Shao and Ramaswamy, 2011; Somerville, 2009). As a result, PATS could meet the demand for more healthy, nutritious, and convenient shelf-stable food (Mújica-Paz et al., 2011). The commercial use of PATS for processing of low-acid foods was approved in 2009 by the US Food and Drug Administration (FDA) (Sablani, 2014; Somerville, 2009). Although there are still no PATS-processed shelf-stable foods commercially available, this corresponds to a viable method for processing of heat-sensitive products.

The widest application of HPP within the food industry is in the context of food processing, but other uses are also foreseen with the progress of research. These include solute diffusion processes (salting, sugaring), assisted freezing-thawing processes, and modification of functional properties of proteins and other macromolecules (San Martin et al., 2002). Recently HPP has been successfully applied in the extraction of some cellular compounds since pressure increases cell permeability and the solubility of the compounds to extract (Galotto et al., 2010; Xi, 2006). HPP can also be used to modulate food fermentations, possibly leading to processes with novel characteristics or even development of new fermentations with interest for the food industry (Mota et al., 2015). Another promising application that arose very recently is the so-called hyperbaric storage, a new food preservation method under high pressure, as a possible energetically costless process. In this case, the goal is to inhibit microbial growth, similarly to freezing and refrigeration, but at naturally variable uncontrolled room temperature (Fernandes et al., 2015a). The studies published so far showed that it is possible to use hyperbaric storage to preserve different types of food products, including different fruit juices, meat, and soup (Bermejo-Prada et al., 2015; Duarte et al., 2015; Fernandes et al., 2015b; Fidalgo et al., 2014; Moreira et al., 2015; Queirós et al., 2014; Santos et al., 2015; Segovia-Bravo et al., 2012).

Pulsed electric field (PEF) is another interesting nonthermal processing technique, which consists of an electrical treatment of short time (from several nanoseconds to several milliseconds) with pulse electric field strength from 100 to 300 V/cm to 20–80 kV/cm applied on food products (Buckow et al., 2014). This treatment results in inactivation of quality-related enzymes, as well as destruction of spoilage and pathogenic microorganisms, possibly due to loss of cell membrane functionality. This can be explained by the formation of hydrophilic pores in the membrane as well as the forced opening of protein channels (Buckow et al., 2014; Sharma et al., 2014; Teissié et al., 1985). Although heat may be generated during the PEF processing treatment, microbiological inactivation is achieved by nonthermal means, ie, due to the applied electrical fields. However, there is a clear synergy between a moderate heating (eg, 40–45°C) and the electric fields (Dunn, 2001). Since PEF processing only causes a small increase of temperature, there is no appreciable loss of flavor, color, and bioactive compounds (Brennan, 2012). Because of that, PEF processing constitutes a potential alternative to traditional thermal processing, with the advantage of retaining or minimally modifying sensorial, nutritional, and health-promoting attributes of the products (Barbosa-Cánovas and Altunakar, 2006).

In parallel with the development of PEF processing, the achievement of high-voltage pulsed power systems led to PEF performance scales from laboratory-scale systems to large production-scale installations (Puértolas et al., 2010). Currently, a successful transfer of processing conditions from lab to industrial scale has been achieved. However, this technology still shows some limitations in the context of food processing. For instance, food products with large electrical conductivity are not suitable for PEF processing because the peak electric field across the chamber is reduced. The ionic strength of a food material directly influences its conductivity and, as the conductivity rises, the lethality of a process decreases. Another problem is the presence of particulates in the liquid because high-energy inputs may be needed to inactivate microorganisms and there is a risk of dielectric breakdown of food (Brennan, 2012). Therefore, it is important to determine not only the optimal conditions to obtain safe and fresh-like products but also to carefully select the food products that are most suitable for PEF processing. Besides the application in food processing, PEF technology can also be used by the food industry to improve the functionality, extractability, and recovery of bioactive compounds by increasing cell permeability (Donsì et al., 2010; Rodríguez-Roque et al., 2015). In addition, it can be used to reduce the allergenicity of certain food products (Johnson et al., 2010) and the formation of processing contaminants (Zhang et al., 2012). Also, PEF is a promising tool to improve drying and freezing processes (Ammar et al., 2010; Jalté et al., 2009; Wiktor et al., 2013).

Ohmic heating is defined as a process where electric currents are applied directly to the food (Sun, 2012). When materials contain sufficient water and electrolytes to allow the passage of electric current, ohmic heating can be used to generate heat within the product, due to electrical resistance of the food to the current (Imai et al., 1995; Knirsch et al., 2010; Sun, 2012). Thus, there is no need to transfer heat through solid–liquid interfaces or inside solid particles (Knirsch et al., 2010).

One of the main advantages of this technology is the much shorter heating times that can be applied (from a few seconds to a few minutes) since the temperature required is achieved very quickly. Therefore, ohmic heating allows the rapid and uniform heating of liquids, allowing the food to maintain its nutritional and sensory characteristics (Sastry, 2005; Sun, 2012). Under certain circumstances, it enables large particulates and carrier fluids to heat at comparable rates, which is particularly useful for aseptic processing of fluids containing particulates and fluids of high viscosity, maintaining the quality of the final products (Palaniappan and Sastry, 2002; Rice, 1995; Wang et al., 2001). Additionally, this technology has low maintenance costs and high-energy conversion efficiencies (Sakr and Liu, 2014). However, there are still several limitations (eg, difficulties in controlling the heat rate during the processes) due to change of electric conductivity of the heating material. As a result, it is necessary to adjust the processing conditions according to the conductivity of each food product (Sakr and Liu, 2014). Contrarily to PEF, the principal mechanisms of microbial inactivation in ohmic heating are thermal in nature. Recent research indicates that ohmic heating may also present mild nonthermal cellular damage due to the presence of the electric field (Cho et al., 1999; Pereira et al., 2007; Sun et al., 2008). However, further research is still needed to completely understand all the effects of ohmic heating in food products (Knirsch et al., 2010).

Besides pasteurization and sterilization, ohmic heating provides several other applications in food industry, including blanching, evaporation, dehydration, fermentation, and extraction (FDA, 2000). Another possibility is the application of this technology for heating of foods to serving temperature, including in the military field or in long-duration space missions (Knirsch et al., 2010).

Microwave corresponds to electromagnetic waves of certain frequencies (300 MHz–300 GHz) that are used to generate heat within products (Chandrasekaran et al., 2013; Sun, 2012). The heating is caused by the ability of the materials to absorb microwave energy and then to convert it into heat. Microwave heating is a convenient way to heat materials since it is considered to be a fast, clean, and easy-to-use technology (Salazar-González et al., 2012). Microwave heating has vast applications in the field of food processing, such as cooking, drying, pasteurization, and preservation of food materials, but herein only microwave pasteurization will be discussed (Chandrasekaran et al., 2013). Microbial destruction by microwave requires lower temperatures than conventional pasteurization due to significant enhancement of the thermal effects (Chandrasekaran et al., 2013). Moreover, application of microwave results in higher and more homogeneous heating rates and subsequent shorter processing times (Sun, 2012). As a result, microwave-pasteurized products are able to better retain taste, color, and nutritional value compared to those pasteurized by other conventional methods (Chandrasekaran et al., 2013). Another advantage relies on the possibility of instantly turning on/off the heating systems (Datta, 2003). However, this technology also has several drawbacks, usually related to the nonuniform temperature distribution, resulting in hot and cold spots in microwave-heated products (Chandrasekaran et al., 2013; Vadivambal and Jayas, 2010). In general, the center region of a food material generates more cold spots than the other regions (Chandrasekaran et al., 2013). In addition, the application of microwave heating greatly depends on the dielectric properties of the food product (Salazar-González et al., 2012).

Microwave pasteurization largely has been applied to fluid foods, such as fruit juices, apple cider, coconut water, vegetable purées, and milk. In terms of solid foods, microwave pasteurization already has been applied to eggs and pickled asparagus. Recently, the FDA approved the microwave sterilization process for mashed potatoes in trays and salmon fillet in sauce in pouches (Brody, 2012; Chandrasekaran et al., 2013). Ultrasound is a nonconventional processing technology that is particularly suitable for preservation of fluid foods. Ultrasound can be defined as inaudible sound waves at frequencies above 20 kHz. For food preservation, low-frequency ultrasound waves (18–100 kHz) are usually the most effective (Dujmić et al., 2013). The main advantages of ultrasound processing over the conventional thermal pasteurization methods include reduced water and energy consumption, minimal flavor and nutrient losses, and increased homogeneity of the product (Soria and Villamiel, 2010).

Data on inactivation of food microorganisms by ultrasound alone are scarce because the effects of ultrasound are usually not severe enough for a sufficient lethal effect (Zinoviadou et al., 2015). However, this technology may be used in combination with other preservation techniques, such as mild heat temperatures, high pressures, and antimicrobials. The application of ultrasound combined with mild heat treatments (<100°C) has given rise to the procedure known as thermoultrasonication. The combination with pressure (<600 MPa) is known as manosonication, while the use of the three strategies together is called manothermosonication (Knorr et al., 2004). Manosonication and manothermosonication are particularly effective for sterilization of jams and liquid eggs and, in general, to extend the shelf life of liquid foods (Zinoviadou et al., 2015).

Isolated ultrasonication is effective for the decontamination of raw vegetables and whole eggs submerged in liquid medium. For purposes other than preservation, it has been used successfully to tenderize meat and for the recovery of valuable compounds from different vegetal and algae matters (Barba et al., 2015; Deng et al., 2014; Roselló-Soto et al., 2015). Better known and more widely practiced is the use of ultrasound in emulsification and homogenization systems and in the cleaning of certain equipment (Knorr et al., 2004; Zinoviadou et al., 2015).

In addition to the technologies described and discussed in this section, there are several other emerging food processing techniques, including cold plasma, pulsed light, ultraviolet light, ozone, supercritical carbon dioxide, and radio frequency, that are currently being studied by food science researchers. However, the application of these technologies is still in a less-advanced stage, and little if any information is available on their industrial utilization, associated costs, and environmental impact.

7.2 COMMERCIALIZATION, SAFETY DATA, AND ENERGY

The quality and safety of food products are a top priority for authorities, influencing current consumers' demands and hence becoming a prime goal for food processors and distributors (Ohlsson and Bengtsson, 2002). Besides these requests, energetic optimization and heat recovery are gaining focus in the food industry since the implementation of emerging food processing technologies is showing a high potential to provide products with improved quality and a reduced environmental footprint, while reducing operation costs and improving the added-value of the products (Pereira and Vicente, 2010). Furthermore, not only economic benefit but also environmental protection, social sustainability, energy supply security, and industrial competitiveness can be provided with an improvement of energy efficiency (Wang, 2014).

7.2.1 HIGH-PRESSURE PROCESSING

The treatment of food using HPP dates back to 1899 (in West Virginia, USA) when Hite performed core research with milk. However, milk products have only recently been commercially treated with HPP, mainly due to the complex changes that milk and derived products undergo during HPP (Hite, 1899; Huppertz et al., 2006). Despite the early pioneering studies of Hite, the interest in HPP application revived in the mid-1980s, and the first commercial food products (strawberry, apple, and kiwi jams) that were treated appeared in Japan in 1990, manufactured by Meidi-ya Food Factory Co. (Horie, 1992). The jams were treated in a 50-L Mitsubishi Heavy Industry (Japan) cold isostatic press at 400 MPa for 20 min, which was shown to provide commercial microbial safety while maintaining fresh fruit sensory qualities (Horie, 1992). Over the last decades, significant advances enabled the scale-up of pilot units to commercial systems, in which industrial HPP treatment of food may be carried out using a batch or semicontinuous process, depending on the physical properties of food (Moreau, 1995).

Since the 1990s there has been an exponential growth of HPP industrial applications. In 2010 there were about 145 full-scale units installed worldwide, operating in over 60 companies and producing hundreds of different HPP-treated products. For the industrial vessels, 35% were in use for vegetable products, 31% for meat products, 14% for shellfish production, 12% for juice and beverage production, and the balance for other specialty applications. Fig. 7.1 shows that at the end of 2014, 252 units of HPP industrial equipment were in production worldwide, indicating a fast increase of this technology in the food industry. During the current year, the total number of HPP operating machines has already reached 320 units (personal communication by Hiperbaric, 2015).

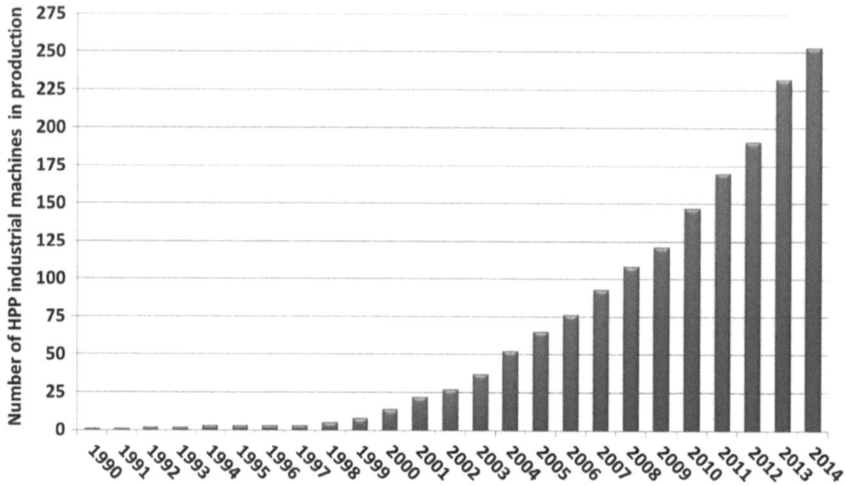

FIGURE 7.1

Evolution of total number of HPP industrial machines.

Courtesy of Hiperbaric, Spain.

Avure Technologies (Västerås, Sweden), Hiperbaric (Burgos, Spain), and UHDE High Pressure Technologies (Hagen, Germany) are major suppliers of commercial-scale pressure equipment, with both horizontal and vertical pressure vessel configurations available for batch HPP equipment. Avure Technologies also produces semicontinuous systems for liquid beverages processing such as juices. The industrial units are characterized by internal volumes ranging from 30 to over 500L and provide pressures levels up to 700MPa at room temperature, with a maximum overall production close to 5000kg/h, depending on the holding time and package size (Van der Berg et al., 2002). Some current applications and products by HPP are shown in Table 7.1.

As an emerging technology for food processing, HPP is continuously being evaluated by the scientific community, which is increasingly concerned with getting a higher product quality and also water and energy savings, and reduced emissions, and consequently, having less impact on the environment. In a high-pressure process energy is required for the compression work. Theoretically, the compression of pure water up to 600MPa, under isothermal conditions, requires 52kJ/kg (Wang, 2008). However, all compressible materials change their temperature during compression due to adiabatic heating. The main ingredient of most foods is water and, thus, it is possible to estimate the temperature increase of food during pressurization based on the thermodynamic properties of water. The temperature increase of pure water per 100MPa increase in pressure level is about 3°C, so the temperature increase of water is then 18°, at 600MPa corresponding to 70kJ/kg of heat. The theoretical total energy input into a high-pressure process at 600MPa for processing pure water is thus about 122kJ/kg (Wang, 2008).

An interesting technique is to combine high pressure with conventional heating for food sterilization (Furukawa et al., 2003; Koutchma et al., 2010). This combined treatment can be utilized to achieve an inactivation of spores of *Clostridium* spp. similar to that of conventional sterilization. However, the specific energy input required for sterilization of cans can be reduced from 300 to 270kJ/kg when applying the HHP treatment.

Table 7.1 Current Applications and Marketed Food Products Treated by High-pressure Processing (Galanakis et al., 2012)

			Processing Conditions			
Country	**Company**	**Product**	***P* (MPa)**	***T* (°C)**	***t* (min)**	**Shelf Life**
Czech Republic	Beskyd Fryc̆ovice	Broccoli and apple or carrot juice	500	20	10	21 days
France	Pampryl	Fruit juices	400	20	10	18 days (at 4°C)
	Ultrafruit	Orange juice	500	20	5–10	–
Italy	Ortogel	Fruit juices and desserts	600	17	3–5	1–2 months
Japan	Meidi-Ya Co.	Jams, sauces, jellies	400	20	10–30	2–3 months (at 4°C)
	Pokka Co.	Grape juice	120–400	23	2–20	–
	Wakayama Food Industries	Mandarin juice	300–400	23	2–20	–
	Chiyonosono	Sake	400	15	30	6–12 months (at 4°C)
Lebanon	–	Fruit juices	500	–	–	1 month
Mexico	Jumex	Smoothies, citrus juices	500	–	–	–
Portugal	Frubaça	Fruit juices	450	12	0.33–1.5	28 days
Spain	–	Prepared vegetable dishes	500	–	–	–
Sweden	Västerås	Fruit juices	500–600	–	–	–
UK	Orchard house foods	Orange juice	500	20	–	–
USA	Avomex	Avocado purée	700	–	10	–
	Odwalla	Apple juice	–	–	–	>2–3 times than fresh

7.2.2 PULSED ELECTRIC FIELD

Similarly as high-hydrostatic pressure processing, PEF has received considerable attention within the last decades to improve or replace existing processes for food preservation, tissue disintegration, and food modification (Toepfl et al., 2006). In fact, when a food product is submitted to electrical pulses, several events occur that contribute to the inactivation of microorganisms, such as resistance heating (Sastry and Barach, 2000), electrolysis (Hülsheger and Niemann, 1980), and disruption of cell membranes (Sitzmann, 1995).

During the 1980s the interest in PEF technology increased among scientific communities for the commercialization of PEF application, and their investigations were followed by several patents focusing on microbial inactivation, tissue response to electric fields, enzyme inactivation, engineering aspects, inactivation modeling, and scale-up (Sepulveda and Barbosa-Cánovas, 2005). Maxwell Laboratories (Food Corporation, San Diego, CA, USA) patented the concept of applying very intensive electric field pulses to inactivate microorganisms in foods and developed it to pilot-plant scale in cooperation with Tetra Pak, the US Army Natick Research Center, and others (Dunn and Pearlman, 1987). To date, PurePulse

Technologies Co., a subsidiary of Maxwell Laboratories, owns several patents to preserve dairy fluid food products, fruit juices, and fluid eggs by treatment with high-intensity electric discharges of 5–100 kV/cm with flat-topped exponentially decaying pulse shapes. Pulse duration is controlled to prevent electrical breakdown of the food product; the typical pulse duration is 1–100 µs with repetition rates between 0.1 and 100 Hz. The patents describe both a batch and continuous processing system, whereas they recommend that PEF treatments could be applied to preheated liquid foods, which enhance microbial inactivation and shelf life stability (Bushnell et al., 1993; Dunn and Pearlman, 1987).

The effectiveness of the technology has been studied on spoilage and pathogenic vegetative microorganisms that may represent a public health risk under commercial conditions, specifically those such as *Listeria monocytogenes*, *Staphylococcus aureus*, *Escherichia coli*, and *Salmonella* Typhimurium, achieving log reductions of 0.6–1.5, 0.5–4.0, 1.5–3.3, and 2–4.2, respectively, depending on processing conditions and food type (Saldaña et al., 2014). Earlier, Dunn and Pearlman (1987) reported more than five logarithmic cycles of microbial count reduction of naturally occurring microorganisms in orange juice after 35 pulses of 100 µs at a voltage intensity of 33.6–35.7 kV/cm and a process temperature of 42–65°C. The shelf life of orange juice was increased from 3 days to 1 week with no significant change in odor or taste.

In spite of all the research and development, most of the PEF applications lie at the experimental level and in the pilot plant range, with limited but increasing commercialization. A great research effort has been placed on the new chamber designs that make it possible to apply a uniform electric field in large-scale equipment (Huang and Sites, 2007; Toepfl et al., 2006). Several pilot-plant-scale units have been developed by different research groups with flow rates of 400–2000 L/h (Min et al., 2003a; Min et al., 2003b) and commercial-scale PEF units for food processing with an overall flow rate of 400–6000 L/h (Kempkes et al., 2010). Major suppliers of commercial-scale equipment are Elea (Quakenbrück, Germany), a spin-off of the German Institute of Food Technologies (DIL), Pulsemaster (Ladel, The Netherlands), and PurePulse from the Dutch Cool Wave Processing (Wageningen, the Netherlands), a subsidiary of the technology and a product development company, specializing in PEF processing for mild preservation.

The first commercial PEF application was implemented in 2005 by the Genesis Juice Corporation (Eugene, OR, USA) for nonthermal pasteurization of fruit juices in a 200 L/h processing unit (Clark, 2013). The company-cited motivations were the avoidance of loss of flavor and the shelf life increase up to 4 weeks. However, the firm subsequently switched from PEF to HPP for undisclosed reasons (Sampedro et al., 2014). In recent years new market opportunities have enlarged the number of commercially available foods processed by pulsed electric fields. Table 7.2 summarizes current commercial applications of PEF for food processing.

Table 7.2 Marketed Food Products Processed by Pulsed Electric Field

Country	Company	Product	Application	Shelf-Life
The Netherlands	Hoogesteger	Fruit juices and smoothies	Microbial inactivation (at 12–15 kV/cm)	21 days
The Netherlands	The fruit lab	Fruit juices and smoothies	Microbial inactivation	–
The Netherlands	Fruity line	Vegetable/fruit juices and smoothies	Microbial inactivation	–
Germany	Beckers Bester	Apple juice	Cell disintegration	–
Germany	Wernsing Feinkost GmbH	Potatoes	Cell disintegration	–

Besides hundreds of studies published showing the benefits of PEF on the safety and quality of foods, its potential to improve the sustainability of food processing and/or to reduce energy requirements while maintaining or improving food quality and safety have also been highlighted.

A study by Heinz et al. (2003) estimated that a specific energy input of 100 kJ/kg or more is required for the application of pulsed electric fields for processing of liquid foods at 30°C. It has also been stated that a PEF application for microbial decontamination of liquid food requires a higher input of energy than a thermal treatment, as heat recovery rates of thermal energy are in the range of 90–95% (Toepfl et al., 2006). However, the impact of pulse energy dissipation and the simultaneous resistive heating of the suspending medium must be taken into account, and energy efficiency of PEF processing may take advantage of synergistic effects of mild heat (Toepfl et al., 2006). When operating at elevated treatment temperatures and making use of synergetic heat effects, the PEF energy input might be reduced close to the amount of 20 kJ/kg required for conventional thermal pasteurization, assuming 95% of heat recovery (Toepfl et al., 2006).

Apart from the reduction of energy input, when operating at elevated temperatures the need to preheat the media to the initial treatment temperature provides a potential to recover the electrical energy dissipated into the product in a heat exchanger (Toepfl et al., 2006). When operating at ambient temperatures, there is no need for preheating and, therefore, high cooling efforts are required, thus causing additional energy costs (Toepfl et al., 2006). Although possible, cooling systems should be avoided, since if cooling is necessary it means that energy dissipation is too high due to high conductivity, indicating that the product may not be suitable for the PEF process (Pereira and Vicente, 2010). Despite the increase of the delivered electrical power, PEF pasteurization seems to be less energy intensive than traditional pasteurization methods, leading to annual savings of 791.2–1055 TJ per year of fossil fuel equivalents while also contributing to the reduction of CO_2 emissions (Lelieveld, 2005).

An important issue with PEF is the potential nonuniform distribution of the field strength that can lead to over- or underprocessing or dielectric breakdowns (Gerlach et al., 2008) caused by the geometry of the chamber and/or by the presence of air bubbles or other dielectric materials/particles such as fat globules (Saldaña et al., 2014). According to Saldaña et al. (2014), possible ways to improve the treatment uniformity are as follows: (1) generating a turbulent flow by modifying the treatment chamber geometry, and (2) introducing a grid before the treatment zone.

7.2.3 OHMIC HEATING

The concept of ohmic heating (OH) of foods first appeared in the 19th century, when processes were patented for the use of electrical currents to heat pumpable liquids (Getchell, 1935). In the early 20th century, the so-called "electropure process" was used in over 50 electric sterilizers for milk pasteurization, serving about 50,000 consumers globally (Anderson and Finkelstein, 1919; Moses, 1938). In 1988 PV Baker licensed a continuous-flow ohmic heater, patented by the Electricity Council of Great Britain, with new and improved electrode material and were, for many years, the principal suppliers of the technology (Skudder, 1988).

Current commercial OH systems are available from SPX (formerly APV, Crawley, UK), Raztek Corp. (Sunnyvale, CA, USA), Emmepiemme SRL (Piacenza, Italy), and Simaco (Piacenza, Italy). APV produces two commercial systems with power outputs of 75 and 300 kW, corresponding to product capacities of 750 and 3000 kg/h, respectively, for a temperature rise of 75°C in water. This ohmic system has been installed for pasteurization and sterilization of a number of food products with excellent quality, particularly when plug flow is being achieved (Tempest, 1996). Emmepiemme SRL produces systems in the range of 60–480 kW for production throughputs of 1000–6500 kg/h (FDA, 2015b).

This technology appears to be particularly suitable for the processing of liquid and viscous products and also pumpable multiphase foods because the heat is generated inside the food matrix (Goullieux and Pain, 2005) and, therefore, overheating problems are avoided. However, the direct application of this technology on solid food products is still limited to the laboratory environment (Stratakos and Koidis, 2015).

The commercialization of food products processed by ohmic heating is currently happening in several countries. Its major successful application has been in the area of fruit and vegetable processing, but it has been reported that it can also be used for cooking and sterilizing meat products (Ghnimi and Fillaudeau, 2014). Cold spot occurrence in complex or multicomponent foods needs to be taken into account by food producers (Stratakos and Koidis, 2015). Table 7.3 shows examples of processed food products and respective industrial plants installed by Emmepiemme SRL.

Since ohmic heating is fundamentally a thermal process, temperature and time are the principal critical factors influencing microbial inactivation. Despite the thermal effect, recent research also

Table 7.3 List of Industrial Ohmic Heating Plants Installed Globally by Emmepiemme SRL (as of 2007)

Installation Year	Country	Product	Heating Power (Hz)
1994	Italy	Tomato sauces and pastes[a]	50
1996	Ivory Coast	Tomato pastes and mango purées[a]	64
	Italy (through Simarco)	Fruit slices and dices[a]	100
1998	Greece (through Simarco)	Peach and apricot slices and dices[a]	150
1999/2000	Greece (through Simarco)	Peach and apricot slices, dices and halves[a]	200, 150, 240, respectively
2000	Italy	Pears and apples dices[a]	150
2001	Italy	Low-acid vegetable purées[a]	100
2002	Mexico	Strawberries	250
	France	Fruit preparations	100
2003	France	Meat recipes (processing line)	50
2004	Italy	Plum peeled tomato and tomato dices[a]	480
2005	Italy	New vegetable sauces and special recipes	60
2006	Italy	Mushrooms and tomato dices[b]	240
	Italy	Tomato dices, pulp and sauces[b]	480
	Italy	Tomato sauces and other derivatives[b]	100
	Greece	Peach and apricot slices and dices[c]	120
2007	Spain	Onion sauces[c]	60

[a]*Aseptic process and filling.*
[b]*Aseptic processing line.*
[c]*Processing and aseptic filling line.*
Adapted from Ghnimi, S., Fillaudeau, L., 2014. Tubular and jet fluid units. In: Ramaswamy, H.S., Marcotte, M., Sastry, S., Abdelrahim, K. (Eds.), Ohmic Heating in Food Processing. CRC Press, Taylor & Francis Group, Boca Raton, US, pp. 183–196.

indicates that ohmic heating may present mechanical cellular damage due to the presence of the electric field, creating pores (electroporation) that lead to leakage of the cell contents (eg, amino acids, proteins, nucleic acids) (Knirsch et al., 2010). Moreover, chemical inactivation can cause microbial death due to the formation of free radicals and metal ions (Guillou and El Murr, 2002). The chemical effect may, however, be a threat to the safety of the food product. Indeed, the formation of radicals may have a negative impact on the food antioxidant activity, and free radicals can cause adverse effects on the human body (Stratakos and Koidis, 2015).

In a study by Cho et al. (1999), ohmic heating was shown to be more effective at killing *Bacillus subtilis* spores than conventional heating, leading to a reduction of the D-value for microbial inactivation. This reduction has also been observed for *B. subtilis* (Cho et al., 1999), *Bacillus licheniformis* and *E. coli* (Pereira et al., 2007), *Byssochlamys fulva* (Castro, 2007), and *Streptococcus thermophilus* (Sun et al., 2008).

One of the key advantages of ohmic heating technology has been its energy efficiency, which typically is in the order of 90% and above (Nguyen et al., 2013). A small amount of heat is lost from the heating chamber to surrounding air, which may slightly reduce the overall efficiency of the system (Nguyen et al., 2013). The efficacy of ohmic processing can be influenced by the behavior and conductivities of individual components within the food and interactions during the heating process (Shirsat et al., 2004). For example, De Halleux et al. (2005) concluded that ohmic heating provided 82–97% of energy saving while reducing the heating times by 90–95% compared to conventional heating. The authors suggested that it could be possible to obtain efficiencies greater than 90% in an industrial process in which these losses were controlled by the wall insulation.

In studies with meat products, conducted at Agri-Food Canada's Food Research and Development Center (FRDC), where traditional smokehouse cooking was replaced by an ohmic process, energy savings of at least 70% could be achieved (Vicente et al., 2006). Moreover, at the Louisiana State University Agricultural Center, sweet potato samples were processed using ohmic heating prior to freeze-drying. Ohmic heating increased the rate of freeze-drying up to 25%, which led to significant savings in both processing time and energy use (Lima et al., 2002; Masanet, 2008).

7.2.4 MICROWAVE HEATING

The development of microwave heating technology started in the beginning of the 20th century with the desire for higher energy transfer rate enhancement (Püschner, 1966). The first patent, describing an industrial conveyor belt microwave system, was issued in 1952 (Spencer, 1952), however, its first application started 10 years later owing to the need for high-power microwave sources to be developed. Over the years this novel thermal technology has established different applications in the field of food processing, such as tempering and thawing, drying, blanching, cooking, baking, pasteurization, and sterilization.

The fast popularity of microwave technology among researchers, food producers, and consumers is related to the ability of achieving high-heating rates, significant reduction in cooking time, more uniform heating, safe handling, ease of operation, and low maintenance (Salazar-González et al., 2012; Zhang et al., 2006). The promise of a rapid and volumetric heating has prompted many research studies to deal with the microbiological safety of food products.

In general, microbial inactivation curves obtained with microwave heating are similar to those of conventional heating. Despite the heating effect, potential nonthermal factors have been addressed as possible additional inactivation effects: selective heating of microorganisms, electroporation, cell membrane rupture, and cell lysis due to electromagnetic energy coupling (Kozempel et al., 1998b). A general consensus stated that the nonthermal effect is likely to be due to the lack of precise measurements of the time–temperature history and its spatial variations (Heddleson and Doores, 1994).

The industrial setting up of microwave heating processes is currently facing two major obstacles related to the nonuniform temperature distribution inside a food product and the high energy costs associated with this technology (Ryynänen et al., 2001). The changes of dielectric properties of food products during the heating processes are not yet fully understood or modeled, and, consequently, the validation of the processes has to be done almost individually for each food product, slowing down the dissemination of microwave industrial lines. The problem of nonuniform heating appears to be even more pronounced in multicomponent ready-to-eat meals (Chandrasekaran et al., 2013; James et al., 2002). Bearing in mind the increased demand in ready-to-eat foods in recent years, significant safety issues may arise. According to Ahmed et al. (2007), temperature uniformity of multicomponent foods depends on the food component place as well as the geometry of products and packages, and it is possible to partially balance the nonuniformity by taking advantage of the increased temperatures occurring in the edges of the foods during microwave heating (Vicente and Castro, 2007).

In 1998 Top's Foods (Olen, Belgium), a company that develops in-package microwave sterilized foods, produced over 13 million ready meals, and 1 year later, the company installed a commercial new microwave heating system (using a frequency of 2450 MHz) that consists of microwave tunnels with several launchers distributed across a number of ready-meal packages (George et al., 2003). The importance of exact positioning of the package within the microwave chamber is recognized by the company; using guide rails within the conveying system, precise positioning of the package ensures that each package receives a precalculated, spatially varying microwave power profile. It is reported that the central areas of the product are provided with higher power levels, in an attempt to direct more heating toward areas likely to contain the slowest heating point (George et al., 2003).

More recently a few companies released novel continuous microwave systems into the market (AMT, 2014; IMS, 2012), which have been applied to several food matrices, such as juices, meat emulsions, and dairy products. This type of continuous microwave processing can potentially offer a solution to the inability to heat-treat food products in a predictable and uniform manner, enabling enhanced process control as well as increased throughput and retention of nutrient compounds and flavors.

Over the years some alternative solutions have been proposed regarding the nonuniform temperature distribution (Table 7.4), and, consequently, the decrease of energetic efficiency of the processes. Since in microwave technology thermal energy is generated directly inside the food, it allows overcoming excessive cooking times and, consequently, may have direct implications in terms of both energetic and heating efficiency. A study by Lakshmi et al. (2007) reported that although the absorption of microwave energy by water in foods is 86–89%, the conversion efficiency of electrical energy to microwave energy is only about 50%. The total thermal efficiency was only around 44%. During microwave heating, moisture loss may occur due to the evaporation of moisture at a high temperature. In this case, part of the thermal energy delivered to the foods releases from moisture evaporation rather than increases the temperature of the foods.

Table 7.4 Possible Ways to Decrease Nonuniform Temperature Distribution During Microwave Heating

Food Product	Solution to Reduce Nonuniform Heating	References
n.a.	Combination of conventional and microwave heating	Fung and Cunningham (1980)
Poultry meat	Treatment with reduced microwave power for a long duration	Goksoy et al. (1998)
Multicomponent ready-to-eat meal	Adjusting food geometry	Ohlsson and Thorsell (1984) and Ryynänen and Ohlsson (1996)
Multicomponent ready-to-eat meals	Adjusting food formulation. Changing design of heating chamber	Buffler (1992) and Ryynänen and Ohlsson (1996)
Cylindrical food model made of agar	Shielding using metallic bands for cylindrical samples	Ho and Yam (1992)
Cylindrical food model made of agar	Pulsed microwave heating delivered in pulses and not continuously	Gunasekaran and Yang (2007)
Multiphase porous media transport model	Combination of microwave heating with (1) air jet impingement or (2) infrared heating	Datta and Ni (2002)
Meat	Combination of water as heating medium (as an intermediate heating step) with direct exposure of food to microwaves	Barbosa-Cánovas et al. (2014)

n.a., *not applicable.*
Adapted from Stratakos, A.C., Koidis, A., 2015. Suitability, efficiency and microbiological safety of novel physical technologies for the processing of ready-to-eat meals, meats and pumpable products. International Journal of Food Science & Technology 50, 1283–1302.

7.2.5 ULTRASOUND

Ultrasonic treatment is rapidly growing in the field of novel nonthermal food processing technologies due to its wide range of applications (Patist and Bates, 2008). From the beginning of the 20th century, ultrasound has been established as a laboratory-based prototype technology, and the mechanism of microbial inactivation is generally attributed to intracellular cavitation (Hughes and Nyborg, 1962).

Although at laboratory scale the efficiency of inactivation approaches 100%, in industrial food processing less control of critical factors would prevent inactivation efficiencies from reaching such high levels (FDA, 2015a). In addition, the protective nature of foods to the ultrasonic inactivation of bacterial cells or spores is quite evident when comparing results from microorganisms suspended in buffer (FDA, 2015a). Although these limitations reduce the current probability of commercial development, combination of ultrasound with other preservation processes, such as heat (thermosonication), pressure (manosonication), or both (manothermosonication), appears to have the greatest potential for industrial applications (Stratakos and Koidis, 2015).

Recent studies have focused on these combinations as they show promising results against pathogens such as *Salmonella senftenberg* 775W and *L. monocytogenes* (Álvarez et al., 2006; Pagán et al., 1999). Ultrasound has also been tested for microbial inactivation on solid foods, but this technology is usually not suitable for this type of product. For instance, Seymour et al. (2002) used ultrasound to remove microorganisms from the surface of vegetable pieces, which afterward were deactivated by chemical sanitizers. However, only a low increase in cleaning efficiency was achieved. Although the applications of ultrasound for food preservation purposes are not commercially feasible at the present time (Stratakos and Koidis, 2015), this technology has established efficient tools for other commercial applications, such as emulsification, homogenization, extraction, crystallization, dewatering, degassing, and defoaming (Patist and Bates, 2008). The scaling-up of ultrasound technologies from the laboratory to commercial production was possible thanks to an improved equipment design and higher efficiencies of large-scale continuous flow-through systems (Patist and Bates, 2008). Manufacturers of high-power ultrasound equipment have been focusing on the design of large continuous-flow treatment chambers (flow cells) causing the cost per volume material treated to be reduced. A typical large flow chamber provides 16 kW for flows ranging from 5 to 500 L/min, depending on the application. Larger flow rates would require multiple systems in series or parallel.

The energetic efficiency of ultrasonic generators and transducers has also been improved over the years, thereby reducing internal heating (and subsequent expensive cooling systems), which often caused system failure. Current systems have an energy efficiency around 85%, which simply means that most of the power sent to the transducer is transferred into the medium (Patist and Bates, 2008). Depending on the application, the amount of energy required per liter material treated (often defined as kWh/L) is comparable to any other unit operation in the industry. Depending on the amplitude and the abrasiveness of the medium, the lifetime of a sonotrode ranges from 1 to 18 months (Patist and Bates, 2008).

7.3 IMPLEMENTATION OF EMERGING TECHNOLOGIES IN THE FOOD INDUSTRY

Cost is an extremely important factor that needs to be taken into account before fully implementing any emerging processing technology into an industrial setting. However, systematic cost analysis studies concerning these technologies are still scarce in the literature (Sampedro et al., 2014). This can be changed by the establishment of academia–industry relationships (eg, through the creation of joint projects), thus placing the industry in a central research role and allowing the performance of reliable cost analyses.

Additionally, the environmental impact is also an important factor for the implementation of novel technologies in food industry, since the concern about wastewater and gas emissions, as well as energy consumption, is increasing (Mattson and Sonesson, 2003; Pereira and Vicente, 2010).

7.3.1 THE CASE OF ORANGE JUICE

Sampedro et al. (2013, 2014) analyzed the cost and environmental impact of orange juice processing by PEF and HPP, respectively, and compared both with thermal processing. The total commercial processing costs were estimated through the development of a cost model for PEF associated with heat recovery (30 kV, 60°C) and nonthermal HPP (550 MPa, 90 s, room temperature), comparing with thermal pasteurization with heat recovery (85°C for 5 s). Additionally, a medium-sized facility was used, while it was assumed that 16,500,000 L of orange juice are pasteurized per year with a throughput of 3000 L/h, 20 h of operation/day (4 h used for sanitizing and maintaining the process equipment) and 5500 h/year. The energy costs (electricity and steam charges) were estimated concerning specifically the United States, but it is known that energy prices may change according to country, region, and energy source. Nevertheless, the relative cost difference among processes (% of variance) should remain relatively constant.

Process schemes of PEF, HPP, and thermal systems were designed for juice pasteurization. Both PEF and thermal pasteurization were attached to a heat recovery system, which was proposed as an energy-saving strategy (Meneses et al., 2011; Toepfl et al., 2006). Regarding to PEF system, the refrigerated incoming juice (7°C) is preheated to 48°C in a heat recovery exchanger by the hot juice (60°C) leaving the PEF processing unit. Subsequently, juice enters in the PEF unit, which has six processing chambers connected in three pair series with a heat exchanger after each pair to remove the ohmic heating generated. The hot juice (60°C) that leaves the last pair of chambers enters the heat exchanger where it is cooled down to 19°C. Then, before it is packed and stored, the juice is cooled down to 7°C. In the HPP system, juice is firstly packaged in a suitable container that will fill the plastic basket. The filling ratio on the plastic basket for bottled orange juice was assumed to be about 60–75%. Subsequently, the plastic baskets are moved into the high-pressure vessel (350–420L, 380–386-mm diameter, eight pressure intensifiers, 12cycles/h) for commercial pasteurization, being the conditions recommended by Hiperbaric (Burgos, Spain) for a pasteurized orange juice with a shelf life of 2 months under refrigeration conditions (4°C). The packaged juice is then moved to a cooling system, where it is cooled down to 7°C, being subsequently placed in a cold storage room. The thermal system used in the study was very common, being used by almost all thermal processors of juices. Heat exchangers are used to heat up the raw incoming juice (7°C) to 79.5°C with hot juice (85°C) that is leaving the pasteurization process, which in turn is cooled down to 15.4°C. Subsequently, the pasteurized juice is chilled to 7°C, packaged, and stored.

Table 7.5 shows the detailed cost models for PEF, HPP, and thermal processing of orange juice, namely the respective capital, utility, labor, and facility-related costs, for the conditions described.

The capital costs of PEF processing were estimated taking into account the several factors needed for orange juice pasteurization, namely PEF equipment ($988,000), pumps ($12,000), heat exchangers ($12,000), and cooling system at the end of the process ($38,000). On the other hand, the estimated cost of HPP equipment was $2,500,000, which was based on the average value of two commercial equipment models (320 and 420L vessels) available in the market that reached the desired throughput. Additionally, the supplemental pumps, coolers, and heat exchangers were estimated to cost $45,000. Others capital costs for installing the equipment and commissioning it were added, too. However, some costs could only be determined after the final design and engineering were completed. Following the instructions of Bauman (1964), Jelen et al. (1983), and AACE (1990), total capital costs were estimated to be twice the equipment cost, applying an installation factor of 200%, which may vary depending on the company and the project. Therefore, the estimated total capital costs of PEF, HPP, and thermal processing were $2,100,000 (1.3 ¢/L), $5,090,000 (3.1 ¢/L) and $132,000 (0.08 ¢/L), respectively. Comparing the novel processing technologies to thermal pasteurization, increases of 16-fold and 38-fold in capital costs were achieved for PEF and HPP, respectively.

Utility costs (electricity) were estimated based on an 85% energy efficiency of PEF system with heat recovery and adding a 15% loss factor due to the energy consumed by the PEF unit. Thus, the estimated total energy of the PEF process was 176kJ/L (147kW for 3000L/h). Adding the other industrial components involved (eg, pumping and cooling processes), the total energy consumed by PEF was 865,000kW/year ($69,000 per year), corresponding to 11% of the pasteurization costs (0.4 ¢/L). In the case of the HPP system, the estimated total energy consumed was 1,020,000kW/year ($70,000 per

Table 7.5 Cost of Commercial PEF, HPP, and Thermal Pasteurization Processes for Orange Juice Pasteurization

Process Parameters	Unit of Measure	PEF	HPP	Thermal
Process flow	kg/year	16,500,000	16,500,000	16,500,000
Annual hours of production	h/year	5500	5500	5500
Capital costs				
Heat exchangers	$	12,000	2,000	18,000
PEF/HPP equipment	$	988,000	2,495,000	–
Holding tube	$	–	–	5000
Process chilling	$	38,000	31,000	31,000
Process pumps	$	12,000	12,000	12,000
Other capital costs	$	1,050,000	2,545,000	66,000
Total capital costs	$	2,100,000	5,090,000	132,000
Utility costs				
Process electricity	kW/year	864,000	1,020,000	38,100
Steam	kg/year	–	–	284,000
Cooling water	kg/year	79,100,000	–	–
Electric charges	$/year	60,000	70,000	3000
Steam charges	$/year	–	–	5000
Cooling water charges	$/year	9000	–	–
Total utility costs	$/year	69,000	70,000	8000
Labor costs				
Plant operators per shift		1	3	1
Labor costs	$/h	40.00	40.00	40.00
Total labor costs	$/year	220,000	660,000	220,000
Facility-related costs				
Estimated plant life	year	10	10	10
Maintenance charges	% capital	3.0%	8.0%	2.0%
Other administration charges	% capital	2.5%	2.5%	2.5%
Depreciation	$/year	210,000	508,000	13,000
Maintenance and other administration charges	$/year	117,000	533,000	6000
Total annual costs	$/year	616,000	1,771,000	247,000
Unit pasteurization cost	$/L	0.037	0.107	0.015

Adapted from Sampedro, F., McAloon, A., Yee, W., Fan, X., Zhang, H., Geveke, D., 2013. Cost analysis of commercial pasteurization of orange juice by pulsed electric fields. Innovative Food Science & Emerging Technologies 17, 72–78, Sampedro, F., McAloon, A., Yee, W., Fan, X., Geveke, D., 2014. Cost analysis and environmental impact of pulsed electric fields and high pressure processing in comparison with thermal pasteurization. Food and Bioprocess Technology 7, 1928–1937.

year), corresponding to 4% of the pasteurization costs (0.42 ¢/L). The energy consumed by PEF and HPP systems was similar, but the energy requirements for cooling in the PEF system were higher. In contrast, thermal systems reused thermal energy (≈90%) through a heat recovery system, which reduced the need of energy (Kozempel et al., 1998a). Thus, the total energy consumed by these systems accounts only for $8000 per year, corresponding to 3% of the total costs (0.05 ¢/L). Taking into account that the energy costs contribute in a small proportion to the overall costs of these processing technologies, the reduction in these costs will influence slightly the overall production cost. For instance, a 50% reduction on energy costs would only reduce 2% of the overall production costs.

Regarding the labor costs, one full-time operator was assigned to both PEF and thermal systems, as they are continuous and fairly automated processes. The total labor costs were estimated to be $220,000 per year, corresponding to 35% (1.3¢/L) and 89% (1.33¢/L) of PEF and thermal pasteurization costs, respectively. However, this may be a conservative estimate since some operators may also have other operations, which reduces significantly the labor costs and subsequently the overall ones. Therefore, it may be interesting to combine PEF processes with other ones, in order to reduce the labor costs. In contrast, HPP works in a batch mode and therefore additional labors are required for product load and unload operations. Thus, the authors considered that three operators were needed in this case for the same throughput, increasing the labor costs to $660.000 per year (corresponds to 37% [4.0¢/L] of the total costs). This value can be reduced through the reduction of number of operators, but then a costly automatic line would be required.

Facility-related costs were estimated taking into account a 10-year plant life. The equipment depreciation estimation by a straight line was based on that life period. The annual depreciation costs accounted for $210,000, $508,000, and $13,000 in the case of PEF, HPP, and thermal, respectively. Maintenance and other administration charges were applied to the capital costs, as represented in Table 7.5. The higher percentages of maintenance charges for a HPP system were due to the higher cost of spare parts and labor, and the need of vessel replacement after a certain number of cycles (200,000–500,000 cycles). Therefore, the overall depreciation, maintenance, and other administration charges for PEF, HPP, and thermal systems accounted for 54%, 30%, and 2% of the overall production costs (2.0, 3.2, and 0.04¢/L), respectively.

In conclusion, the overall cost of pasteurizing 1 L of orange juice by the three processing technologies was calculated by adding the capital, labor, and energy costs. For PEF, HPP, and thermal processes, the calculated values were 3.7, 10.7, and 1.5¢/L, respectively. Therefore, the estimated costs of nonthermal technologies were 2.5- and 7-fold (PEF and HPP, respectively) higher than thermal pasteurization. Others authors also studied the costs of PEF and HPP processing for pasteurization of other juices and food products and obtained similar results (Huang and Wang, 2009; Jin and Zhang, 2002; Ludikhuyze et al., 2002; Mújica-Paz et al., 2011; Thakur and Nelson, 1998; Tonello, 2011).

The environmental impact of these processing technologies was also evaluated, being expressed as the production of greenhouse gas levels, eg, carbon dioxide equivalents (kg of CO_2 equivalents). Sampedro et al. (2014) was the first one to study this parameter based on the equivalent CO_2 emissions (kg/year) from electricity (equipment, steam, and cooling water) and natural gas consumption by the processes analyzed (Table 7.6).

The equivalent CO_2 annual emission was similar for PEF and HPP systems (700,000 kg CO_2/year and 773,000 kg CO_2/year, respectively). However, these values represented an increase of 777–858% in the CO_2 emissions compared to the thermal systems, which is mainly due to the higher electricity consumption. In a final stage of the study, Sampedro et al. (2014) performed a sensitivity analysis in order to estimate the effect of the processing volume (500–5500 L/h) on the total cost of pasteurizing 1 L of orange juice. For PEF it was verified that the increase of two-, three- and fivefold in production size (from 1000 to 2,000, 3000 and 5000 L/h) reduced the total costs by nearly 50%, 60%, and 75%, respectively. In the case of HPP, an increase in production size by two-, four-, and sixfold (from 500 to 1000, 2000 and 3000 L/h) reduced the overall costs by 43–73%. In conclusion, the implementation of PEF and HPP in food industry was demonstrated for orange juice processing and, comparing with traditional thermal pasteurization, these nonthermal processing technologies had higher costs and higher environmental impact (in terms of CO_2 production).

Table 7.6 Environmental Impact of PEF, HPP, and Thermal Processes (Sampedro et al., 2014)

Process Parameters	Unit of Measure	PEF	HPP	Thermal
Steam consumption	kg/year			284,000
Cooling water consumption	kg/year	79,131,000		
Natural gas requirements[a]	kg/year			19,000
Electricity requirements (steam)[b]	kW/year			125
Electricity requirements (cooling water)[b]	kW/year	55,000		
Electricity requirements (equipment)	kW/year	865,000	1,020,000	38,100
Total electricity consumption	kW/year	920,000	1,020,000	38,200
Subtotal CO_2 equivalent emissions (electricity)[b]	kg CO_2/year	700,000	773,000	29,000
Subtotal CO_2 equivalent emissions (natural gas)[b]	kg CO_2/year			70,000
Total CO_2 equivalent emissions	kg CO_2/year	700,000	773,000	90,000

[a] *With 80% conversion efficiency factor.*
[b] *Conversion factors obtained from the US Department of Energy.*

7.3.2 THE CASE OF MILK

The agricultural emissions attributed to the dairy sector are approximately 1.9% of the total US greenhouse gas emissions, being derived from both off-farm and on-farm activities (Thoma et al., 2013). The production of fluid milk with extended shelf life contributes to reduction of the milk loss and waste, thus reducing the energy use and the greenhouse emissions. Milk processing by ultra-high temperature (UHT) is the most common technique to extend its shelf life, but other processing technologies may also be used for milk pasteurization. Tomasula et al. (2014) applied a computer simulation model to calculate the use of energy and water to analyze the environmental impact and to perform cost analysis of the milk processing with PEF. The obtained results were compared with thermal processes usually applied in industry: high temperature, short time (HTST), and UHT processing (Tomasula et al., 2013).

The simulations were based on a medium-size plant with processing capacity of 113.6 million L/year (30.0 million gal/year) or 27,300 L/h of raw whole milk to produce 3.25% milk (whole milk) and 40% cream (40% fat content). The simulations took into account the entire process, from the raw milk storage in silos to the temporary cold storage of the packaged product. In addition, it was assumed that the plants operate continuously, with milk processing during 260 days/year, 16 h/day, with 8 h used for equipment cleaning (Tomasula et al., 2013).

The simulation models were designed according to the process schemes for HSTS pasteurization, UHT and PEF processing. In the case of HTST pasteurization, it was assumed that raw milk is firstly pumped from the milk silos to the balance tank, where it is preheated in the first heat exchanger. Thereafter the milk is deaerated and separated into skim milk and cream streams, which is processed into heavy cream. Milk is then homogenized, before entering in the second preheat regenerating heat exchanger. Later, milk is pasteurized at 77°C for 22 s, followed by cooling, packaging, and storing at 1.8°C. On the other hand, the UHT processing was modeled similarly to the first four steps in the HTST process followed by a tubular heat exchanger and an aseptic packaging line replacing the HTST pasteurizers and the packaging equipment. For milk pasteurization using PEF processing, the raw milk was firstly preheated to 55°C in a heat exchanger and then entered into the PEF system at 50°C. The

Table 7.7 Energy and Water Use and Greenhouse Gas Emissions Associated With Electricity and Natural Gas Used in Fluid Milk Process Model (FMPM), UHT, and PEF Processing for Whole Milk Production

Process Parameters	Unit of Measure	FMPM, HTST	UHT	PEF
Process flow	L/year	113,600,000	113,600,000	113,600,000
Energy use				
Total energy	MJ	30,178,030	30,589,790	60,878,110
SEC[a]	MJ/kg of RMP[c]	0.267	0.270	0.538
SEC[a] electrical	MJ/kg of RMP[c]	0.14	0.10	0.44
SEC[a] natural gas	MJ/kg of RMP[c]	0.13	0.17	0.10
Water use	L of water/kg of RMP[c]	0.245	0.273	0.245
Environmental impact				
Total GHG emissions	kg of CO_2e	4,268,060	3,564,979	11,234,118
GHG[b]	g of CO_2e/kg of RMP[c]	37.6	31.4	99.1
GHG[b] natural gas	g of CO_2e/kg of RMP[c]	7.9	10.9	6.1
GHG[b] electricity	g of CO_2e/kg of RMP[c]	29.7	20.6	93.0

[a]*Specific energy consumption.*
[b]*Greenhouse gas emissions.*
[c]*Raw milk processed.*
Adapted from Tomasula, P., Datta, N., Yee, W., McAloon, A., Nutter, D., Sampedro, F., Bonnaillie, L., 2014. Computer simulation of energy use, greenhouse gas emissions, and costs for alternative methods of processing fluid milk. Journal of Dairy Science 97, 4594–4611.

PEF system corresponds to three PEF units in parallel, with each unit handling up to 10,000 L/h of milk. The electric field strength used in this processing ranged from 30 to 40 kV/cm and the treatment time in each unit ranged from 50 to 150 μs. After leaving the PEF units, the hot milk entered in the heat exchanger to be cooled and then was once more cooled to 1.7°C, in a final cooling heat exchanger. The use of energy and water by the different processing technologies as well as the processing costs and their environmental impact were analyzed and is presented in Table 7.7.

For the base fluid milk process model (FMPM, pasteurized by HTST), the energy use was higher for the cleaning-in place step (29.4%) and lower for the wastewater treatment (6.3%), with a total energy consumption of 0.267 MJ/kg of raw milk processed. In this case, the thermal energy (natural gas) used for cleaning and pasteurization operations (0.13 MJ/kg of raw milk processed) and the electrical energy used for milk packaging and homogenization (0.14 MJ/kg of raw milk processed) were mainly responsible for the total energy consumption in this process. Regarding the energy use of UHT milk production, the natural gas used in the process increased to 0.17 MJ/kg of raw milk processed, due to the higher temperatures required when compared to HTST. However, the electrical energy decreased to 0.10 MJ/kg of raw milk processed, since the aseptic packaging has lower energy consumption relatively to HTST packaging, and, in the end, UHT-treated milk does not require refrigeration. Thus, the total energy consumption of the UHT process was 0.270 MJ/kg of raw milk processed. On the other hand, the PEF process has a total energy consumption of 0.538 MJ/kg of raw milk processed, with the PEF units accounting for 51.7% of the total energy used in this process. However, while the electrical energy was 0.44 MJ/kg of raw milk processed, the natural gas use was only 0.10 MJ/kg of raw milk processed, being the highest and the

Table 7.8 Cost of Fluid Milk Process Model (FMPM), UHT, and PEF Processing for Whole Milk Production (Tomasula et al., 2014)

Process Parameters	Unit of Measure	FMPM, HTST	UHT	PEF
Process flow	L/year	113,600,000	113,600,000	113,600,000
Capital costs	$\times$1,000	28,496	36,617	35,712
Operating costs	$\times$1,000	60,347	70,565	62,091
Utilities	$\times$1000	429	470	1123
Waste treatment	$\times$1000	12	13	12
Raw materials	$\times$1,000	46,775	55,768	46,775
Labor	$\times$1000	8986	8986	8986
Other facility costs	$\times$1000	1297	1666	1625
Capital depreciation	$\times$1000	2850	3662	3571
Unit products costs	$/MP entity[a]	1.92	0.60	1.98

[a]*MP entity = 1 gallon of whole milk packaged = 3.785 L of whole milk packaged for all processes except UHT, where MP entity = 1 L of whole milk packaged.*

lowest value obtained in this study, respectively. This lowest value can be attributed to more efficient energy recovery during the process. Based on the energy consumption data, the environmental impact of these processing technologies was also calculated, namely in terms of greenhouse gas emissions, which were expressed in CO_2 equivalents (CO_2e). The lower energy consumption by the processing technologies usually leads to lower greenhouse gas emissions and consequently lower environmental impact. However, it is important to take into account that this lower environmental impact is not necessarily related to a reduction in technology cost. Therefore, the authors also performed an economic analysis of whole milk production, using the three processing technologies. The capital, operating costs, and unit production costs were calculated for each technology and are presented in Table 7.8.

HTST was demonstrated to be the most energy efficient of the three technologies examined, with the lowest energy consumption and the lowest unit production costs. However, UHT processing was the technology with the lowest environmental impact but the highest unit production cost. As previously observed with the orange juice case, the PEF technology applied to milk processing had higher costs compared to the thermal technologies, which is explained by the fact that this technology is still under development.

7.3.3 THE CASE OF OYSTERS

The FDA established that Gulf of Mexico oysters harvested during the warm-weather months require a postharvest processing (FDA, 2009b) due to the association of *Vibrio vulnificus* illnesses to raw oyster consumers (FDA, 2009a). *V. vulnificus* is a bacterium found in warm seawater that can be transmitted to humans through consumption of raw shellfish harvested from waters containing this organism (CDC, 2009). Usually it does not affect healthy individuals; those who are immunocompromised have a greater risk of being affected (CDC, 2009).

One of the processing technologies that can be applied to oyster postharvest processing is HPP. In addition to microbial inactivation, HPP can also be used for both half-shell and shucked oysters, since pressure helps to release the adductor muscle from the shell, making it easy to remove the oyster from the shell. Therefore, Muth et al. (2013) applied HPP to oysters harvested in the summer and analyzed the costs and economic feasibility of this technology. For that, the resource requirements for installing and operating the HPP equipment were determined and subsequently estimated the costs of postharvest processing on a per-oyster basis. Regarding the HPP equipment, the authors used four sizes of pressure vessel in this study: 100L horizontal machine operating at 11 cycles per hour processing 120 shell-weight pounds per cycle; 320L vertical machine operating at 12 cycles per hour processing 450 shell-weight pounds per cycle; 350L horizontal machine operating at 12 cycles per hour processing 500 shell-weight pounds per cycle; and 687L horizontal machine operating at 10 cycles per hour processing 700 shell-weight pounds per cycle. The estimation of the throughput, total costs, and per-unit costs for each HPP piece of equipment are represented in Table 7.9. Two operating schedules were studied (2000 and 4800h per year), assuming the same processing time for both half-shell and shucked oysters.

For 2000 operating hours per year, the per-oyster costs (including the amortized capital equipment costs and annual operating costs) ranged from 5.5 to 7.8 cents per half-shell oyster and 1.8–0.5 cents per shucked oyster. On the other hand, when the operating hours increased to 4800h, the per-oyster costs reduced (4.3–5.3 cents per half-shell oyster and 3.0–2.0 cents per shucked oyster). This small reduction in the costs resulted from the yields increasing for postharvest processed oysters and the shucking time reduction due to the separation of oyster muscle from the shell during HPP process.

In this study the costs of establishing a central facility for postharvest processing using HPP (eg, for land, site development, construction, process validation, postharvest processing equipment, and its installation) were estimated. The cost values achieved were $2.4 million for 100L equipment and $5.3 million for 320L equipment. Despite the other HPP equipment in the study meeting the volume requirements for the central facilities, the authors considered the 320L equipment, due to its lower initial investment costs.

In conclusion, HPP can be successfully used as a postharvest processing technology for oysters harvested in the summer, eliminating the illness risk due to *V. vulnificus*. However, the success of the implementation of this postharvest technology to oysters will depend on the consumer acceptance of this product, which is affected by safety, sensorial properties, and other quality characteristics of these processed oysters. Until now, several sensorial studies were performed and no differences were detected between traditional and processed oysters by raw oyster consumers and trained sensory panels (Andrews et al., 2002; Andrews, 2003; Andrews and Coggins, 2004; Balthrop, 2001; Coggins, 2004; Otwell et al., 2010). However, although HPP-processed oysters are available, the cost is relatively high compared to traditional oysters (Balthrop, 2001; Hanson et al., 2003; Morgan et al., 2009; Posadas and Posadas, 2004).

7.3.4 MEASURES FOR IMPLEMENTATION INCREASING

The great drawback for the implementation of these novel processing technologies in food industry seems to be the higher price of the final product when compared to products thermally processed. Subsequently, the consumer acceptance of products processed by emerging technologies is reduced since most of them are not willing to pay more for these products. Thus, developers and sponsors of these technologies must take several measures to facilitate their implementation in food industry. Consequently, it is expected that implementation increase will reduce the final product price by decreasing the total processing costs (Sampedro et al., 2014). For instance, several authors think that capital costs of PEF may be reduced by changing several aspects of equipment design and a higher demand for HPP units may also result in its capital cost reduction over time (Gaudreau et al., 2005; Kempkes et al., 2010; Sampedro et al., 2014; Yin et al., 1997).

Table 7.9 Throughput Assumptions and Costs for HPP Application to Oyster Postharvest Process, Assuming 2000 and 4800 Operating Hours per Year (Muth et al., 2013)

Process Parameters	100 L Horizontal	320 L Vertical	350 L Horizontal	687 L Horizontal
2000 operating hours per year				
Annual throughput assumptions				
Half-shell oysters	3,960,000	16,200,000	18,000,000	21,000,000
Shucked oysters	2,640,000	10,800,000	12,000,000	14,000,000
Total oysters	6,600,000	27,000,000	30,000,000	35,000,000
Total shell-weight pounds	2,640,000	10,800,000	12,000,000	14,000,000
Total sacks	26,400	108,000	120,000	140,000
Total cost estimation				
Total capital costs	\$1,295,000	\$2,065,000	\$2,421,500	\$3,125,000
Total annual operating costs	\$323,198	\$713,512	\$784,455	\$973,755
Per-unit cost estimation				
Per half-shell oyster	\$0.078	\$0.056	\$0.055	\$0.057
Per shucked oyster[a]	−\$0.005	−\$0.017	−\$0.018	−\$0.016
Per sack	\$12.24	\$6.61	\$6.54	\$6.96
4800 operating hours per year				
Annual throughput assumptions				
Half-shell oysters	15,840,000	38,880,000	43,200,000	50,400,000
Shucked oysters	6,336,000	25,920,000	28,800,000	33,600,000
Total oysters	22,176,000	64,800,000	72,000,000	84,000,000
Total shell-weight pounds	8,870,400	25,920,000	28,800,000	33,600,000
Total sacks	88,704	259,200	288,000	336,000
Total cost estimation				
Total capital costs	\$1,295,000	\$2,065,000	\$2,421,500	\$3,125,000
Total annual operating costs	\$383,390	\$905,752	\$1,079,655	\$1,169,755
Per-unit cost estimation				
Per half-shell oyster	\$0.053	\$0.043	\$0.044	\$0.043
Per shucked oyster[a]	−\$0.020	−\$0.030	−\$0.029	−\$0.030
Per sack	\$5.88	\$3.49	\$3.75	\$3.48

[a]*Negative cost values for shucked oysters mean that processors incur "savings" resulting from increased yields for shucked oysters.*

However, additional corporative measures should be considered as possible encouragements for implementation of these novel technologies. One example is the application of the open-innovation model, where companies use external ideas, technology, and solutions in addition to the internal ones to innovate, opening channels for knowledge access (Saguy and Sirotinskaya, 2014). This can be performed by the promotion of win–win collaborations between universities and industries, where universities provide the technical knowledge about food products and technology for the implementation of processing technology in food industry. On the other hand, this partnership can work the other way around, with the universities providing pilot-scale processors for treatment and companies the technical information about food products and consumers' needs. This case may work for small and medium-sized enterprises (SMEs), since they have several limitations and barriers concerning the high capital costs of the emergent processing technologies. For these partnerships to succeed, all participants of open-innovation model have to be proactive, not only universities and industry but also government and private sector. For instance, some governmental measures must be taken to support innovation and the private sector must provide funds to help the occurrence of open innovation in SMEs (Saguy, 2013). These measures altogether may reduce the final product price and subsequently increase consumer acceptance.

Additionally, there are other possible reasons for the lower acceptance of these novel products, since despite new food-processing technologies usually being well-received by food scientists consumers may have a more conservative position and do not always readily accept the benefits of new processing methods. However, providing more information about the technologies seems to be the key to increasing consumer acceptance of these products. This is especially important in the case of introducing PEF-processed products since many consumers associate the technology name with electricity and are skeptical about the possible side effects on food (Sonne et al., 2012). Therefore, in some cases it would be important to have information about the processing technologies described on the package, allowing the consumer to recognize the advantages of these products over the thermally processed ones. In addition, the establishment of a new label specifically for food products processed by nonconventional technologies would help to demonstrate the value added to these products, and then the consumer may be willing to pay more for them.

ACKNOWLEDGMENTS

F. J. Barba was supported from the Union by a postdoctoral Marie Curie Intra-European Fellowship (Marie Curie IEF) within the 7th European Community Framework Program (http://cordis.europa.eu/fp7/mariecurieactions/ief_en.html) (project number 626524—HPBIOACTIVE—Mechanistic modeling of the formation of bioactive compounds in high pressure processed seedlings of Brussels sprouts for effective solution to preserve healthy compounds in vegetables).

REFERENCES

AACE, 1990. Conducting Technical and Economic Evaluations in the Process and Utility Industries. Association for the Advancement of Cost Engineering, Morgantown, WV.

Ahmed, J., Ramaswamy, H.S., Alli, I., Raghavan, V.G.S., 2007. Protein denaturation, rheology, and gelation characteristics of radio-frequency heated egg white dispersions. International Journal of Food Properties 10, 145–161.

Alemán, G.D., Walker, M., Farkas, D.F., Torres, J.A., Ting, E.Y., Mordre, S.C., Hawes, A.C., 1996. Pulsed ultra high pressure treatments for pasteurization of pineapple juice. Journal of Food Science 61, 388–390.

Álvarez, I., Mañas, P., Virto, R., Condón, S., 2006. Inactivation of *Salmonella senftenberg* 775W by ultrasonic waves under pressure at different water activities. International Journal of Food Microbiology 108, 218–225.

Ammar, J.B., Lanoisellé, J.-L., Lebovka, N.I., Van Hecke, E., Vorobiev, E., 2010. Effect of a pulsed electric field and osmotic treatment on freezing of potato tissue. Food Biophysics 5, 247–254.

AMT, 2014. Advanced Microwave Technologies.

Anderson, A.K., Finkelstein, R., 1919. A study of the electropure process of treating milk. Journal of Dairy Science 2, 374–406.

Andrews, L.S., Coggins, P., 2004. Consumer acceptability of post-harvest processed and value added oysters – year 2. In: Jamir, T.V.C., Jamison, J., Posadas, R., Smith, E., Mitchell, T., McNeely, J., Balthrop, P. (Eds.), Integrated Oyster Market Research, Product Development, Evaluation, Promotion and Consumer Education for the Gulf of Mexico's Oyster Industry. Gulf and South Atlantic Fisheries Foundation, Inc., Tampa, US. Sea Grant Contract #NA16RG2195 (GSAFF # 88) Project R/LR-Q-23 Year II.

Andrews, L., Posadas, B., Jahncke, M., 2002. Oyster irradiation: pathogenic *Vibrio* response and consumer difference testing. In: 26th Annual Tropical and Subtropical Fisheries Technological Conference of the Americas, Orlando, US.

Andrews, L.S., 2003. Consumer acceptability of post harvest processed and value added oysters. In: 27th Annual Tropical and Subtropical Fisheries Technological Conference of the Americas, Gainesville, US.

Balthrop, P., 2001. New oyster product: processing and market research. In: U.S. Department of Commerce – National Oceanic and Atmospheric Administration. University of Florida, Gainesville, US. Cooperative agreement NA76RG-0120.

Barba, F.J., Esteve, M.J., Frígola, A., 2012. High pressure treatment effect on physicochemical and nutritional properties of fluid foods during storage: a review. Comprehensive Reviews in Food Science and Food Safety 11, 307–322.

Barba, F.J., Grimi, N., Vorobiev, E., 2015. New approaches for the use of non-conventional cell disruption technologies to extract potential food additives and nutraceuticals from microalgae. Food Engineering Reviews 7, 45–62.

Barbosa-Cánovas, G.V., Altunakar, B., 2006. Pulsed Electric Fields processing of foods: an overview. In: Raso, J., Heinz, V. (Eds.), Pulsed Electric Fields Technology for the Food Industry. Springer, New York, US, pp. 153–194.

Barbosa-Cánovas, G.V., Juliano, P., 2008. Food sterilization by combining high pressure and thermal energy. In: Gutiérrez-Lopez, G.F., Barbosa-Cánovas, G.V., Welti-Chanes, J., Parada-Arias, E. (Eds.), Food Engineering: Integrated Approaches. Springer Science, New York, US, pp. 9–46.

Barbosa-Cánovas, G.V., Medina-Meza, I., Candoğan, K., Bermúdez-Aguirre, D., 2014. Advanced retorting, microwave assisted thermal sterilization (MATS), and pressure assisted thermal sterilization (PATS) to process meat products. Meat Science 98, 420–434.

Bauman, H.C., 1964. Fundamentals of Cost Engineering in the Chemical Industry. Reinhold Pub. Corp., London, UK.

Bermejo-Prada, A., Vega, E., Perez-Mateos, M., Otero, L., 2015. Effect of hyperbaric storage at room temperature on the volatile profile of strawberry juice. LWT – Food Science and Technology 62, 906–914.

Brennan, J.G., 2012. Food Processing Handbook. Wiley-VCH, Weinheim, Germany.

Brody, A.L., 2012. The coming wave of microwave sterilization and pasteurization. Food Technology Magazine 66, 78–80.

Buckow, R., Chandry, P.S., Ng, S.Y., McAuley, C.M., Swanson, B.G., 2014. Opportunities and challenges in pulsed electric field processing of dairy products. International Dairy Journal 34, 199–212.

Buffler, C.R., 1992. Microwave Cooking and Processing. Van Nostrand Reinhold, New York, US.

Bushnell, A.H., Dunn, J.E., Clark, R.W., Pearlman, J.S., 1993. High pulsed voltage systems for extending the shelf life of pumpable food products. Foodco Corporation US Patent 5235905.

Carlez, A., Rosec, J.-P., Richard, N., Cheftel, J.-C., 1994. Bacterial growth during chilled storage of pressure-treated minced meat. LWT – Food Science and Technology 27, 48–54.

Castro, I., 2007. Ohmic Heating as an Alternative to Conventional Thermal Treatment. Universidade do Minho, Portugal.

CDC, 2009. *Vibrio vulnificus*: General Information. Centers for Disease Control and Prevention, Atlanta, US.

Chandrasekaran, S., Ramanathan, S., Basak, T., 2013. Microwave food processing – a review. Food Research International 52, 243–261.

Chawla, R., Patil, G.R., Singh, A.K., 2011. High hydrostatic pressure technology in dairy processing: a review. Journal of Food Science and Technology 48, 260–268.
Cheftel, J.C., 1995. High-pressure, microbial inactivation and food preservation. Food Science and Technology International 1, 75–90.
Cho, H.-Y., Yousef, A.E., Sastry, S.K., 1999. Kinetics of inactivation of *Bacillus subtilis* spores by continuous or intermittent ohmic and conventional heating. Biotechnology and Bioengineering 62, 368–372.
Clark, J.P., 2013. Processing and engineering highlights. Food Technology 67, 117–120.
Coggins, P., 2004. Sensory differences of Gulf post harvest processed oysters. In: Jamir, T.V.C., Jamison, J., Posadas, R., Smith, E., Mitchell, T., McNeely, J., Balthrop, P. (Eds.), Integrated Oyster Market Research, Product Development, Evaluation, Promotion and Consumer Education for the Gulf of Mexico's Oyster Industry. Gulf and South Atlantic Fisheries Foundation, Inc., Tampa, US. Sea Grant Contract #NA16RG2195 (GSAFF # 88) Project R/LR-Q-23 Year II.
Datta, A., Ni, H., 2002. Infrared and hot-air-assisted microwave heating of foods for control of surface moisture. Journal of Food Engineering 51, 355–364.
Datta, A.K., 2003. Microwave food preservation. In: Heldman, D.R. (Ed.), Encyclopedia of Agricultural, Food, and Biological Engineering. Marcel Dekker, New York, US, pp. 657–661.
De Halleux, D., Piette, G., Buteau, M., Dostie, M., 2005. Ohmic cooking of processed meats: energy evaluation and food safety considerations. Canadian Biosystems Engineering 47, 41–47.
De Vleeschouwer, K., Van der Plancken, I., Van Loey, A., Hendrickx, M.E., 2010. The effect of high pressure-high temperature processing conditions on acrylamide formation and other maillard reaction compounds. Journal of Agricultural and Food Chemistry 58, 11740–11748.
Deng, Q., Zinoviadou, K.G., Galanakis, C.M., Orlien, V., Grimi, N., Vorobiev, E., Lebovka, N., Barba, F.J., 2014. The effects of conventional and non-conventional processing on glucosinolates and its derived forms, isothiocyanates: extraction, degradation, and applications. Food Engineering Reviews 7 (3), 357–381.
Donsì, F., Ferrari, G., Pataro, G., 2010. Applications of pulsed electric field treatments for the enhancement of mass transfer from vegetable tissue. Food Engineering Reviews 2, 109–130.
Duarte, R.V., Moreira, S.A., Fernandes, P.A.R., Fidalgo, L.G., Santos, M.D., Queiros, R.P., Santos, D.I., Delgadillo, I., Saraiva, J.A., 2015. Preservation under pressure (hyperbaric storage) at 25 °C, 30 °C and 37 °C of a highly perishable dairy food and comparison with refrigeration. CyTA – Journal of Food 13, 321–328.
Dujmić, F., Brnčić, M., Karlović, S., Bosiljkov, T., Ježek, D., Tripalo, B., Mofardin, I., 2013. Ultrasound-assisted infrared drying of pear slices: textural issues. Journal of Food Process Engineering 36, 397–406.
Dunn, J.E., Pearlman, J.S., 1987. Methods and Apparatus for Extending the Shelf-life of Fluid Food Products. Maxwell Laboratories, Inc.. US Patent 4695472.
Dunn, J., 2001. Pulsed electric field processing: an overview. In: Barbosa-Cánovas, G.V., Zhang, Q.H. (Eds.), Pulsed Electric Fields in Food Processing: Fundamental Aspects and Applications. Technomic Publishing Company, Lancaster, UK, pp. 1–30.
FDA, 2000. Kinetics of Microbial Inactivation for Alternative Food Processing Technologies. Institute of Food Technologists.
FDA, 2009a. Backgrounder on Measures to Eliminate Risk Caused by *Vibrio vulnificus* Infection from Consumption of Raw Molluscan Shellfish. (Silver Spring).
FDA, 2009b. In: Letter to Interstate Shellfish Sanitation Conference, Washington, DC.
FDA, 2015a. Kinetics of Microbial Inactivation for Alternative Food Processing Technologies – Ultrasound.
FDA, 2015b. Safe Practices for Food Processes – Kinetics of Microbial Inactivation for Alternative Food Processing Technologies – Ohmic and Inductive Heating.
Fernandes, P.A., Moreira, S.A., Fidalgo, L.G., Santos, M.D., Queirós, R.P., Delgadillo, I., Saraiva, J.A., 2015a. Food preservation under pressure (hyperbaric storage) as a possible improvement/alternative to refrigeration. Food Engineering Reviews 7, 1–10.

Fernandes, P.A.R., Moreira, S.A., Duarte, R., Santos, D.I., Queirós, R.P., Fidalgo, L.G., Santos, M.D., Delgadillo, I., Saraiva, J.A., 2015b. Preservation of sliced cooked ham at 25, 30 and 37°C under moderated pressure (hyperbaric storage) and comparison with refrigerated storage. Food and Bioproducts Processing 95, 200–207.

Fidalgo, L.G., Santos, M.D., Queirós, R.P., Inácio, R.S., Mota, M.J., Lopes, R.P., Gonçalves, M.S., Neto, R.F., Saraiva, J.A., 2014. Hyperbaric storage at and above room temperature of a highly perishable food. Food and Bioprocess Technology 7, 2028–2037.

Fung, D.Y., Cunningham, F., 1980. Effect of microwaves on microorganisms in foods. Journal of Food Protection 43, 641–650.

Furukawa, S., Shimoda, M., Hayakawa, I., 2003. Mechanism of the inactivation of bacterial spores by reciprocal pressurization treatment. Journal of Applied Microbiology 94, 836–841.

Galanakis, C.M., Barba, F.J., Mitropoulou, V.S., Martinez-Saez, N., del Castillo, M.D., 2012. Patented and commercialized applications. In: Galanakis, C.M. (Ed.), Food Waste Recovery: Processing Technologies and Industrial Techniques. Elsevier – Academic Press, London, UK, pp. 339–362.

Galotto, M.J., Ulloa, P., Escobar, R., Guarda, A., Gavara, R., Miltz, J., 2010. Effect of high-pressure food processing on the mass transfer properties of selected packaging materials. Packaging Technology and Science 23, 253–266.

Gaudreau, M., Hawkey, T., Petry, J., Kempkes, M., 2005. Solid-state power systems for pulsed electric field (PEF) processing. In: Pulsed Power Conference. IEEE, Monterey, US, pp. 1278–1281.

George, R.M., Leadley, C.E., Emond, S.P., Saraiva, C., Smith, S., Turner, J., 2003. New Technologies Bulletin 26. Campden and Chorleywood Food Research Association, Chipping Campden, Gloucestershire, UK.

Gerlach, D., Alleborn, N., Baars, A., Delgado, A., Moritz, J., Knorr, D., 2008. Numerical simulations of pulsed electric fields for food preservation: a review. Innovative Food Science & Emerging Technologies 9, 408–417.

Getchell, B., 1935. Electric pasteurization of milk. Agriculture Engineering 16, 408–410.

Ghnimi, S., Fillaudeau, L., 2014. Tubular and jet fluid units. In: Ramaswamy, H.S., Marcotte, M., Sastry, S., Abdelrahim, K. (Eds.), Ohmic Heating in Food Processing. CRC Press, Taylor & Francis Group, Boca Raton, US, pp. 183–196.

Goksoy, E.O., James, C., James, S.J., 1998. Non-uniformity of surface temperatures after microwave heating of poultry meat. The Journal of Microwave Power and Electromagnetic Energy 34, 149–160 A Publication of the International Microwave Power Institute.

Goullieux, A., Pain, J.-P., 2005. Ohmic heating. In: Sun, D.-W. (Ed.), Emerging Technologies for Food Processing. Academic Press, London, UK, pp. 47–65.

Guillou, S., El Murr, N., 2002. Inactivation of *Saccharomyces cerevisiae* in solution by low-amperage electric treatment. Journal of Applied Microbiology 92, 860–865.

Gunasekaran, S., Yang, H.-W., 2007. Effect of experimental parameters on temperature distribution during continuous and pulsed microwave heating. Journal of Food Engineering 78, 1452–1456.

Hanson, T., House, L., Sureshwaran, S., Posadas, B., Liu, A., 2003. Opinions of U.S. Consumers Toward Oysters: Results of a 2000–2001 Survey. Mississippi Agricultural and Forestry Experiment Station Bulletin 1133. Mississippi State University, Mississippi, US.

Heddleson, R.A., Doores, S., 1994. Factors affecting microwave heating of foods and microwave induced destruction of foodborne pathogens – a review. Journal of Food Protection 57, 1025–1037.

Heinz, V., Toepfl, S., Knorr, D., 2003. Impact of temperature on lethality and energy efficiency of apple juice pasteurization by pulsed electric fields treatment. Innovative Food Science & Emerging Technologies 4, 167–175.

Hite, B.H., 1899. The Effects of Pressure in the Preservation of Milk, vol. 58. Bulletin West Virginia University Agricultural Experiment Station, pp. 15–35.

Ho, Y., Yam, K., 1992. Effect of metal shielding on microwave heating uniformity of a cylindrical food model. Journal of Food Processing and Preservation 16, 337–359.

Hoogland, H., Heij, W.d., Schepdael, L.v, 2001. High pressure sterilization: novel technology, new products, new opportunities. New Food 4.

Horie, Y., 1992. Fruit preparations – development of a new pressurizing process. In: Charalambous, G. (Ed.), Off-flavors in Foods and Beverages. Elsevier Science, Amsterdam, The Netherlands, pp. 313–328.
Huang, L., Sites, J., 2007. Automatic control of a microwave heating process for in-package pasteurization of beef frankfurters. Journal of Food Engineering 80, 226–233.
Huang, K., Wang, J., 2009. Designs of pulsed electric fields treatment chambers for liquid foods pasteurization process: a review. Journal of Food Engineering 95, 227–239.
Hughes, D.E., Nyborg, W.L., 1962. Cell disruption by ultrasound streaming and other activity around sonically induced bubbles is a cause of damage to living cells. Science 138, 108–114.
Hülsheger, H., Niemann, E.-G., 1980. Lethal effects of high-voltage pulses on *E. coli* K12. Radiation and Environmental Biophysics 18, 281–288.
Huppertz, T., Fox, P.F., de Kruif, K.G., Kelly, A.L., 2006. High pressure-induced changes in bovine milk proteins: a review. Biochimica et Biophysica Acta (BBA)-Proteins and Proteomics 1764, 593–598.
Imai, T., Uemura, K., Ishida, N., Yoshizaki, S., Noguchi, A., 1995. Ohmic heating of Japanese white radish *Rhaphanus sativus* L. International Journal of Food Science & Technology 30, 461–472.
IMS, 2012. Industrial Microwave Systems.
Jalté, M., Lanoisellé, J.-L., Lebovka, N.I., Vorobiev, E., 2009. Freezing of potato tissue pre-treated by pulsed electric fields. LWT – Food Science and Technology 42, 576–580.
James, C., Swain, M.V., James, S.J., Swain, M.J., 2002. Development of methodology for assessing the heating performance of domestic microwave ovens. International Journal of Food Science & Technology 37, 879–892.
Jelen, F.C., Black, J.H., Engineers, A.A.o.C., 1983. Cost and Optimization Engineering. McGraw-Hill Book Company, New York, US.
Jin, T., Zhang, Q., 2002. Cost evaluation of a commercial scale PEF system. In: 2002 IFT Annual Meeting, Technical Program 91E-21, Anaheim, US.
Johnson, P.E., Van der Plancken, I., Balasa, A., Husband, F.A., Grauwet, T., Hendrickx, M., Knorr, D., Mills, E., Mackie, A.R., 2010. High pressure, thermal and pulsed electric-field-induced structural changes in selected food allergens. Molecular Nutrition & Food Research 54, 1701–1710.
Kempkes, M.A., Doona, C.J., Kustin, K., Feeherry, F.E., 2010. Pulsed electric field (PEF) systems for commercial food and juice processing. In: Doona, C.J., Kustin, K., Feeherry, F.E. (Eds.), Case Studies in Novel Food Processing Technologies: Innovations in Processing, Packaging, and Predictive Modelling. Woodhead Publishing, Cambridge, UK, pp. 73–102.
Knirsch, M.C., Dos Santos, C.A., de Oliveira Soares, A.A.M., Penna, T.C.V., 2010. Ohmic heating – a review. Trends in Food Science & Technology 21, 436–441.
Knoerzer, K., Juliano, P., Gladman, S., Versteeg, C., Fryer, P.J., 2007. A computational model for temperature and sterility distributions in a pilot-scale high-pressure high-temperature process. AIChE Journal 53, 2996–3010.
Knorr, D., Zenker, M., Heinz, V., Lee, D.-U., 2004. Applications and potential of ultrasonics in food processing. Trends in Food Science & Technology 15, 261–266.
Knorr, D., 1994. Effect of high hydrostatic pressure on food safety and quality. Food Technology 47, 156–161.
Knorr, D., 1995. Hydrostatic pressure treatment of food microbiology. In: Gould, G.W. (Ed.), New Methods of Food Preservation. Blackie Academic and Professional, New York, US, pp. 159–175.
Koutchma, T., Song, Y., Setikaite, I., Juliano, P., Barbosa-Cánovas, G.V., Dunne, C.P., Patazca, E., 2010. Packaging evaluation for high-pressure high-temperature sterilization of shelf-stable foods. Journal of Food Process Engineering 33, 1097–1114.
Kozempel, M., McAloon, A., Yee, W., 1998a. The cost of pasteurizing apple cider. Food Technology 52, 50–52.
Kozempel, M.F., Annous, B.A., Cook, R.D., Scullen, O.J., Whiting, R.C., 1998b. Inactivation of microorganisms with microwaves at reduced temperatures. Journal of Food Protection 61, 582–585.
Lakshmi, S., Chakkaravarthi, A., Subramanian, R., Singh, V., 2007. Energy consumption in microwave cooking of rice and its comparison with other domestic appliances. Journal of Food Engineering 78, 715–722.
Ledward, D.A., 1995. High pressure processing – the potential. In: Ledward, D.A., Johnston, D.E., Earnshaw, R.G., Hasting, A.P.M. (Eds.), High Pressure Processing of Foods. Nottingham University Press, Nottingham, UK, pp. 1–5.

Lelieveld, H., 2005. Pef – a food Industry's view. In: Barbosa-Cánovas, G.V., Tapia, M.S., Cano, M.P., Martín-Belloso, O., Martínez, A. (Eds.), Novel Food Processing Technologies. CRC Press, Boca Raton, US, pp. 145–156.

Lima, M., Zhong, T., Lakkakula, N.R., 2002. Ohmic Heating: A Value-added Food Processing Tool. (A technical report from Louisiana Agriculture Magazine).

Ludikhuyze, L., Van Loey, A., Indrawati, Hendrickx, M., 2002. High pressure processing of fruit and vegetables. In: Jongen, W.M.F. (Ed.), Fruit and Vegetable Processing. Woodhead Publishing, Cambridge, UK, pp. 346–362.

Masanet, E., 2008. Energy Efficiency Improvement and Cost Saving Opportunities for the Fruit and Vegetable Processing Industry. An Energy Star Guide for Energy and Plant Managers (Technical report: Environmental Energy Technologies Division. U.S. Environmental Protection Agency – Environmental Energy Technologies Division).

Matser, A.M., Krebbers, B., van den Berg, R.W., Bartels, P.V., 2004. Advantages of high pressure sterilisation on quality of food products. Trends in Food Science & Technology 15, 79–85.

Mattson, B., Sonesson, U., 2003. Introduction. In: Mattson, B., Sonesson, U. (Eds.), Environmentally-friendly Food Processing. Woodhead Publishing Limited, Cambridge, UK, pp. 1–2.

Meneses, N., Jaeger, H., Knorr, D., 2011. Minimization of thermal impact by application of electrode cooling in a co-linear PEF treatment chamber. Journal of Food Science 76, E536–E543.

Min, S., Jin, Z.T., Min, S.K., Yeom, H., Zhang, Q.H., 2003a. Commercial-scale pulsed electric field processing of orange juice. Journal of Food Science 68, 1265–1271.

Min, S., Jin, Z.T., Zhang, Q.H., 2003b. Commercial scale pulsed electric field processing of tomato juice. Journal of Agricultural and Food Chemistry 51, 3338–3344.

Moreau, C., 1995. Semi-continuous high pressure cell for liquid processing. In: Ledward, D.A., Johnston, D.E., Earnshaw, R.G., Hasting, A.P.M. (Eds.), High Pressure Processing of Foods. Nottingham University Press, Nottingham, UK, pp. 181–197.

Moreira, S.A., Fernandes, P.A.R., Duarte, R., Santos, D.I., Fidalgo, L.G., Santos, M.D., Queirós, R.P., Delgadillo, I., Saraiva, J.A., 2015. A first study comparing preservation of a ready-to-eat soup under pressure (hyperbaric storage) at 25°C and 30°C with refrigeration. Food Science & Nutrition 3 (6), 467–474.

Morgan, O.A., Martin, G.S., Huth, W.L., 2009. Oyster demand adjustments to counter-information and source treatments in response to *Vibrio vulnificus*. Journal of Agricultural and Applied Economics 41, 683–696.

Moses, B.D., 1938. Electric pasteurization of milk. Agriculture Engineering 19, 525–526.

Mota, M.J., Lopes, R.P., Delgadillo, I., Saraiva, J.A., 2015. Probiotic yogurt production under high pressure and the possible use of pressure as an on/off switch to stop/start fermentation. Process Biochemistry 50, 906–911.

Mújica-Paz, H., Valdez-Fragoso, A., Samson, C.T., Welti-Chanes, J., Torres, J.A., 2011. High-pressure processing technologies for the pasteurization and sterilization of foods. Food and Bioprocess Technology 4, 969–985.

Muth, M.K., Viator, C.L., Karns, S.A., Cajka, J.C., O'Neil, M., 2013. Analysis of the costs and economic feasibility of requiring postharvest processing for raw oysters. Comprehensive Reviews in Food Science and Food Safety 12, 652–661.

Nguyen, L.T., Choi, W., Lee, S.H., Jun, S., 2013. Exploring the heating patterns of multiphase foods in a continuous flow, simultaneous microwave and ohmic combination heater. Journal of Food Engineering 116, 65–71.

Oey, I., Lille, M., Van Loey, A., Hendrickx, M., 2008a. Effect of high-pressure processing on colour, texture and flavour of fruit-and vegetable-based food products: a review. Trends in Food Science & Technology 19, 320–328.

Oey, I., Van der Plancken, I., Van Loey, A., Hendrickx, M., 2008b. Does high pressure processing influence nutritional aspects of plant based food systems? Trends in Food Science & Technology 19, 300–308.

Ohlsson, T., Bengtsson, N., 2002. Minimal processing of foods with non-thermal methods. In: Ohlsson, T., Bengtsson, N. (Eds.), Minimal Processing Technologies in the Food Industries. Woodhead Publishing Limited, Cambridge, UK, pp. 34–60.

Ohlsson, T., Thorsell, U., 1984. Problems in microwave reheating of chilled foods. Foodservice Research International 3, 9–16.

Olsson, S., 1995. Production equipment for commercial use. In: Ledward, D.A., Johnston, D.E., Earnshaw, R.G., Hasting, A.P.M. (Eds.), High Pressure Processing of Foods. Nottingham Univeristy Press, Nottingham, UK, pp. 167–180.

Otwell, S., Garrido, L., Garrido, V., Sims, C., 2010. Sensory Assessment Study for Post-harvest Processed (PHP) Oysters. University of Florida, Gainesville, US.

Pagán, R., Manas, P., Raso, J., Condón, S., 1999. Bacterial resistance to ultrasonic waves under pressure at nonlethal (manosonication) and lethal (manothermosonication) temperatures. Applied and Environmental Microbiology 65, 297–300.

Palaniappan, S., Sastry, S., 2002. Ohmic heating. In: Juneja, V.K., Sofos, J.N. (Eds.), Control of Foodborne Microorganisms. Marcel Dekker, New York, US, pp. 451–460.

Patist, A., Bates, D., 2008. Ultrasonic innovations in the food industry: from the laboratory to commercial production. Innovative Food Science & Emerging Technologies 9, 147–154.

Pereira, R.N., Vicente, A.A., 2010. Environmental impact of novel thermal and non-thermal technologies in food processing. Food Research International 43, 1936–1943.

Pereira, R., Martins, J., Mateus, C., Teixeira, J.A., Vicente, A.A., 2007. Death kinetics of *Escherichia coli* in goat milk and *Bacillus licheniformis* in cloudberry jam treated by ohmic heating. Chemical Papers 61, 121–126.

Posadas, B.C., Posadas, R.A., 2004. Consumer preferences for postharvest processed raw oyster products in coastal Mississippi. In: Jamir, T.V.C., Jamison, J., Posadas, R., Smith, E., Mitchell, T., McNeely, J., Balthrop, P. (Eds.), Integrated Oyster Market Research, Product Development, Evaluation, Promotion and Consumer Education for the Gulf of Mexico's Oyster Industry. Gulf and South Atlantic Fisheries Foundation, Inc., Tampa, FL, US. Sea Grant Contract #NA16RG2195 (GSAFF # 88) Project R/LR-Q-23 Year II.

Puértolas, E., López, N., Condón, S., Álvarez, I., Raso, J., 2010. Potential applications of PEF to improve red wine quality. Trends in Food Science & Technology 21, 247–255.

Püschner, H.A., 1966. Heating with Microwaves. Philips Technical Library, Berlin, Germany.

Queirós, R.P., Santos, M.D., Fidalgo, L.G., Mota, M.J., Lopes, R.P., Inácio, R.S., Delgadillo, I., Saraiva, J.A., 2014. Hyperbaric storage of melon juice at and above room temperature and comparison with storage at atmospheric pressure and refrigeration. Food Chemistry 147, 209–214.

Ramaswamy, H.S., Chen, C., Marcotte, M., 1999. Novel processing technologies in food preservation. In: Barrett, D.M., Somogyi, L.P., Ramaswamy, H.S. (Eds.), Processing Fruits: Science and Technology. CRC, Boca Raton, US, pp. 201–220.

Ramirez, R., Saraiva, J., Lamela, C.P., Torres, J.A., 2009. Reaction kinetics analysis of chemical changes in pressure-assisted thermal processing. Food Engineering Reviews 1, 16–30.

Raso, J., Barbosa-Cánovas, G.V., 2003. Nonthermal preservation of foods using combined processing techniques. Critical Reviews in Food Science and Nutrition 43, 265–285.

Rice, J., 1995. Ohmic adventures. Food Processing 56, 87–91.

Rodríguez-Roque, M.J., de Ancos, B., Sánchez-Moreno, C., Cano, M.P., Elez-Martínez, P., Martín-Belloso, O., 2015. Impact of food matrix and processing on the in vitro bioaccessibility of vitamin C, phenolic compounds, and hydrophilic antioxidant activity from fruit juice-based beverages. Journal of Functional Foods 14, 33–43.

Roselló-Soto, E., Galanakis, C.M., Brnčić, M., Orlien, V., Trujillo, F.J., Mawson, R., Knoerzer, K., Tiwari, B.K., Barba, F.J., 2015. Clean recovery of antioxidant compounds from plant foods, by-products and algae assisted by ultrasounds processing. Modeling approaches to optimize processing conditions. Trends in Food Science & Technology 42, 134–149.

Ryynänen, S., Ohlsson, T., 1996. Microwave heating uniformity of ready meals as affected by placement, composition, and geometry. Journal of Food Science 61, 620–624.

Ryynänen, S., Tuorila, H., Hyvönen, L., 2001. Perceived temperature effects on microwave heated meals and meal components. Food Service Technology 1, 141–148.

Sablani, S.S., 2014. 75 Years of IFT: food engineering and physical properties, and nanoscale food science, engineering and technology in JFS – 1936 to present. Journal of Food Science 79, iii–v.

Saguy, I.S., Sirotinskaya, V., 2014. Challenges in exploiting open innovation's full potential in the food industry with a focus on small and medium enterprises (SMEs). Trends in Food Science & Technology 38, 136–148.

Saguy, I.S., 2013. Academia-industry innovation interaction: paradigm shifts and avenues for the future. In: Yannniotis, S., Taoukis, P., Stoforos, N.G., Karathanos, V.T. (Eds.), Advances in Food Process Engineering Research and Applications. Springer, New York, US, pp. 645–656.

Sakr, M., Liu, S., 2014. A comprehensive review on applications of ohmic heating (OH). Renewable and Sustainable Energy Reviews 39, 262–269.

Salazar-González, C., San Martín-González, M.F., López-Malo, A., Sosa-Morales, M.E., 2012. Recent studies related to microwave processing of fluid foods. Food and Bioprocess Technology 5, 31–46.

Saldaña, G., Álvarez, I., Condón, S., Raso, J., 2014. Microbiological aspects related to the feasibility of PEF technology for food pasteurization. Critical Reviews in Food Science and Nutrition 54, 1415–1426.

Sampedro, F., McAloon, A., Yee, W., Fan, X., Zhang, H., Geveke, D., 2013. Cost analysis of commercial pasteurization of orange juice by pulsed electric fields. Innovative Food Science & Emerging Technologies 17, 72–78.

Sampedro, F., McAloon, A., Yee, W., Fan, X., Geveke, D., 2014. Cost analysis and environmental impact of pulsed electric fields and high pressure processing in comparison with thermal pasteurization. Food and Bioprocess Technology 7, 1928–1937.

San Martin, M., Barbosa-Canovas, G., Swanson, B., 2002. Food processing by high hydrostatic pressure. Critical Reviews in Food Science and Nutrition 42, 627–645.

Santos, M.D., Queiros, R.P., Fidalgo, L.G., Inacio, R.S., Lopes, R.P., Mota, M.J., Sousa, S.G., Delgadillo, I., Saraiva, J.A., 2015. Preservation of a highly perishable food, watermelon juice, at and above room temperature under mild pressure (hyperbaric storage) as an alternative to refrigeration. LWT – Food Science and Technology 62, 901–905.

Sastry, S.K., Barach, J.T., 2000. Ohmic and inductive heating. Journal of Food Science 65, 42–46.

Sastry, S.K., 2005. Advances in ohmic heating and moderate electric field (MEF) processing. In: Barbosa-Cánovas, G.V., Tapia, M.S., Cano, M.P. (Eds.), Novel Food Processing Technologies. CRC Press, Boca Raton, US, pp. 491–500.

Segovia-Bravo, K., Guignon, B., Bermejo-Prada, A., Sanz, P., Otero, L., 2012. Hyperbaric storage at room temperature for food preservation: a study in strawberry juice. Innovative Food Science & Emerging Technologies 15, 14–22.

Sepulveda, D.R., Barbosa-Cánovas, G.V., 2005. Present status and the future of PEF technology. In: Barbosa-Cánovas, G.V., Tapia, M.S., Cano, M.P. (Eds.), Novel Food Processing Technologies. CRC Press, Boca Raton, US, pp. 1–45.

Sevenich, R., Bark, F., Crews, C., Anderson, W., Pye, C., Riddellova, K., Hradecky, J., Moravcova, E., Reineke, K., Knorr, D., 2013. Effect of high pressure thermal sterilization on the formation of food processing contaminants. Innovative Food Science & Emerging Technologies 20, 42–50.

Seymour, I.J., Burfoot, D., Smith, R.L., Cox, L.A., Lockwood, A., 2002. Ultrasound decontamination of minimally processed fruits and vegetables. International Journal of Food Science & Technology 37, 547–557.

Shao, Y., Ramaswamy, H.S., 2011. *Clostridium sporogenes*-ATCC 7955 spore destruction kinetics in milk under high pressure and elevated temperature treatment conditions. Food and Bioprocess Technology 4, 458–468.

Sharma, P., Oey, I., Everett, D.W., 2014. Effect of pulsed electric field processing on the functional properties of bovine milk. Trends in Food Science & Technology 35, 87–101.

Shirsat, N., Lyng, J.G., Brunton, N.P., McKenna, B., 2004. Ohmic processing: electrical conductivities of pork cuts. Meat Science 67, 507–514.

Sitzmann, W., 1995. High-voltage techniques for food preservation. In: Gould, G.W. (Ed.), New Methods of Food Preservation. Blackie Academic and Professional, London, UK, pp. 236–252.

Skudder, P.J., 1988. Ohmic heating: new alternative for aseptic processing of viscous foods. Journal of Food Engineering 988, 99–101.

Somerville, J.A., 2009. The Effects of Pressure-assisted Thermal Processing on the Quality Attributes of Black Beans (*Phaseolus vulgaris* L.). Food Science and Nutrition. The Ohio State University, US.

Sonne, A.-M., Grunert, K.G., Veflen Olsen, N., Granli, B.-S., Szabó, E., Banati, D., 2012. Consumers' perceptions of HPP and PEF food products. British Food Journal 114, 85–107.

Soria, A.C., Villamiel, M., 2010. Effect of ultrasound on the technological properties and bioactivity of food: a review. Trends in Food Science & Technology 21, 323–331.

Spencer, P., 1952. Means for Treating Foodstuffs Raytheon Mfg Co.. US Patent 2605383.

Stratakos, A.C., Koidis, A., 2015. Suitability, efficiency and microbiological safety of novel physical technologies for the processing of ready-to-eat meals, meats and pumpable products. International Journal of Food Science & Technology 50, 1283–1302.

Sun, H., Kawamura, S., Himoto, J.-I., Itoh, K., Wada, T., Kimura, T., 2008. Effects of ohmic heating on microbial counts and denaturation of proteins in milk. Food Science and Technology Research 14, 117–123.

Sun, D.-W., 2012. Thermal Food Processing: New Technologies and Quality Issues. CRC Press, Boca Raton, US.

Teissié, J., Prats, M., Soucaille, P., Tocanne, J., 1985. Evidence for conduction of protons along the interface between water and a polar lipid monolayer. Proceedings of the National Academy of Sciences 82, 3217–3221.

Tempest, P., 1996. Electroheat technologies for food processing. Bulletin of APV Processed Food Sector, England.

Thakur, B., Nelson, P., 1998. High pressure processing and preservation of food. Food Reviews International 14, 427–447.

Thoma, G., Popp, J., Nutter, D., Shonnard, D., Ulrich, R., Matlock, M., Kim, D.S., Neiderman, Z., Kemper, N., East, C., 2013. Greenhouse gas emissions from milk production and consumption in the United States: a cradle-to-grave life cycle assessment circa 2008. International Dairy Journal 31, S3–S14.

Toepfl, S., Mathys, A., Heinz, V., Knorr, D., 2006. Review: potential of high hydrostatic pressure and pulsed electric fields for energy efficient and environmentally friendly food processing. Food Reviews International 22, 405–423.

Tomasula, P., Yee, W., McAloon, A., Nutter, D., Bonnaillie, L., 2013. Computer simulation of energy use, greenhouse gas emissions, and process economics of the fluid milk process. Journal of Dairy Science 96, 3350–3368.

Tomasula, P., Datta, N., Yee, W., McAloon, A., Nutter, D., Sampedro, F., Bonnaillie, L., 2014. Computer simulation of energy use, greenhouse gas emissions, and costs for alternative methods of processing fluid milk. Journal of Dairy Science 97, 4594–4611.

Tonello, C., 2011. Case studies on high-pressure processing of foods. In: Zhang, Q.H., Barbosa-Cánovas, G.V., Balasubramiam, V.M., Dunne, C.P., Farkas, D.F., Yuan, J.T.C. (Eds.), Nonthermal Processing Technologies for Food. Wiley-Blackwell, New Jersey, US, pp. 36–50.

Torres, J.A., Velazquez, G., 2005. Commercial opportunities and research challenges in the high pressure processing of foods. Journal of Food Engineering 67, 95–112.

Vadivambal, R., Jayas, D., 2010. Non-uniform temperature distribution during microwave heating of food materials – a review. Food and Bioprocess Technology 3, 161–171.

Van der Berg, R.W., Hoogland, H., Lelieveld, H.L.M., Van Schepdael, L., 2002. High pressure equipment designs for food processing applications. In: Hendrickx, M., Knorr, D. (Eds.), Ultra High Pressure Treatments of Foods. Kluwer Academic, New York, US, pp. 297–313.

Vicente, A., Castro, I., 2007. Novel Thermal processing technologies. In: Tewari, G., Juneja, V. (Eds.), Advances in Thermal and Non-Thermal Food Preservation. Blackwell Publishing, Oxford, UK, pp. 99–130.

Vicente, A., Teixeira, J., Castro, I., 2006. Ohmic heating for food processing. In: Sun, D.-W. (Ed.), Thermal Food Processing: New Technologies and Quality Issues. Taylor and Francis, Boca Raton, US, pp. 459–500.

Wang, C.S., Kuo, S.Z., Kuo-Huang, L.L., Wu, J.S.B., 2001. Effect of tissue infrastructure on electric conductance of vegetable stems. Journal of Food Science 66, 284–288.

Wang, L., 2008. Energy efficiency and conservation in high-pressure food processing. In: Wang, L. (Ed.), Energy Efficiency and Management in Food Processing Facilities. CRC Press, Boca Raton, US, pp. 323–332.

Wang, L., 2014. Energy efficiency technologies for sustainable food processing. Energy Efficiency 7, 791–810.
Wiktor, A., Iwaniuk, M., Śledź, M., Nowacka, M., Chudoba, T., Witrowa-Rajchert, D., 2013. Drying kinetics of apple tissue treated by pulsed electric field. Drying Technology 31, 112–119.
Xi, J., 2006. Effect of high pressure processing on the extraction of lycopene in tomato paste waste. Chemical Engineering & Technology 29, 736–739.
Yin, Y., Zhang, Q.H., Sastry, S.K., 1997. High Voltage Pulsed Electric Field Treatment Chambers for the Preservation of Liquid Food Products. Ohio State University. US Patent 5690978.
Zhang, M., Tang, J., Mujumdar, A.S., Wang, S., 2006. Trends in microwave-related drying of fruits and vegetables. Trends in Food Science & Technology 17, 524–534.
Zhang, Y., Hou, Y., Zhang, Y., Chen, J., Chen, F., Liao, X., Hu, X., 2012. Reduction of diazinon and dimethoate in apple juice by pulsed electric field treatment. Journal of the Science of Food and Agriculture 92, 743–750.
Zimmerman, F., Bergman, C., 1993. Isostatic high-pressure equipment for food preservation. Food Technology 47, 162–163.
Zinoviadou, K.G., Galanakis, C.M., Brnčić, M., Grimi, N., Boussetta, N., Mota, M.J., Saraiva, J.A., Patras, A., Tiwari, B., Barba, F.J., 2015. Fruit juice sonication: implications on food safety and physicochemical and nutritional properties. Food Research International 77 (4), 743–752.

CHAPTER

SUSTAINABLE INNOVATION IN FOOD SCIENCE AND ENGINEERING

8

E. Betoret[1], L. Calabuig-Jiménez[2], N. Betoret[2], C. Barrera[2], L. Seguí[2], P. Fito[2]
[1]*University of Bologna, Cesena, Italy;* [2]*Universitat Politècnica de Valencia, Valencia, Spain*

8.1 INTRODUCTION

Sustainability means meeting the needs and aspirations of the present without compromising the ability of future generations to meet theirs. To achieve food and agricultural needs, traditionally, the system has been directed toward promotion of organic and local food. As explained by Spiertz (2010), there is another possibility that suggests to continue the production hegemony, emphasizing biotechnology and technological panaceas.

Global food production attempts to obtain safer and higher quality foods than ever before, but at a great cost. Current food supply systems not only deliver insufficient foods but are economically and environmentally unsustainable, lacking in resilience, inequitable, and risk a human health disaster. A food system in which nearly 1 billion people are undernourished and 1.5 billion are overweight is at the very least testament to a massive system failure. The need for urgent action is widely recognized (Godfray et al., 2010; Royal Society, 2009; MacMillan and Benton, 2014). A host of environmental problems such as greenhouse gas emissions, deforestation, desertification, eutrophication, and biodiversity loss are exacerbated through current food system activities (Garnett, 2011).

Nowadays the sustainability of a product, a process, or a system is assessed according to three dimensions: environmental, social, and economic. Sustainability challenges occur at all stages in the food system from production through processing, distribution, and retailing to consumption and waste disposal. There are increasing demands from policy makers, stakeholders, and public interest groups for research to adopt more integrated perspectives in pursuit of more holistic solutions (Defra, 2007; Kates et al., 2001). Integrated perspectives are particularly called for to improve understanding of the mutual interaction between technological change and the economic, social, and environmental contexts in which it occurs. The promise is held out for holistic solutions combining adaptations in socio-technical systems, rather than single-minded technological responses. As explained by OECD (2004), the development of a sustainable agri-food system places responsibilities on both the natural and the social sciences. While advances in basic and strategic biological research have greatly expanded the potential to produce nutritious food in an efficient and environmentally sustainable manner, social and economic factors will determine the uptake and value of this research, as well as its future direction (Lowe et al., 2008).

Sustainable food production stands at the intersection of several growing needs. First, there are the needs of consumers for improved food security and safety, as well as more sophisticated needs. Second,

Innovation Strategies in the Food Industry. http://dx.doi.org/10.1016/B978-0-12-803751-5.00008-8

there is the quest for economic sustainability of food production, based on cost reduction and increased product differentiation. And third, there is the growing concern for reversing the overexploitation of natural resources, waste generation, and the contribution to climate change (Fava et al., 2013).

Functional foods are foods that beneficially affect one or more target functions in the body, beyond adequate nutritional effects, in a way that is relevant to either an improved state of health and well-being and/or a reduction of health risk, and they are consumed as a part of a normal food pattern (not a pill, a capsule, or any form of dietary supplement) (European Commission, 2010). Many diseases are closely related to diet as well as lifestyle and are of concern to society because of their prevalence. Functional foods can help to prevent or improve those diseases, thus contributing directly to public health. In Betoret et al. (2011) the technologies used to develop functional foods are classified into three groups. The first group consists of the technologies traditionally used in food processing, formulation and blending, and cultivation and breeding. The second group consists of the technologies forming a structure to prevent the deterioration of physiologically active compounds, ie, microencapsulation, edible films and coatings, and vacuum impregnation are part of this group. Finally, the third group includes recent technologies aimed to design personalized functional foods. This chapter provides an overview of the technologies described in Betoret et al. (2011), emphasizing the sustainability innovations achieved in each case as well as their industrial applications.

8.2 FORMULATION AND BLENDING

Formulation and blending constitutes the most common, traditionally used, simple, and cheap methodology to develop functional foods and has been widely used in food processing. Its use in functional food development has a long history for the successful control of deficiencies (Burgi et al., 1990; FAO & WHO, 2006; Betoret et al., 2011). In more recent years, the emergence of dietary compounds with health benefits has offered an excellent opportunity to improve public health, and thus this category of compounds has received considerable attention from the scientific community, consumers, and food manufacturers. The list of dietary active compounds (vitamins, probiotics, bioactive peptides, antioxidants, etc.) is endless and the types of final products obtained are growing steadily (Wildman, 2006).

All these natural compounds used in the formulation of new foods or functional products can be obtained from food-producing industries' by-products (which differ from waste in that they have not been widely exposed to environmental contamination) (Fava et al., 2013). Vegetable, cereals, and fruit-processing by-products and waste are typically rich in proteins, sugars, and lipids and contain particular aromatic and aliphatic complex compounds. Thus they are cheap, abundant sources of value-added biobased chemicals and materials. Indeed, after specific pretreatments with physical and biological agents followed by tailored recovery procedures, they might provide specific natural antioxidants, antimicrobial agents, and vitamins, among others, along with macromolecules (eg, soluble fibers), bioactive oligosaccharides, oligopeptides, and pigments.

A sustainable agro-food industry recognizes that waste prevention, minimization, and valorization, rather than "end of pipe," are the required solutions for waste management. The legislation in Europe is promoting the use of these solutions.

8.3 CULTIVATION AND BREEDING

Plant and animal breeding techniques have been used since the beginning of human history. The evaluation and selection of different breeds started with the domestication of animal and plant species around 12,000 years ago, which was led by the wish to obtain desired traits, dictated by social, nutritional, and environmental needs with no understanding of the molecular processes involved (National Research Council, 1989). In cases where agronomic and breeding approaches cannot achieve significant improvement of food products, biotechnology offers a useful alternative (Zhao and Shewry, 2011). In recent decades, with the use of molecular biology tools and the development of genetically modified plants, biotechnology turns into a useful technology that offers an additional way to improve the traits of foods (a genetically modified organism, or GMO). A GMO is "an organism in which the genetic material has been altered in a way that does not occur naturally by mating and/or natural recombination" (EU Directive, 2001/18). In the plant sciences, attention was directed at increasing yields by enhancing soil fertility, reducing pests, and developing new genetic varieties. By contrast, in animal science genetic modifications are mainly focused on health, nutrition, and breeding (Lyson, 2002).

In plant biotechnology, the focus has been mostly on producing genetically modified crops that are resistant to insects, viral pathogens, and herbicides (Barling et al., 1999; Hails, 2000; Deisingh and Badrie, 2005). In food manufacture, biotechnology has been mostly applied to enzyme production by microorganisms (Barling et al., 1999). Even so, there are experiments to produce crops with enhanced nutritional and health benefits to obtain enriched crop products as "functional foods" and "nutraceuticals" (Zhao and Shewry, 2011; Betoret el at., 2011), and with the capacity to produce pharmaceuticals ("pharming"). The metaphor of "crops becoming factories, producing vaccines, plastics, industrial starches, and feed supplements and enzymes" captures the trajectory of this type of research (Vergrart and Brown, 2008).

In recent decades population has grown considerably in the world. Projections for human population growth suggest that by 2050, more than 9 billion people will inhabit the earth (United Nations, 2015), hence there is the need to produce more food to satisfy food security to the world population. As Ban Ki-moon, Secretary-General of the United Nations, remarked at the Food and Agriculture Organization of the United Nations High-Level Conference on World Food Security, "Food production needs to rise by 50 percent by the year 2030 to meet the rising demand" (Ki-moon, 2008). This context shows a big challenge to the food production systems and to agriculture, which can be faced using plant and animal breeding techniques such as biotechnology. The challenge of global food security in a sustainable way requires the intensification of knowledge-intensive approaches and the use of modern agrotechnologies and biotechnologies (Spiertzt et al., 2010). Experts assert that biotechnology innovations will triple crop yields without requiring any additional farmland, saving valuable rain forests and animal habitats; can reduce or eliminate reliance on pesticides and herbicides that might contribute to environmental degradation; and can preserve precious ground soils and water resources (Lyson, 2002).

The integration of biotechnology in the sustainable framework has been shown by Ervin et al. (2010), who suggested that production and environmental benefits are the two legs of the sustainability stool. Genetically modified crops generally showed progress in reducing agriculture's environmental footprint and improving farmers' profits (herbicide-resistant and insect-resistant crop varieties). But this option of biotechnology has paid insufficient attention to the integrated and systemic requirements of sustainable agriculture. In particular, the consideration of socioeconomic distributive or equity must be accounted for in any assessment of sustainability (Ervin et al., 2010). Fundamental changes are required in the way public and private research, and technology development and commercialization should be structured (Kvakkestad, 2009). Strong efforts and coordination between stakeholders are needed and combined with science to obtain good results in the consideration of sustainability (Ervin et al., 2010; Spierzt, 2010).

The plea is to maximize the potential of the plant genome sciences in contributing significantly to human health, energy security, and environmental stewardship. Strikingly, food production is not listed as one of the major challenges; however, climate change and the world food crisis bring a "sense of urgency" in the debate on meeting the demands of a growing global and wealthier population.

Biotechnology and plant breeding, agriculture crop management, and government policies contribute to counteract the growth demand (Tester and Langridge, 2010). To satisfy the growing demand it will be necessary to make use of the best science and technology to raise crop productivity (Pardue, 2010). So far, a combination of advanced plant breeding, systems innovations, development of best practices, and legislation have demonstrated to be effective in developing more environmentally friendly agricultural systems that are profitable, ecologically safe, and socially acceptable (Spierzt, 2010).

The issue of genetically modified crops has been highly controversial since the introduction of recombinant DNA technology in the 1970s (Singer and Soll, 1973; Berg et al., 1974; Devos et al., 2007). It is more intense in the European Union, based on their precautionary principles (Vergragt and Brown, 2008; Ervin et al., 2010; Spiertz et al., 2010). The public objections have been numerous, including concerns about risk assessments, ethics and equity issues, power relations, and the mistrust of technocrats and public authorities (Vergragt and Brown, 2008).

Biotechnologies have the potential to revolutionize virtually all aspects of society, from how, where, when, and by whom food is produced, processed, and consumed to how dietary changes might be used to treat illness and disease. In the field of food science and agriculture, biotechnologies have the potential to satisfy the worldwide growing demand for food with sustainable agriculture. Biotechnologies will participate in farm and agriculture changes improving the sector in a sustainable manner (social, economic, and environmental approach), to guarantee food security in the world.

8.4 MICROENCAPSULATION

Microencapsulation is the envelopment of small solid particles, liquid droplets, or gases in a coating (Thies, 1987). Microencapsulation is based on the embedding effect of a polymeric matrix, which creates a microenvironment in the capsule able to control the interactions between the internal part and the external one (Borgogna et al., 2010). This technology allows the protection of a wide range of materials of biological interest, from small molecules and proteins (enzymes, hormones, etc.) to cells of bacterial, yeast, and animal origin (Thies, 2005). Thanks to microencapsulated ingredients, many products that were considered technically unfeasible are now possible (Gharsallaoui et al., 2007).

Such versatile technology is widely studied and exploited in high-technology fields for applications ranging from cell therapy to drug delivery (Smidsrød and Skjak-Braek, 1990). There are numerous industrial applications of microencapsulation. Some examples are carbonless paper, "scratch-and-sniff" fragrance samples, "intelligent" textiles, controlled release of drugs, pesticides, cosmetic active agents, and functional foods. Martins et al. (2014) illustrate the distribution, in percentages, of microencapsulation over different fields of applications showing that the sector with the highest level is the drug sector (68%), followed by the food (13%) and cosmetic sectors (8%).

In the food industry, Shahidi and Han (1993) proposed three reasons for applying microencapsulation:

1. to reduce the core reactivity with environmental factors; to decrease the transfer rate of the core material to the outside environment;
2. to promote easier handling; to control the release of the core material;
3. to mask the core taste; and finally to dilute the core material when it should be used in only very small amounts.

Various applications for the production of high-value foods and nutraceuticals are quite interesting and are being developed quickly. Some examples are described in Betoret et al. (2011). In this regard the main challenge of technology is to maintain the active form in the food during preparation and processing, until the time of consumption, and deliver it in the appropriate specific site of the organism.

It is absolutely necessary that the technologies used for functional food development are sustainable. Research studies related to microencapsulation and sustainability are growing significantly although they are still scarce. In this regard, the published studies are focused mainly in four main research directions:

1. wall (matrix) materials for microencapsulation
2. microencapsulating materials
3. processes for microencapsulation
4. innovative applications

The selection of wall material for microencapsulation is an important part of the process. Each material possesses unique emulsifying and film-forming properties and from its macro- and microstructural properties this will depend on the capacity to form small-sized, physically stable microcapsules. There is a need for selection, development, and characterization of biodegradable, biocompatible, safe, and environmentally friendly materials suitable for utilization as encapsulating agents.

In recent years, researchers have focused on the utilization of secondary plant materials in coatings. Plant materials are mainly made up of three types of biopolymers: cellulose, lignin, and hemicellulose. Various agricultural residues such as corn fiber, corn peel, and sugar beet contain about 20–40% hemicellulose, making it the second most abundant polysaccharide in nature (Tatar et al., 2014). It is now agreed that hemicellulose is valuable due to its adhesive, thickening, stabilizing and emulsifying (Yadav et al., 2009), and film-forming properties (Hansen and Plackett, 2008). Ebringerovà (2006) demonstrated that hemicellulose produces stable foams and oil/water type emulsions due to the presence of small amounts of lignin and proteins, acting as the hydrophobic centers, and due to film-forming effects. Yadav et al. (2009) deduced that the corn fiber gum is a better emulsifier than gum arabic in an oil-in-water emulsion system. On the other hand, McPherson et al. (2006) evaluated the application of hemicellulose hydrolyzate from corn hull as coating. They found that the hemicellulose was more efficient than the gum arabic. Tatar et al. (2014) tested the effectiveness of hemicellulose-based coating isolated from corn wastes and showed that hemicellulose can be used in combination with gum arabic in coatings to be microencapsulated by spray drying method. Although the chemical and physical properties of hemicellulose-based or derived products were studied (Ebringerovà, 2006; Hansen and Plackett, 2008), there is still need to investigate the potential of hemicellulose in encapsulation or coating applications (Celebioglu et al., 2012). The use of plant proteins as encapsulating materials represents another trend for renewability and sustainability in microencapsulation technology. In this regard, vegetable proteins used as a wall material in microencapsulation include soy protein isolate, pea protein isolate, and cereal proteins (Gharsallaoui et al., 2007; Nesterenko et al., 2012; Tang and Li, 2013a,b), among which soy protein isolate was most frequently applied, possibly due to its widely commercial availability. Many previous studies have indicated that relative to sodium caseinate or whey proteins, the microencapsulated products with vegetable proteins as the encapsulating materials exhibit comparable or even better encapsulation efficiency, and higher stability against oxidation (Charve and Reineccius, 2009; Kim and Morr, 1996; Rascón et al., 2011). Gharsallaoui et al. (2007) reported a novel system using pea protein isolate-stabilized emulsions, to encapsulate lipophilic ingredients by spray drying. Liu et al., 2014 compared the microencapsulating potential of *Phaseolus* legumes, including red bean, kidney bean, and mung bean with soy protein isolate and showed that all the tested proteins exhibited similar emulsifying properties but the interfacial properties noting that more studies should be conducted on these proteins to warrant their application as wall materials in spray-drying microencapsulation.

In the case of microencapsulated materials, the sustainability of the technology is based on the use of bioactive compounds extracted from food processing by-products. To date certain classes of phenolics such as procyanidins of grape seeds (Zhang et al., 2007), polyunsaturated fatty acids ω-3 from marine by-products (Ferraro et al., 2010), phenolics from pomegranate peel (Çam et al., 2014), polyphenols from star fruit (*Averrhoa carambola*) pomace (Saikia et al., 2015), and polyphenols from *Vitis vinifera* grape wastes (Aizpurua-Olaizoa et al., 2016) are some examples of bioactive compounds extracted from food by-products that have been microencapsulated. Polyphenols have a poor long-term stability, as they are affected by pH variation, presence of metal ions, light, temperature, oxygen, and enzymatic activities (Bakowska et al., 2003). Moreover, due to low water solubility, they often present poor bioavailability (Munin and Edwards-Lévy, 2011) and they are unstable in alkaline conditions encountered in biological fluids (Dube et al., 2010).

Microencapsulation can be achieved by a wide range of methods or techniques, providing isolation, entrapment, protection, or controlled release of sensitive or reactive materials from/across the surrounding matter (Martins et al., 2014). The adequate microencapsulation method depends on the specific molecular structure of the compound to be encapsulated and on the specific characteristics of the wall material, as well as on the desired functionality of the obtained ensemble in each case. The different types of microcapsules and microspheres are produced from a wide range of wall materials (monomers and/or polymers) and by a large number of different microencapsulation processes such as spray-drying, spray-cooling, spray-chilling, air suspension coating, extrusion, centrifugal extrusion, freeze-drying, coacervation, rotational suspension separation, co-crystallization, liposome entrapment, interfacial polymerization, molecular inclusion, etc (Desai and Park, 2005; Gibbs et al., 1999; Gouin, 2004; King, 1995; Shahidi and Han, 1993). The current industrial scenery is founded on compromises based on the needs of the industrial processes developed to satisfy both the increasing market requirements and the mandatory rules in sustainable production such as raw material/energy savings, respect of environmental constraints of industrial-scale processes (Charpentier, 2007). New microencapsulation technologies are relentlessly devised and invented by academics and industrial researchers; in 2002 over 1000 patents were filed concerning various microencapsulation processes. Some of these new processes have very little industrial relevance because of the extremely high cost-in-use, difficult scale-up, and/or narrow applicability range. However, some of these processes stand out as being promising, sensible, and likely to be scaled up in the near future for the encapsulation of active ingredients (Tran et al., 2011; Gouin, 2004). Dalmoro et al. (2012) reviewed the new approaches to the microencapsulation processes focusing on the emerging ultrasonic atomization technique and presented fundamentals and novel aspects of the technology emphasizing the advantages in terms of intensification and low-energy request. Amongst the various techniques developed to encapsulate food ingredients, spray-drying is the most common technology due to its low cost and the availability of equipment (Gharsallaoui et al., 2007). Aghbashlo et al. (2012) applied the exergy analysis using the second law of thermodynamics to obtain a quantitative measure regarding the efficiency, losses, and performance for microencapsulation of fish oil using different wall materials and drying air temperatures. The authors concluded that the use of exergy analyses is a good way to manage and improve the sustainability of the process.

The microcapsules produced can be used in the development of functional foods but other interesting applications regarding the sustainability of the field are emerging. For example, Lamppa et al. (2014) reviewed the nutritional delivery systems emerging from the drug delivery field aimed at reducing waste in food and beverage and eliminating waste in food and beverage packaging. The authors presented the microencapsulation as one technology able to functionalize nutrient powders by decreasing the dissolution times and improving the consumer experience by modulating particle size, porosity, and hydrophilicity, and by preventing degradation of the bioactive compound before it reaches the desired functional location. Takei et al. (2008) used the microencapsulated *Lactobacillus delbrueckii* ssp. *bulgaricus* NBRC 13953 as soil bioamendments and evaluated the effect of preparation parameters in the emulsion system on the survival activity of the encapsulated bacteria. Furthermore, the soil application demonstrated that the microcapsules are effective in the removal of root-knot nematode.

8.5 EDIBLE FILMS AND COATINGS

As pointed out by Attila E. Pavlath at the 24th National Meeting and Exposition of the American Chemical Society (ACS, 2013), the use of edible films and coatings has grown in the last 30 years from 10 to more than 1000 companies in the business, with annual sales exceeding $100 million. The great success of these products is due to the fact that they can be applied in almost any sector of the food industry to meet requirements related to the marketing of safe foods, with high quality and enhanced nutritional value, as well as being cost effective and environmentally friendly (Gennadios et al., 1997). Because coatings are applied and formed directly on the food product, whereas films are applied after being formed separately, their composition, manufacture, properties, and uses are not exactly the same (Bourtoom, 2008). Regarding edible films and coatings' contribution to sustainable economic and social progress, it takes place in different ways.

Edible coatings date from the 12th century, consisting of layers of molten wax applied on the surface of citrus fruits with the aim of preserving them for later consumption (Pavlath and Orts, 2009). Extending shelf life is still one of the main objectives of scientific research and industrial application of edible films and coatings on the surfaces of foods. As stated in a recent study (FAO, 2011), about one-third of the edible parts of food produced for human consumption gets lost or wasted, basically for not meeting food safety and/or quality standards. Food losses involve a waste of land, water, energy, and several inputs used in production, so any technique that effectively reduces these losses will also contribute to the more efficient use of natural resources.

In the particular case of fruits and vegetables, postharvest losses are estimated to range between 5% and 25% in developed countries, and between 20% and 50% in developing countries, depending on the type of product (Pérez-Gago et al., 2008). Cold and storage in controlled and/or modified atmosphere are usually employed in slowing down the senescence process. However, the result of applying these techniques is not always homogeneous, which, in addition to its relatively high cost, makes it necessary to search for new storage alternatives. In general terms, it is more usual for this kind of product to apply an edible coating than a film, and formulations combining hydrocolloids (polysaccharides or proteins) and lipid compounds are reported to be the most appropriate (Olivas et al., 2008). On one hand, hydrocolloids provide selective permeability to CO_2 and O_2 while lipids confer resistance to water vapor migration. Although emulsions based on natural waxes are the most commercially applied products due to their effectiveness with citrus fruits and apples, there is increasing use of cellulose ethers (such as carboxymethylcellulose, hydroxypropyl cellulose, and methylcellulose) as ingredients in coatings for fruits and vegetables (Han, 2009; Pérez-Gago and Rhim, 2014). In recent years, research has focused on the development of edible coatings with antimicrobial activity for the control of microorganisms that cause postharvest diseases. Apart from the use of hydrocolloids with certain antimicrobial properties, such as chitosan or aloe vera, incorporating ingredients with antifungal properties, such as certain food-grade additives, natural extracts and antagonistic to fungal pathogens, have been reported to be good alternatives to the use of synthetic chemical fungicides in whole fruits (Pérez-Gago, 2015). In addition, packaged or coated products are more protected from losses of volatile compounds and nutrients, as well as from physical damage caused by physical impact, pressure, vibrations, and other mechanical factors, which are especially prevalent when handling grapes, cherries, stone fruits, berries, and fresh-cut products. In the particular case of the latter ones, food-grade additives are usually included in the film or coating formulation in order to prevent browning or tissue softening.

Although losses and waste of meats, poultry, and seafood account for half of the ones for fruits and vegetables (FAO, 2011), using edible coatings on such highly perishable products has intensified recently. Indeed, weight losses taking place during storage under refrigeration, freezing, or vacuum conditions have been successfully reduced by applying edible coatings with good moisture barrier properties as is the case of lipid-based coatings (Gennadios et al., 1997). In the same way, rancidity and brown coloration, respectively, caused by lipid and myoglobin oxidation are reduced by using edible coatings of low-oxygen permeability.

According to their ability to extend the shelf life of several foods, edible films and coatings contribute in the food globalization process that involves a larger supply of products to vary and improve consumers' diets. Beyond the conservation of each food's nutritional properties, new edible coatings arise with the aim of improving them. For example, coatings are applied on the surface of several foods to reduce oil uptake during frying (García et al., 2002; Suárez et al., 2008). Moreover, edible films and coatings are used as physiologically active ingredients carriers (Salgado et al., 2015). In fact, a wide range of naturally occurring antioxidants such as essential oils and plant extracts (basil, thyme, cinnamon leaf, and tea tree essential oil among others), as well as pure compounds (like ascorbic acid and α-tocopherol) have been incorporated into edible films and coatings to improve their bioactive properties (Bonilla et al., 2012, 2013; Perdones et al., 2014; Sánchez-González et al., 2009). To a lesser extent, the incorporation of probiotics into functional edible films and coatings has been studied for the development of probiotic breads (Soukoulis et al., 2014) and probiotic fresh-cut fruits (Tapia et al., 2007).

Finally, edible films and coatings play an important role in reducing the environmental impact of the feeding process. On the one hand, they contribute to the revalorization of those industrial by-products that meet the necessary conditions to be included in their formulation. This is the case for starch and cellulose obtained from different vegetables, chitosan from crustaceans, carrageenan and protein extracted from seaweed, whey protein from the dairy industry, gelatin from slaughterhouses and tanneries, soybean and sunflower proteins from oil cakes, and keratin from feathers (Salgado et al., 2015). On the other hand, the use of edible films and coatings as primary packaging may simplify the total packaging structure and, therefore, the overall utilization of synthetic materials (Han, 2009). Apart from food applications, the use of edible films for agricultural purposes, as grocery bags or in place of cushioning foams, could also reduce the accumulation of nonrenewable and nonbiodegradable materials derived from petroleum.

8.6 VACUUM IMPREGNATION

Vacuum impregnation is a mass transfer operation between a porous solid matrix immersed in a liquid media. The creation of a pressure gradient between the porous and the liquid media produces a degasification of the porous structure in the first step of the process, and secondly an important input of liquid from the external media to the solid matrix (Fito et al., 1996). A great amount of research work has been reported in the last 15 years, and some of the results have been transferred to the food-processing industry (Zhao and Xie, 2004; Radziejewska-Kubzdela et al., 2014). Because of the perishability and the short shelf life of vegetal tissues that compromise the sustainability of natural resources, vacuum impregnation is of relevant interest.

By means of vacuum impregnation, it has been possible to include inside a food-porous structure, mainly a vegetal tissue, cryoprotectants, stabilizers, enzymes, inhibitors of enzymatic browning, and physiologically active compounds (eg, minerals, vitamins, probiotics, or antioxidants among others). This way, it has been possible to enhance freezing tolerance; to increase stability and shelf life of minimally processed fruits or vegetables; to improve kinetics and efficiency of processes in which a solute, like sugar or salt, has to be homogenously distributed inside a food; and to develop nutritionally fortified products. In any case, vacuum impregnation makes possible food quality and safety without high costs, contributing to the sustainability of the food processing system. Additionally, some results have proved the possibility to effectively reduce the incidence of chronic disease, including vacuum-impregnated foods specifically formulated for specific dietary needs, with resulting reduction of costs to public medical systems.

Xie and Zhao (2004) used a high-fructose corn syrup (50%) or high-methylated pectin (3%) to impregnate strawberries prior to the freezing process. They found that vacuum impregnation resulted in a strengthened structure and reduced water drip in thawed strawberries. High-fructose corn syrup decreased the amount of frozen water in tissue, while high-methylated pectin penetrated into intracellular spaces, protecting fruit against freezing damage. An example for the application of vacuum impregnation to inhibit enzymatic browning is shown in a study by Perez-Cabrera et al. (2011). During the impregnation of pears, the authors used an isotonic solution containing enzymatic browning inhibitors (ascorbate; 4-hexylresorcinol; EDTA; citrate) with or without the addition of calcium lactate. In the study reported by Derossi et al. (2010), it was observed that the reduction of pH values was greater during vacuum impregnation of pepper slices than in the case of blanching at atmospheric pressure.

Related to the improvement in the distribution of a solute within a food, Tamer et al. (2013) reported the duration decrease of debittering process in olive fruits impregnated with NaOH (1.5%) and NaCl (3%). In the case of drying, vacuum osmotic dehydration in a hypertonic solution may result in a reduction of raw material moisture content and a shortened drying time, thus contributing to an improvement of dried material quality. Pallas et al. (2013) reported a significant reduction of drying time as a result of vacuum osmotic dehydration of rabbiteye blueberries in a sucrose solution (60 Brix).The use of vacuum impregnation allows the introduction of other structure-forming compounds, eg, polyamines, into the plant tissue. Polyamines exhibit properties similar to those of calcium ions. They may bind with cell walls and pectins found in the middle lamella (Kramer et al., 1991). On the other hand, they may inhibit the synthesis of ethylene in damaged tissue and reduce the activity of enzymes responsible for its softening (Kramer et al., 1989). The use of vacuum-impregnation technique to introduce these compounds to the tissue of strawberries was tested by Ponappa et al. (1993).

Regarding the impregnation with physiologically active compounds such as probiotics or antioxidants, Betoret et al. (2003) applied vacuum impregnation of apples with apple juice supplemented with *Saccharomyces cerevisiae* or milk inoculated with *S. cerevisiae* and *Lactobacillus casei*. Impregnation facilitated the effective introduction of probiotics to apple tissue, providing the content of microorganisms in the product after convection drying (air-drying) at 10^6–10^7 CFU/g. This is equivalent to the level of bacteria in dairy products. Similarly, Krasaekoopt and Suthanwong (2008) obtained the level of microorganisms in fruit after air-drying at 10^7 CFU/g during the vacuum impregnation of guava and papaya fruits using *L. casei*, which makes this product a probiotic food.

The current state of knowledge on the biological activity of flavonoid compounds clearly indicates that their positive effect on the human organism results mainly from antioxidant properties. The capacity of flavonoids to scavenge reactive oxygen species and chelate transition metals may have a significant role in pathological conditions (eg, inflammations, atherosclerosis, diabetes, neurodegenerative diseases, or cancer), which are accompanied by oxidative stress (Hanasaki et al., 1994; Yao et al., 2004). Betoret et al. (2012) applied vacuum impregnation to introduce homogenized mandarin juice with low-pulp content to apple snacks. The authors obtained the content of hesperidin in 40 g of the enriched product equivalent to that in 250 mL of mandarin juice. In a continuation of studies by Betoret et al. (2012), apple snacks vacuum impregnated with mandarin juice were administered to obese children in order to alleviate inflammatory conditions and improve the antioxidant capacity of the organism (Betoret et al., 2012; Codoñer-Franch et al., 2013). A considerable improvement was observed in systolic blood pressure and the lipid profile following the treatment period. The authors concluded that the addition of the product to diet contributed to an alleviation of oxidative stress and the inflammatory condition, as well as several other risk factors connected with atherosclerosis in the examined obese children.

The most controversial aspect of the application of vacuum impregnation in the food industry is the great amount of residual liquid generated as a consequence of the process. Different solutions can be applied in order to assure environmental and economic sustainability. Some of them are:

1. To use as an impregnation liquid by-product generated in other food processes. Vegetable, cereal, and fruit-processing by-products are typically rich in proteins, sugars, and lipids and contain particular aromatic and aliphatic complex compounds. When it has a suitable composition for vacuum-impregnation purposes, it could be used directly. Otherwise, components of interest could be extracted and incorporated to the impregnation liquid.

2. To reuse the impregnation liquid. Some studies show that it is possible to reuse the same liquid medium in different cycles of impregnation. The impregnation liquid composition is hardly affected by the output of native liquid during the first stage of the operation (Castagnini et al., 2015). Regarding the microbiological quality of impregnation liquid along reuse, it can be controlled by a cold treatment like a high-pressure homogenization.
3. To use the impregnation liquid waste to formulate new food products. Ingredients obtained from fruit and cereal by-products might have great potential and market opportunities in modern society where the consumption of "ready-to-eat" products with health-promoting properties is increasing. The possibility of using the liquid waste to produce new food products, including fruit juice beverages and snacks, self-stable fillers for bakery products, and fiber-enriched bakery products is being investigated (Fava et al., 2013).

Additionally, after specific pretreatments with physical and biological agents followed by tailored recovery procedures, they might provide specific natural antioxidants, antimicrobial agents, vitamins, among others, along with macromolecules (eg, soluble fibers), bioactive oligosaccharides, oligopeptides, and pigments. Furthermore, some of the compounds occurring in the hydrolyzates resulting from the by-product pretreatment can be transformed into more sophisticated molecules like flavors and fermentation products, through tailored biotechnological processes (Wynan, 2003; Laufenberg et al., 2003). All these natural compounds can be combined in the formulation of new food products with the attempt to close the circle within large fruit and cereal food industries and/or to create new synergies between fruit/cereal processing industries and food-producing industries.

8.7 NUTRIGENOMICS

Nutrigenomics, also called nutritional genomics, considers the interactions between foods or dietary supplements, proteome (the sum total of all proteins), and a metabolome (the sum of all metabolites) on an individual's genome, and the consequent downstream effects on their phenotype (Debusk et al., 2005; Ferguson et al., 2010). Consequently, nutritional genomics has the potential to provide tailored nutrition advice to populations or to individuals. It recognizes that what is appropriate dietary advice for one individual may be inappropriate, or actually harmful, to another.

In nutritional terms, few will question the repeated documented link between diet, an environmental factor, and the risk and incidence of a number of the chronic diseases of today. The acceptance of such associations has contributed significantly to the formulation of current public health nutrition policies, which, although based on credible evidence, have not been shown to benefit equally those who follow them.

Functional foods may be a mechanism through which optimal dietary advice can be tailored to a population's needs. But to do this link in a sustainable way, it is essential that individuals from all the relevant professions have active input integrating the new knowledge into appropriate training programs. Input through the food industry into new functional foods development will also be necessary (Ferguson, 2009). Food and nutrition professionals have the responsibility to ensure that all groups of society are given appropriate nutrition advice and the resources are used in an optimal way.

Experts and other stakeholders predict that nutrigenomics will deliver improvements in public health by identifying genetically determined differences in how diet impacts on chronic disease, both in terms of food and food components as a cause of disease, and as a preventative or curative agent, although there is uncertainty regarding the concrete forms that such developments will take (Komduur et al., 2007). There is little evidence that societal benefits and sustainability (eg, reduced health service costs) increase acceptance of nutrigenomics, though some stakeholders have assumed this (Frewer et al., 2011).

8.8 CONCLUSIONS

Current food supply systems and practices are economically and environmentally unsustainable and the need for urgent action is widely recognized. Traditionally, regarding sustainability, the agri-food system has been directed toward promotion of organic and local foods, but there is the other possibility that implies emphasizing biotechnology and technological panaceas. Management of food processes in an adequate way can contribute to achieve a full sustainability concept with three dimensions implied: environmental, social, and economic. Environmental because the processes and the raw materials used in the development of functional foods are more and more environmentally friendly with less waste production, enhancing reuse and biodegradable products. Social because functional foods contribute clearly to improve public health, and economic because the improvement of public health implies reducing the rising costs of the health care system.

REFERENCES

ACS, 2013. Edible coatings for ready-to-eat fresh fruits and vegetables. In: Research Presented at the 24th National Meeting & Exposition of the American Chemical Society Held in Indianapolis in September 2013 [On line resource] http://www.acs.org/content/acs/en/pressroom/newsreleases/2013/september/edible-coatings-for-ready-to-eat-fresh-fruits-and-vegetables.html.

Aghbashlo, M., Mobli, H., Rafiee, S., Madadlou, A., 2012. Energy and exergy analyses of the spray drying process of fish oil microencapsulation. Biosystems Engineering 111, 229–241.

Aizpurua-Olaizola, O., Navarro, P., Vallejo, A., Olivares, M., Etxebarria, N., Usobiaga, A., 2016. Microencapsulation and storage stability of polyphenols from *Vitis vinifera* grape wastes. Food Chemistry 190, 614–621.

Bakowska, A.M., Kucharska, A.Z., Oszmianski, J., 2003. The effects of heating, UV irradiation and storage on stability of anthocyanin–polyphenol copigment complex. Food Chemistry 81 (3), 349–355.

Barling, D., De Vriend, H., Cornelese, J.A., Ekstrand, B., Hecker, E.F.F., Howlett, J., Jensen, J.H., Lang, T., Mayer, S., Staer, K.B., Top, R., 1999. The social aspects of food biotechnology: a European view. Environmental Toxicology and Pharmacology 7, 85–93.

Berg, P., Baltimore, D., Boyer, H.W., Cohen, S.N., Davis, R.W., Hogness, D.S., Zinder, N.D., 1974. Potential biohazards of recombinant DNA molecules. Science 185, 303.

Betoret, E., Betoret, N., Vidal, D., Fito, P., 2011. Functional foods development: trends and technologies. Trends in Food Science & Technology 22, 498–508.

Betoret, E., Sentandreu, E., Betoret, N., Codoñer-Franch, P., Valls-Bellés, V., Fito, P., 2012. Technological development and functional properties of an apple snack rich in flavonoid from mandarin juice. Innovative Food Science and Emerging Technologies 16, 298–304.

Betoret, N., Puente, L., Díaz, M.J., Pagán, M.J., García, M.J., Gras, M.L., Martínez-Monzó, J., Fito, P., 2003. Development of probiotic-enriched dried fruits by vacuum impregnation. Journal of Food Engineering 56, 273–277.

Bonilla, J., Atarés, L., Vargas, M., Chiralt, A., 2012. Effect of essential oils and homogenization conditions on properties of chitosan-based films. Food Hydrocolloids 26 (1), 9–16.

Bonilla, J., Talón, E., Atarés, L., Vargas, M., Chiralt, A., 2013. Effect of the incorporation of antioxidants on physicochemical and antioxidant properties of wheat starch–chitosan films. Journal of Food Engineering 118 (3), 271–278.

Borgogna, M., Bellich, B., Zorzin, L., Lapasin, R., Cesaro, A., 2010. Food microencapsulation of bioactive compounds: rheological and thermal characterisation of non-conventional gelling system. Food Chemistry 122, 416–423.

Bourtoom, T., 2008. Edible films and coatings: characteristics and properties. International Food Research Journal 15 (3), 237–248.

Burgi, H., Supersaxo, Z., Selz, B., 1990. Iodine deficiency diseases in Switzerland one hundred years after Theodor Kocher's survey: a historical review with some new goitre prevalence data. Acta Endocrinologica 123, 577–590.

Çam, M., Cihatiçyer, N., Erdogan, F., 2014. Pomegranate peel phenolics: microencapsulation, storage stability and potential ingredient for functional food development. LWT – Food Science and Technology 55, 117–123.

Castagnini, J.M., Betoret, N., Betoret, E., Fito, P., 2015. Vacuum impregnation and air drying temperature effect on individual anthocyanins and antiradical capacity of blueberry juice included into an apple matrix. LWT – Food Science and Technology 64 (2), 1289–1296.

Celebioglu, H.Y., Cekmecelioglu, D., Dervisoglu, M., Kahyaoglu, T., 2012. Effect of extraction conditions on hemicellulose yields and optimisation for industrial processes. International Journal of Food Science & Technology 47 (12), 2597–2605.

Charpentier, J.C., 2007. In the frame of globalization and sustainability, process intensification, a path to the future of chemical and process engineering (molecules into money). Chemical Engineering Journal 134, 84–92.

Charve, J., Reineccius, G.A., 2009. Encapsulation performance of proteins and traditional materials for spray dried flavors. Journal of Agricultural and Food Chemistry 57, 2486–2492.

Codoñer-Franch, P., Betoret, E., Betoret, N., López-Jaén, A.B., Valls-Belles, V., Fito, P., 2013. Dried apples enriched with mandarin juice by vacuum impregnation improve antioxidant capacity and decrease inflammation in obese children. Nutrición Hospitalaria 28, 1177–1183.

Dalmoro, A., Barba, A.A., Lamberti, G., d'Amore, M., 2012. Intensifying the microencapsulation process: ultrasonic atomization as an innovative approach. European Journal of Pharmaceutics and Biopharmaceutics 80 (3), 471–477.

Debusk, R., Fogarty, C., Ordovas, J., Kornman, K., 2005. Nutritional genomics in practice: where do we begin? Journal of the American Dietetic Association 105 (4), 589–598. http://dx.doi.org/10.1016/j.jada.2005.01.002.

Defra Science Advisory Council, 2007. Social research in Defra. Science Advisory Council Paper, SAC 33 (07).

Deisingh, A.K., Badrie, N., 2005. Detection approaches for genetically modified organisms in foods. Food Research International 38 (6), 639–649.

Derossi, A., de Pilli, T., Severini, C., 2010. Reduction in the pH of vegetables by vacuum impregnation: a study on pepper. Journal of Food Engineering 99, 9–15.

Desai, K.G.H., Park, H.J., 2005. Recent developments in microencapsulation of food ingredients. Drying Technology 23, 1361–1394.

Devos, Y., Maesele, P., Reheul, D., Van Speybroeck, L., De Waele, D., 2007. Ethics in the societal debate on genetically modified organisms: a (re)quest for sense and sensibility. Journal of Agricultural and Environmental Ethics 21, 29–61.

Dube, A., Ng, K., Nicolazzo, J.A., Larson, I., 2010. Effective use of reducing agents and nanoparticle encapsulation in stabilizing catechins in alkaline solution. Food Chemistry 122, 662–667.

Ebringerová, A., 2006. Structural diversity and application potential of hemicelluloses. Macromolecular Symposia 232 (1), 1–12.

Ervin, D.E., Glenna, L.L., Jussaume, R.A., 2010. Are biotechnology and sustainable agriculture compatible? Renewable Agriculture and Food Systems 25 (02), 143–157.

EU Directive. 2001/18/EC of the European Parliament and of the Council on the Deliberate Release into the Environment of Genetically Modified Organisms and Repealing Council Directive 90/220/EEC.

European Commission, 2010. Directorate-General for Research, FP7 Cooperation-Food. Functional Foods, Brussels, Belgium, pp. 1–28.

FAO, 2011. Global Food Losses and Food Waste: Extent, Causes and Prevention. (Rome).

FAO, WHO, 2006. Guidelines on Food Fortification with Micronutrients. WHO Press, Geneva.

Fava, F., Zanaroli, G., Vannini, L., Guerzoni, E., Bordoni, A., Viaggi, D., Robertson, J., Waldron, K., Bald, C., Esturo, A., Talens, C., Tueros, I., Cebrián, M., Sebő k, A., Kuti, T., Broeze, J., Macias, M., Brendle, H.G., 2013. New advances in the integrated management of food processing by-products in Europe: sustainable explotation of fruit and cereal processing by-products with the production of new food products (NAMASTE EU). New Biotechnology 30 (6), 647–655.

Ferguson, L.R., Philpott, M., Barnett, M.P.G., 2010. Nutrigenomics: integrating genomic approaches into nutrition research. Molecular Diagnosis and Therapy 347–363.

Ferguson, L.R., 2009. Nutrigenomics approaches to functional foods. Journal of the American Dietetic Association 109 (3), 452–458.

Ferraro, V., Cruz, I.B., Jorge, R.F., Malcata, X., Pintado, M.E., Castro, P.M.L., 2010. Valorisation of natural extracts from marine source focused on marine by-products: a review. Food Research International 43, 2221–2233.

Frewer, L.J., Bergmann, K., Brennan, M., Lion, R., Meertens, R., Rowe, G., Vereijken, C., 2011. Consumer response to novel agri-food technologies: implications for predicting consumer acceptance of emerging food technologies. Trends in Food Science & Technology 22 (8), 442–456. http://dx.doi.org/10.1016/j.tifs.2011.05.005.

Fito, P., Andrés, A., Chiralt, A., Pardo, P., 1996. Coupling of hydrodynamic mechanism and deformation-relaxation phenomena during vacuum treatments in solid porous food-liquid systems. Journal of Food Engineering 27, 229–240.

García, M.A., Ferrero, C., Bértola, N., Martino, M., Zaritzky, N., 2002. Edible coatings from cellulose derivatives to reduce oil uptake in fried products. Innovative Food Science & Emerging Technologies 3 (4), 391–397.

Garnett, T., 2011. Where are the best opportunities for reducing greenhouse gas emissions in the food system (including the food chain)? Food Policy 36, S23–S32.

Gennadios, A., Hanna, M.A., Kurth, L.B., 1997. Application of edible coatings on meats, poultry and seafoods: a review. LWT – Food Science and Technology 30 (4), 337–350.

Gharsallaoui, A., Roudaut, G., Chambin, O., Voilley, A., Saurel, R., 2007. Applications of spray-drying in microencapsulation of food ingredients: an overview. Food Research International 40 (9), 1107–1121.

Gibbs, B.F., Kermasha, S., Alli, I., Mulligan, C.N., 1999. Encapsulation in the food industry: a review. International Journal of Food Sciences and Nutrition 50, 213–224.

Godfray, H.C.J., Beddington, J.R., Crute, I.R., Haddad, L., Lawrence, D., Muir, J.F., Pretty, J., Robinson, S., Thomas, S.M., Toulmin, C., 2010. Food security: the challenge of feeding 9 billion people. Science 327, 812–818.

Gouin, S., 2004. Micro-encapsulation: industrial appraisal of existing technologies and trends. Trends in Food Science and Technology 15, 330–347.

Hails, R.S., 2000. Genetically modified plants–the debate continues. Trends in Ecology & Evolution 15 (1), 14–18.

Han, J.H., 2009. Edible films and coatings: a review. In: Embuscado, M.E., Huber, K.C. (Eds.), Edible Films and Coatings for Food Applications. Springer Science + Business Media, LLC, New York, pp. 213–255.

Hanasaki, Y., Ogawa, S., Fukui, S., 1994. The correlation between active oxygens scavenging and antioxidative effects of flavonoids. Free Radical Biology & Medicine 16, 845–850.

Hansen, N.M.L., Plackett, D., 2008. Sustainable films and coatings from hemicelluloses: a review. Biomacromolecules 9 (6), 1493–1505.

Kates, R.W., Clark, W.C., Corell, R., Hall, M.J., Jaeger, C.C., Lowe, I., McCarthy, J.J., Schellnhuber, H.J., Bolin, B., Dickson, N.M., Faucheus, S., Gallopin, G.C., Grübler, A., Huntley, B., Jäger, J., Jodha, N.S., Kasperson, R.E., Mabogunje, A., Matson, P., Mooney, H., More III, B., O´Riordan, T., Svedin, U., 2001. Sustainability science. Science 292, 641–642.

Kim, Y.D., Morr, C., 1996. Microencapsulation properties of gum Arabic and several food proteins: spray-dried orange oil emulsion particles. Journal of Agricultural and Food Chemistry 44, 1314–1320.

Ki-moon, B., June 2008. Rome. The High-level Conference on World Food Security: The Challenges of Climate Change and Bioenergy, vol. 3.

King, A.H., 1995. Encapsulation of food ingredients: a review of available technology, focusing on hydrocolloids. In: Risch, S.J., Reineccius, G.A. (Eds.), Encapsulation and Controlled Release of Food Ingredients. ACS Symposium Series, vol. 590. American Chemical Society, Washington, DC, pp. 26–39.

Kramer, G.F., Wang, C.Y., Conway, W.S., 1989. Correlation of reduced softening and increased polyamine levels during low-oxygen storage of McIntosh apples. Journal of the American Society for Horticultural Science 114, 942–946.

Kramer, G.F., Wang, C.Y., Conway, W.S., 1991. Inhibition of softening by polyamine application in Golden Delicious and McIntosh apples. Journal of the American Society for Horticultural Science 116, 813–817.

Krasaekoopt, W., Suthanwong, B., 2008. Vacuum impregnation of probiotics in fruit pieces and their survival during refrigerated storage. Kasetsart Journal 42, 723–731.

Komduur, R.H., Korthals, M., te Molder, H., 2007. The good life: living for health and life without risks? on a prominent script of nutrigenomics. Nutrition Reviews 65, 301e315.

Kvakkestad, V., 2009. Institutions and the R&D of GM-crops. Ecological Economics 68 (10), 2688–2695.

Lamppa, J.W., Horn, G., Edwards, D., 2014. Toward the redesign of nutrition delivery. Journal of Controlled Release 190, 201–209.

Laufenberg, G., Kunz, B., Nystroem, M., 2003. Transformation of vegetable waste into value added products: (A) the upgrading concept; (B) practical implementations. Bioresource Technology 87, 167–198.

Liu, F., Chen, Z., Tang, C.H., 2014. Microencapsulation properties of protein isolates from three selected *Phaseolus* legumes in comparison with soy protein isolate. LWT – Food Science and Technology 55, 74–82.

Lowe, P., Phillipson, J., Lee, R.P., 2008. Socio-technical innovation for sustainable food chains: roles for social science. Trends in Food Science & Technology 19, 226–233.

Lyson, T., 2002. Advanced agricultural biotechnologies and sustainable agriculture. Trends in Biotechnology 20 (5), 193–196.

MacMillan, T., Benton, T.G., 2014. Engage farmers in research. Nature 509, 25–77.

Martins, I.M., Barreiro, M.F., Coelho, M., Rodrigues, A.E., 2014. Microencapsulation of essential oils with biodegradable polymeric carriers for cosmetic applications. Chemical Engineering Journal 245, 191–200.

McPherson, R.E., Olson, R.L., Eads, A., 2006. Emulsifiers for Citrus Oils and Related Products. U.S. World Intellectual Property Organization (WPO 2006/017744 A1).

Munin, A., Edwards-Lévy, F., 2011. Encapsulation of natural polyphenolic compounds: a review. Pharmaceutics 3 (4), 793–829.

National Research Council, 1989. Past Experience with Genetic Modification of Plants and Their Introduction into the Environment. National Academy Press, Washington DC, pp. 16–36.

Nesterenko, A., Alric, I., Silvestre, F., Durrieu, V., 2012. Influence of soy protein's structural modifications on their microencapsulation properties: α-tocopherol microparticle preparation. Food Research International 48, 387–396.

OECD (Organisation for Economic Co-operation and Development), 2004. Agriculture and the Environment: Lessons Learned from a Decade of OECD Work. OECD, Paris.

Olivas, G.I., Dávila-Aviña, J.E., Salas-Salazar, N.A., Molina, F.J., 2008. Use of edible coatings to preserve the quality of fruits and vegetables during storage. Stewart Postharvest Review 4 (3), 1–10.

Pallas, L.A., Pegg, R.B., Kerr, W.L., 2013. Quality factors, antioxidant activity, and sensory properties of jet-tube dried rabbiteye blueberries. Journal of the Science of Food and Agriculture 93, 1887–1897.

Pardue, S.L., 2010. Food, energy, and the environment. Poultry Science 89 (4), 797–802.

Pavlath, A.E., Orts, W., 2009. Edible films and coatings: why, what and how? In: Embuscado, M.E., Huber, K.C. (Eds.), Edible Films and Coatings for Food Applications. Springer Science + Business Media, LLC, New York, pp. 1–24.

Perdones, A., Vargas, M., Atarés, L., Chiralt, A., 2014. Physical, antioxidant and antimicrobial properties of chitosan–cinnamon leaf oil films as affected by oleic acid. Food Hydrocolloids 36, 256–264.

Perez-Cabrera, L., Chafer, M., Chiralt, A., Gonzalez-Martinez, C., 2011. Effectiveness of antibrowning agents applied by vacuum impregnation on minimally processed pear. LWT – Food Science and Technology 44, 2273–2280.

Pérez-Gago, M.B., 2015. Últimos avances en recubrimientos comestibles antimicrobianos para fruta entera. [On line resource] http://www.interempresas.net/Poscosecha/Articulos/144074-Ultimos-avances-en-recubrimientos-comestibles-antimicrobianos-para-fruta-entera.html.

Pérez-Gago, M.B., Rhim, J.W., 2014. Edible coating and film materials: lipid bi-layers and lipid emulsions. In: Han, J.H. (Ed.), Innovations in Food Packaging. Academic Press, Elsevier Ltd., pp. 325–368.

Pérez-Gago, M.B., Del Río, M.A., Rojas-Argudo, C., 2008. Recubrimientos comestibles en frutas y hortalizas. [On line resource] http://www.horticom.com/pd/article.php?sid=69985.

Ponappa, T., Scheerens, J.C., Miller, A.R., 1993. Vacuum infiltration of polyamines increases firmness of strawberry slices under various storage conditions. Journal of Food Science 58, 361–364.

Radziejewska-Kubzdela, E., Biegańska-Marecik, R., Kidoń, M., 2014. Applicability of vacuum impregnation to modify physico-chemical, sensory and nutritive characteristics of plan origin products. A review. International Journal of Molecular Sciences 15, 16577–16610.

Rascón, M.P., Beristain, C.I., García, H.S., Salgado, M.A., 2011. Carotenoid retention and storage stability of spray-dried encapsulated paprika oleoresin using gum Arabic and soy protein isolate as wall materials. LWT – Food Science and Technology 44, 549–557.

Royal Society, 2009. Reaping the Benefits: Science and the Sustainable Intensification of Global Agriculture. Royal Society, UK.

Saikia, S., Kumar, N., Mahanta, C.L., 2015. Optimisation of phenolic extraction from *Averrhoa carambola* pomace by response surface methodology and its microencapsulation by spray and freeze drying. Food Chemistry 171, 144–152.

Salgado, P.R., Ortiz, C.M., Musso, Y.S., Di Giorgio, L., Mauri, A.N., 2015. Edible Films and Coatings Containing Bioactives. COFS. http://dx.doi.org/10.1016/j.cofs.2015.09.004.

Sánchez-González, L., Vargas, M., González-Martínez, C., Chiralt, A., Cháfer, M., 2009. Characterization of edible films based on hydroxypropylmethylcellulose and tea tree essential oil. Food Hydrocolloids 23 (8), 2102–2109.

Shahidi, F., Han, X.Q., 1993. Encapsulation of food ingredients. Critical Reviews In Food Science and Nutrition 33, 501–547.

Singer, M., Soll, D., 1973. Guidelines for DNA hybrid molecules. Science 181, 1114.

Smidsrød, O., Skjak-Braek, G., 1990. Alginate as immobilization matrix for cells. Trends in Biotechnology 8, 71–78.

Soukoulis, C., Yonekura, L., Gana, H.H., Behboudi-Jobbehdar, S., Parmenter, C., Fiska, I., 2014. Probiotic edible films as a new strategy for developing functional bakery products: the case of pan bread. Food Hydrocolloids 39, 231–242.

Spiertz, H., 2010. Food production, crops and sustainability: restoring confidence in science and technology. Current Opinion in Environmental Sustainability 2 (5–6), 439–443.

Suárez, R.B., Campañone, L.A., García, M.A., Zaritzky, N.E., 2008. Comparison of the deep frying process in coated and uncoated dough systems. Journal of Food Engineering 84 (3), 383–393.

Takei, T., Yoshida, M., Hatate, Y., Shiomori, K., Kiyoyama, S., 2008. Lactic acid bacteria-enclosing poly(ε-caprolactone) microcapsules as soil bioamendment. Journal of Bioscience and Bioengineering 106 (3), 268–272.

Tamer, C.E., İncedayı, B., Yıldız, B., Çopur, Ö.U., 2013. The use of vacuum impregnation for debittering green olives. Food and Bioprocess Technology 6, 3604–3612.

Tang, C.H., Li, X.R., 2013a. Microencapsulating properties of soy protein isolate and storage stability of the correspondingly spray-dried emulsions. Food Research International 52, 419–428.

Tang, C.H., Li, X.R., 2013b. Microencapsulating properties of soy protein isolate: influence of preheating and/or blending with lactose. Journal of Food Engineering 117, 281–290.

Tapia, M.S., Roias-Graü, E.J., Rodríguez, J., Ramírez, J., Carmona, A., Martin-Belloso, O., 2007. Alginate and gellan based edible films for probiotic coatings on fresh cut fruits. Journal of Food Science 72 (4), 190–196.

Tatar, F., Tunç, M.T., Dervisoglu, M., Cekmecelioglu, D., Kahyaoglu, T., 2014. Evaluation of hemicellulose as a coating material with gum arabic for food microencapsulation. Food Research International 57, 168–175.

Tester, M., Langridge, P., 2010. Breeding technologies to increase crop production in a changing world. Science 327, 818–822.

Thies, C., 1987. Microencapsulation. In: Mark, H.F., Bikales, N.M., Overberger, C.G., Menges, G., Kroschwitz, J.I. (Eds.), Encyclopedia of Polymer Science and Engineering. John Wiley & Sons, New York, pp. 724–745.

Thies, C., 2005. A survey of microencapsulation processes. In: Benita, S. (Ed.), Microencapsulation. Marcel Dekker Inc, New York, pp. 1–20.

Tran, V.T., Benoît, J.P., Venier-Julienne, M.C., 2011. Why and how to prepare biodegradable, monodispersed, polymeric microparticles in the field of pharmacy? International Journal of Pharmaceutics 407, 1–11.

United Nations. Department of Economic and Social Affairs. Population Division. World Population Prospects, the 2015 Revision, 2015. August 19, 2015. http://esa.un.org/unpd/wpp/Graphs/.

Vergragt, P.J., Brown, H.S., 2008. Genetic engineering in agriculture: new approaches for risk management through sustainability reporting. Technological Forecasting and Social Change 75 (6), 783–798.

Wildman, R.E.C., 2006. Classifying nutraceuticals. In: Wildman (Ed.), Handbook of Nutraceuticals and Functional Foods, (second ed.). CRC Publisher, pp. 13–31.

Wyman, C.E., 2003. Potential synergies and challenges in refining cellulosic biomass to fuels, chemicals, and power. Biotechnology Progress 19, 254–262.

Xie, J., Zhao, Y., 2004. Use of vacuum impregnation to develop high quality and nutritionally fortified frozen strawberries. Journal of Food Processing and Preservation 28, 117–132.

Yadav, M.P., Johnston, D.B., Hicks, K.B., 2009. Corn fiber gum: new structure/function relationships for this potential beverage flavor stabilizer. Food Hydrocolloids 23 (6), 1488–1493.

Yao, L.H., Jiang, Y.M., Shi, J., 2004. Flavonoids in food and their health benefits. Plant Foods for Human Nutrition 59, 113–122.

Zhang, L.F., Mou, D.H., Du, Y.S., 2007. Procyanidins: extraction and microencapsulation. Journal of the Science of Food and Agriculture 87, 2192–2197.

Zhao, F.J., Shewry, P.R., 2011. Recent developments in modifying crops and agronomic practice to improve human health. Food Policy 36, S94–S101.

Zhao, Y., Xie, J., 2004. Practical applications of vacuum impregnation in fruit and vegetable processing. Trends in Food Science & Technology 15, 434–451.

CHAPTER

INNOVATIVE BIOBASED MATERIALS FOR PACKAGING SUSTAINABILITY

9

K.G. Zinoviadou[1], M. Gougouli[1], C.G. Biliaderis[2]

[1]Perrotis College, American Farm School, Thessaloniki, Greece; [2]School of Agriculture, Aristotle University, Thessaloniki, Greece

9.1 INTRODUCTION

After only a small decline due to the financial crisis in 2008, the production of plastics increased continuously and in 2012 it reached the value of 288 Mt. In Europe, packaging applications are the largest application sector for the plastics industry, representing 39.4% of the total plastics demand (Plastics Europe, 2013). Despite the increased environmental concerns, still most of today's synthetic polymers are produced from petrochemicals. Because of their lack of biodegradability, petroleum-based products result in major waste disposal problems in certain areas. Moreover, the continuing use of oil resources implies an inevitable decrease in availability accompanied with increasing costs. Sustainable packaging is a topic that attracts great interest not only from the industry but also from the government and consumers. It can refer to many different aspects like the reduction of packaging material, the use of recycling packaging, or the use of materials with a lower ecological footprint.

According to the European Bioplastics association there are three types of bioplastics: (1) bioderived and biodegradable/compostable, (2) fossil-fuel derived and biodegradable, and (3) bioderived and nonbiodegradable. In this chapter the first category of bioplastics is discussed (Johansson et al., 2012). Use of such materials is advantageous since biobased materials reduce the depletion of finite fossil resources and CO_2 emissions while their compostability offers environmental benefits in the end-of-life phase (Molenveld et al., 2015). In this context, polymers obtained from renewable recourses or utilization of agro-industrial and marine wastes are considered as a promising alternative. Nowadays, their use seems even more attractive since scientific advances over the last decade have diminished the costs of converting organic material into platform chemicals that can supply other manufacturing streams like polymers. The interest is so great that several national and international research projects have been performed in Europe and others are still in progress with a focus on the potentials of biomaterials for packaging applications. Examples are SustainPack, SustainComp, SUNPAP, FlexPakRenew, Bio-mimetic, and n-chitopak (Johansson et al., 2012).

As illustrated in Fig. 9.1 biobased materials used in packaging technology can be divided into three main categories based on their origin, chemical nature, and means of production. It should be noted that the physical properties of packaging play a dominant role in the selection of an appropriate polymer and can be broadly classified in the following four categories (Alavi et al., 2015):

Innovation Strategies in the Food Industry. http://dx.doi.org/10.1016/B978-0-12-803751-5.00009-X

- thermal: heat capacity, thermal conductivity, melting temperature, and glass transition temperature
- mechanical: tensile properties and tear strength
- barrier: diffusion, solubility, and permeability
- optical: gloss, haze, and transparency

The aim of this chapter is to give an overview of innovative uses of biobased packaging in the field of the food industry and provide information on the relevant regulatory aspects and safety considerations.

9.2 NOVEL BIOBASED PLASTICS

According to the plastic industry sector's projections by 2020 the most important biopolymers that are both biobased and biodegradable include starch and its derivatives, poly(lactic acid), (PLA) and polyhydroxyalkanoates (PHAs) with amounts up to 1.3, 0.8, and 0.4 Mt, respectively (Imre and Pukanszky, 2013). Regarding the range of applications of biodegradable polymers used in the food industry, it includes disposable cutlery, cups, plates, trays, overwrap and laminated films, and containers.

9.2.1 STARCH AND STARCH BLENDS

Starch is the one of the most abundant natural (plant-derived) polysaccharides. Its granules can vary in shape, size, structure, and chemical composition depending on its origin, but in all cases it consists of two main macromolecular species, amylose and amylopectin, differing in fine structure and properties

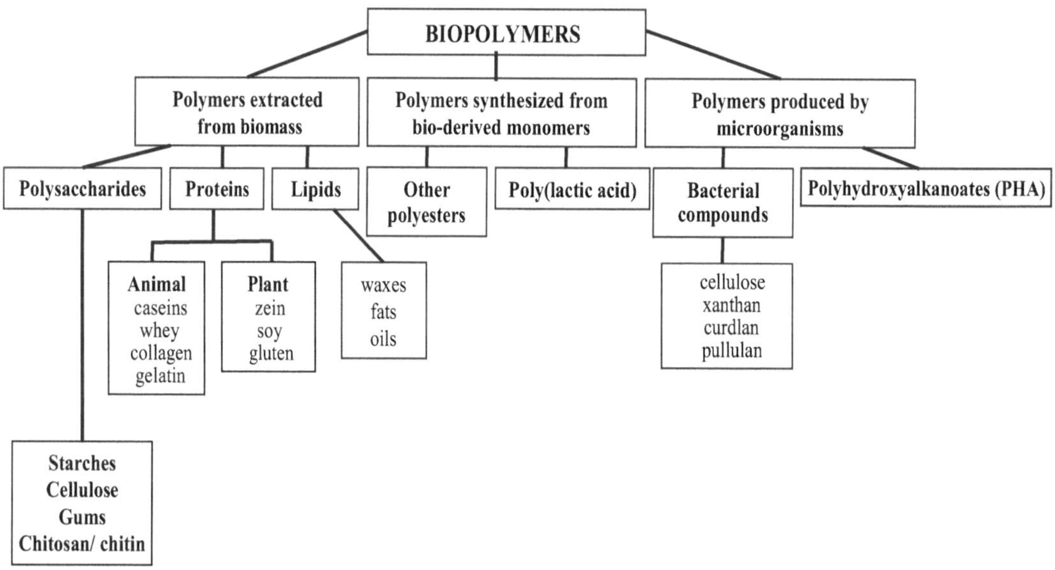

FIGURE 9.1

Biologically derived materials used for biopolymer-based packaging.

Adapted from Cutter, C.N., 2006. Opportunities for bio-based packaging technologies to improve the quality and safety of fresh and further processed muscle foods. Meat Science 74, 131–142.

(Biliaderis, 2009; Jimenez et al., 2012). As a packaging material starch alone cannot form films with adequate mechanical and diffusion-controlled physical properties unless it undergoes certain pretreatment such as plasticization with small molecular weight–compatible constituents, blending, or chemical modification (Alavi et al., 2015). For example, thermoplastic starch can be produced by processing a starch–plasticizer blend in an extruder at temperatures of between 140 and 160°C under high pressure and shear. The presence of plasticizers, apart from water, is essential in order to obtain a rubbery material without brittleness (Jimenez et al., 2012). The Italian company Novamont is the leader in this field with its line Master-Bi (www.novamont.com) while other companies of interest are Plantic (www.plantic.com.au) and Rodenburg biopolymers (www.biopolymers.com) launching the lines Eco Plastic and Solanyl, respectively. Commercially, starch-based packaging has been previously used by companies such as McDonald's or in high profile international events like the Olympic games of 2012. Moreover, outdoor events often feature "green" disposable cups, plates, and cutlery made from starch-based plastics (Johansson et al., 2012).

Interestingly, it has been demonstrated that for the production of starch plastic granules, 25–75% less energy is required and the greenhouse emissions are 20–80% lower than for polyethylene (PE) (Steinbuchel, 2003). However, it should be pointed out that only some of the starch blends are suitable for food-packaging applications. This is due to the fact that the blends may contain additives such as compatibilizers and plasticizers that can migrate out of the matrix.

9.2.2 POLY(LACTIC ACID)

PLA is one of the most promising biodegradable bioplastics since it possesses mechanical properties and transparency comparable to common thermoplastics like polystyrene and polyethylene terephthalate (Benetto et al., 2015). Lactic acid, the single monomer of PLA, can be produced by fermentation of renewable agricultural sources, and upon polymerization the resulting PLA may have a great variability in molecular weight. Various processing techniques can be applied to PLA for the formation of packaging materials such as injection molding, extrusion, blow molding, and thermoforming processing (Peelman et al., 2013). When compared to other hydrocarbon-based polymers, one of the greatest advantages of PLA production is an overall decrease of CO_2 emission. Since large amounts of carbon dioxide are absorbed when corn or other crops used for PLA production are grown, the use of PLA may emit fewer greenhouse gases compared to competitive hydrocarbon-based polymers (Jamshidian et al., 2010). Regarding food-packaging applications, PLA films have been thermoformed into trays for packaging of salads, ready-to-eat meals, and deli products among other uses. NatureWorks Inc (www.natureworksllc.com) owned wholly by Cargill and located in the US Midwest is the leading company in PLA production globally. An overview of different uses of PLA packaging is summarized in the following Table 9.1.

9.2.3 POLYHYDROXYALKANOATES

The PHA family are biodegradable thermoplastic polymers produced by a wide range of microorganisms through a fermentation process (Peelman et al., 2013). They are aliphatic polyesters that accumulate within the microbial cells as granules, constituting up to 90% of the cell mass and exhibiting thermomechanical properties similar to synthetic polymers such as polypropylene (Alavi et al., 2015). A major disadvantage of PHAs in packaging applications is the fact that they are not transparent; currently they are mostly used for the production of carrier bags (Molenveld et al., 2015). Out of the more

Table 9.1 Commercial Applications of PLA-Based Packaging

Product Category	Application	Sources
Milk and dairy	Danone in Germany uses PLA (Ingeo) cup for Activia yogurt	www.danone.com
	Cooperative in Iowa uses PLA (Ingeo) milk bottle	www.nuatreworksllc.com
	Stonyfield Farm replaced all of its petroleum-based multipack yogurt cups with cups made from plant-based Ingeo	www.stonyfield.com
	Sandros Inc. uses PLA-foam thermobox (BioFoam) for ice cream	www.sandros-bio.de
Fruit juice	Noble juice in the United States uses PLA (Ingeo) bottles	www.nuatreworksllc.com
	Polenghi in Italy uses PLA (Ingeo) bottles	www.polenghigroup.it
	Blue Lake Citrus Products uses PLA (Ingeo) bottles	www.nuatreworksllc.com
Fruits and vegetables	Albert Heijn (Dutch supermarket chain) uses PLA films for organic fruit and vegetables	Molevelt et al. (2015)
	Walmart uses Ingeo natural packaging for the fresh organic salads, salsas, and bagged spinach	www.marketside.com
	Compostable fruit net made from BIO-FLEXF1130 and F2110	www.fkur.com
Water	Sant'Anna in Italy uses PLA (Ingeo) bottle for mineral water	www.santanna.it
Coffee/Tea	BASF creates coffee capsules made from compostable bioplastic	www.basf.com
	Beanarella packages coffee in compostable PLA-based cups (Ecovio)	www.beanarella.ch
	Ahlstrom produces a lightweight filament web-based PLA to be used as a tea bag	www.ahlstrom.com

than 100 PHA composites that are known, polyhydroxybutyrate (PHB) is the most common one. This bioderived polymer has high crystallinity and a high melting point and is often blended with PLA, leading to the formation of materials with advanced mechanical, thermal, and physical properties compared to neat PLA (Armentano et al., 2015). Recently it was reported that blending PLA with 25% (w/w) PHB resulted in improved oxygen and barrier properties, whilst reducing the inherent transparency of PLA (Arrieta et al., 2014). The leading producers of PHA-based materials are Metabolix (www.metabolix.com), Tianan (www.tianan-enmat.com), TGBM (www.tjgreenbio.com), Ecoman (www.ecoman.com), and Biomer (www.biomer.de).

9.3 EDIBLE FILMS AND COATINGS

Any type of material used for enrobing (ie, coating or wrapping) various food products to extend their shelf life and maybe eaten together with food with or without further removal is considered an edible film or coating. These films can serve as a physical barrier between food components, can be used directly as a wrap, or can be formed into a pouch that may carry food products (Hernandez-Izquierdo and Krochta, 2008). When applied they can increase the shelf life of products by modulating the transfer of gases, such as oxygen and carbon dioxide, moisture, aroma, and flavor compounds in food systems (Porta et al., 2011). Moreover, edible films and coatings can serve as effective carriers for a wide range of food additives, including various antimicrobials and antioxidant agents (Cagri et al., 2004).

9.3.1 COMPOSITION OF EDIBLE FILMS AND COATINGS

Structuring biopolymers like polysaccharides and proteins have been employed for formulation of edible films and coatings. Polysaccharides such as starch, chitosan, pectin, cellulose, pullulan, and galactomannan have been studied as potential materials for edible packaging, and their properties and uses have been extensively reviewed (Cerqueira et al., 2011; Espitia et al., 2014; Jimenez et al., 2012; Porta et al., 2011). Polysaccharides often render transparent and homogeneous films with moderate mechanical properties. However, the application of polysaccharide-based coatings and films is limited because they are relatively sensitive to moisture and have poor water vapor–barrier properties. On the contrary, they exhibit excellent gas permeability properties, resulting in desirable modified atmospheres. Different methods have been proposed in order to solve this shortcoming such as blending with different biopolymers, the incorporation of hydrophobic substances such as waxes or oils, or chemical modification (Campos et al., 2011). Numerous proteins such as corn zein, soy protein isolates, milk proteins, wheat gluten, and gelatin have been studied as potential film-forming biopolymers. While polysaccharides consist of a few or even one monomer, proteins are based on several amino acids. The secondary, tertiary, and quaternary structures of proteins result in various interactions and bindings differing in position, type, and strength. Upon physical (heating, irradiation), chemical (glutaraldehyde, formaldehyde, glyceraldehyde, glyoxal), or enzymatic (transglutaminase) treatment these structures can be modified, resulting in packaging materials with improved mechanical and barrier properties (Han, 2014).

The flexibility and the processability of the polymers can be improved by the addition of plasticizers, a group of low-molecular weight, nonvolatile compounds that can decrease the intermolecular forces along the polymer chains. However, since adverse effects of plasticizers on edible film attributes have been reported, eg, increased gas and water permeabilities, their impact should always be evaluated and their use (type and amount) should be optimized and controlled (Vieira et al., 2011). Due to their hydrophobic nature, lipids can form films with good moisture barrier properties, but when used alone, the presence of pinholes and cracks on the films' surface results in relatively poor mechanical properties. In this respect, the development of composite or multilayer films can assist in reaching the desirable physical properties (Kristo et al., 2007). The lipids that are commonly used for preparation of lipid-based edible films and coatings include neutral lipids, fatty acids, waxes, and resins. The performance of the lipid materials introduced in a composite packaging matrix depends on the nature of the lipid used (structure, chemical arrangement, hydrophobicity, physical state) as well as its interactions and compatibility with the other components of the film (Han, 2014).

9.3.2 EDIBLE PACKAGING APPLICATIONS

Despite the fact that edible coatings are being commonly used for a multitude of food products, they are not always noticed by consumers. The shiny appearance of certain fruits such as apples and the desired glaze of certain candies are examples of such functions of edible coatings. Application of coatings is also common in the pharmaceutical industry in order to prevent crumbling and stickiness, to mask any undesirable flavor upon swallowing, and to provide controlled release of certain ingredients (Embuscado and Huber, 2009). In Table 9.2 an overview of commercial coatings is provided.

One of the first commercial applications of edible films was in the field of dental care. Listerine PocketPacks are small pieces of edible films that melt instantly on the tongue releasing a breath freshener. They consist of a mixture of hydrocolloids including carrageenan, locust bean gum,

Table 9.2 List of Commercially Used Edible Coatings

Commercial Name	Component	Applications	Sources
FreshSeal	Sucrose esters	Extending shelf life of different fruits and vegetables such as tomatoes, mangoes, and melons	www.basf.com
Semperfresh	Sucrose esters	Controlling weight loss and respiration, retaining moisture and preserving natural color of cherries and pears	www.paceint.com
Pac-Rite SP	Carnauba wax	Providing weight loss control for sweet potatoes	www.paceint.com
NatureSeal	Polysaccharides	Maintaining the taste, texture, and color of sliced apples and pears	www.natureseal.com
Crystalac	Shellac	Glazing hard and soft sugar candy	www.mantrose.com
Cozeen 303 N	Zein protein	Providing a shiny coating to nuts and candies and preventing candies from sticking	www.zeinproducts.com
Fry Shield	Calcium pectinate	Reducing oil uptake during frying of fish, potatoes, and other vegetables	www.kerry.com
Myvacet	Acetylated monoglycerides	Preventing water loss in muscle foods	www.kerry.com

pullulan, and xanthan (www.listerine.com). Moreover, Novartis was the first leading company that used edible film technology in order to deliver an accurate dose of medication against cough in the thin-film format (www.novartis.com). WikiPearls are novel edible forms for consuming transportable foods and drinks without plastic. They consist of a natural food membrane held together by electrostatic forces that is inspired by nature and is water and oxygen impenetrable (www.perfectlyfree.com). Edible films consisting of different fruit and vegetable extracts have been applied as an appealing alternative for seaweed sheets, and are commonly used for sushi, in an attempt to meet consumer demands for increased variety and unique flavors. Additionally, sandwich wraps have been manufactured in a similar way as a replacement for bread and tortillas (www.origamifoods.com). Edible single-use cups consisting of seaweed, organic sweeteners, and natural flavors and colors have been recently developed (www.loliware.com), while the recent launch of an edible single-use spherical water container seems a rather promising innovation (www.skippingrockslab.com). Other relevant innovations in the field of edible packaging include the edible cookie cup (www.lavazza.com), edible wafer cupcake cases (www.oetker.co.uk), and the edible burger wrapper.

Edible packaging that disappears is also an emerging market. MonoSol, LLC is one of the leading companies to create water-soluble bags made from food-grade ingredients for preportioned oatmeal or rice. Vivos Films are transparent, odorless, tasteless films consisting of a blend of food-grade ingredients that also render sufficient mechanical properties, allowing the formation of bags/pouches with existing converting technologies (www.vivosfilms.com). Similarly, Watson, Inc has developed the so-called Sol-u-Pak edible pouches in order to offer premeasured ingredients. The pouches melt in 10–20 s after mixing with the rest of the ingredients (www.watson-inc.com). Recently a design firm presented three quickly biodegrading packages made from edible ingredients such as caramelized sugar coated with wax, seaweed, and beeswax that can be used for serving olive oil, smoothies, and rice, respectively (www.tomorrowmachine.se).

9.4 NANOCOMPOSITES FOR BIOBASED PACKAGING

It has been suggested that inherent shortcomings of biopolymer-based packaging materials such as hydrophilicity, poor barrier properties, low mechanical strength, narrow processing window, or low heat deflection temperatures, may be overcome by nanocomposite technology (Rhim et al., 2013a). Fillers with at least one dimension in the nanometric scale (1–100 nm) can be defined as nanoparticles or nanoreinforcements, and their composites with biopolymers are bionanocomposites (Azeredo, 2009). The potential use of layered clay mineral nanoparticles has received quite a lot of attention due to their commercial availability, low cost, significant mechanical and barrier property enhancements, and relatively simple processability. However, it should be pointed out that although nanoclays are naturally occurring materials, they are not renewable or biodegradable. In this context, biobased nanofillers, derived from partially crystalline biopolymers such as starch, cellulose, or chitin have been gaining considerable interest lately owing to their high availability, low density, and of course biodegradability. For nanocellulosic particles, their performance depends both on the isolation method applied and the origin of the nanofibers (Azizi Samir et al., 2005). Up to now cellulose-based nanoparticles have been derived from various sources of plant materials and demonstrated that they were good fillers to improve the mechanical properties of biopolymer films such as starch (da Silva et al., 2012), alginate (Huq et al., 2012), and chitosan films (Khan et al., 2012). When starch nanocrystals were incorporated into sorbitol-plasticized pullulan films, an increase of the glass transition temperature was noted, and it was attributed to the restriction of the mobility of the polymeric chains (pullulan) as a result of strong interactions between the starch nanocrystals and the filler with the matrix. This effect was accompanied by a large increase of the tensile strength and a drastic decrease of the film's extensibility (Kristo and Biliaderis, 2007). The addition of chitin whiskers in chitosan-based films resulted in great improvement of both the tensile strength and the elongation properties of the composite films; this was attributed to the fact that incorporation of fibrous and rod-like chitin whiskers into the chitosan matrix can form extensive hydrogen bonding due to the numerous hydroxyl groups on the fiber surface (Li et al., 2011).

9.5 BIOPOLYMER-BASED ANTIMICROBIAL PACKAGING

Antimicrobial packaging appears to be a promising form of active food packaging that can be used successfully in order to increase the shelf life of food products and at the same time meet consumer demands regarding safety and high quality. The primary advantage of antimicrobial edible packaging is that inhibitory agents incorporated into these films can be specifically targeted for postprocessing contaminants on the food surface. Moreover, incorporation of the antimicrobial compounds into polymer matrices is advantageous when compared to their direct application to the products' surface, since it allows their controlled release. Consequently, by using antimicrobial films a total reduction of the amount of added active agent can be achieved (Quintavalla and Vicini, 2002). The inherent biodegradability and, in many cases, edibility of these films, could also be maintained by the incorporation of food-grade functional compounds. In this context, natural active agents should be used as prominent alternatives to chemical preservatives. The most common antimicrobial agents used in biopolymer-based films are listed in the following, and in Table 9.2 different applications of the antimicrobial films are summarized.

9.5.1 ORGANIC ACIDS AND SALTS

Organic acids and their salts have a long history as generally recognized as safe (GRAS) food preservatives and are frequently used in the food industry due to their well-evidenced antimicrobial activity against yeasts, molds, and bacteria. The most common food preservatives falling in this category are acetic, benzoic, lactic, citric, malic, tartaric, propionic, fumaric, and sorbic acid and their salts such as potassium sorbate or sodium lactate (Cagri et al., 2004). The antimicrobial activity of the acids is attributed to a decrease of the internal pH of the microbial cells due to the ionization of the otherwise undissociated acid molecules coupled with the disruption of substrate transport (Ramos et al., 2012a). Apart from their antimicrobial activity, some of these compounds, like the lactate salts, are also used as flavor enhancers, emulsifiers, and can increase the water-holding capacity and cooking yields when applied in meat products (Shelef, 1994).

9.5.2 ESSENTIAL OILS

Essential oils (EOs) are volatile, natural compounds that are synthesized as secondary metabolites by aromatic plants and possess characteristic and intense odor and taste. Their chemical composition is complex and depends, apart from the species, on different factors such as the part of the plant used for the extraction, the environmental conditions, the time of harvesting, and the geographical region (Campos et al., 2011). Due to their high antimicrobial activity, as has been reviewed by Burt (2004), the essential oils can be considered as potential alternatives to synthetic preservatives and have already been classified as GRAS. The delivery of such compounds via their incorporation into packaging films is advantageous since lower oil concentrations are required, thus minimizing the occurrence of undesirable flavors and odors (Sanchez-Gonzalez et al., 2011). Moreover, by using films as carriers of the EOs, different problems related to the volatility and low water solubility of the EOs can be alleviated.

9.5.3 ANTIMICROBIAL PEPTIDES

Nisin is a naturally occurring peptide produced by *Lactococcus lactis* subsp. *lactis*. It has already been approved for use in more than 50 countries worldwide, and there is an increasing trend for its application as a natural antimicrobial agent (Han et al., 2015). Nisin exhibits antimicrobial activity against a wide range of gram-positive bacteria such as *Listeria monocytogenes*, *Staphylococcus aureus*, *Clostridium botulinum*, and its spores (FDA, 2001). On the contrary, its effectiveness against gram-negative bacteria is limited due to its inability to cross the lipopolysaccharide outer membrane, but it can be enhanced when used in combination with chelators like EDTA. Due to the fact that it is nontoxic, heat stable, and does not contribute to off-flavors, nisin is commercially used in a variety of food products, whereas its incorporation in biopolymer-based packaging has been investigated (Table 9.3). E-polylysine (ε-PL) is a natural antimicrobial homo-peptide consisting of 25–35 L-lysine units that are

Table 9.3 Applications of Biopolymer-Based Antimicrobial Packaging

	Biopolymer Matrix	Antimicrobial Agent	Microorganisms	Application Medium	References
Organic acids and their salts	Sweet potato starch	Potassium sorbate	*Escherichia coli* *Staphylococcus aureus*	Solid Semi-solid media	Shen et al. (2010)
	Chitosan	Sodium lactate Sodium diacetate Potassium sorbate	*Listeria monocytogenes*	Cold-smoked salmon	Jiang et al. (2011)
	Chitosan	Potassium sorbate	*Spoilage molds*	Butter cake	Sangsuwan et al. (2015)
	Sodium caseinate	Potassium sorbate Sodium lactate	*L. monocytogenes*	Agar media	Kristo et al. (2008)
	Whey proteins	Lactic acid Propionic acid	*E. coli* *S. aureus* *Yarrowia lipolytica*	Agar media	Ramos et al. (2012b)
	Argentine anchovy proteins	Sorbic acid Benzoic acid	*E. coli* O157:H7 *L. monocytogenes* *S. aureus* *Salmonella* Enteritidis	Agar media	Da Rocha et al. (2014)
	Argentine anchovy proteins	Sorbic acid Benzoic acid	*E. coli* O157:H7 *L. monocytogenes*	Meat	Da Rocha et al. (2014)
	Carboxymethyl cellulose Oxidized potato starch Soy Protein Isolate Gelatin	Potassium sorbate	*E. coli* *S. aureus* *Bacillus cereus* *Pectobacterium carotovorum* *Botrytis cinerea* *Monilinia fructigena* *Alternaria alternata* *Rhizopus nigricans*	Agar media	Kowalczyk et al. (2015)
	Carrageenan	Citric acid	*S. aureus* *Proteus mirabilis* *Pseudomonas aeruginosa* *E. coli* *Dickeya chrysanthemi*	Agar media	Ela Fawal (2014)

Continued

Table 9.3 Applications of Biopolymer-Based Antimicrobial Packaging—cont'd

	Biopolymer Matrix	Antimicrobial Agent	Microorganisms	Application Medium	References
Essential oils	Whey protein isolate	Oregano oil	Total viable count *Pseudomonas* spp. Lactic acid bacteria	Beef muscle	Zinoviadou et al. (2009)
		Oregano oil Clove oil Tea tree oil Coriander oil Mastic thyme oil Laurel oil Rosemary oil Sage oil	*Listeria innocua* *S. aureus* *Salmonella* Enteritidis *Pseudomonas fragi*	Agar media	Fernandez-Pan et al. (2012)
		Oregano oil Clove oil	Total aerobic mesophiles *Enterobacteriaceae* *Pseudomonas* spp. Lactic acid bacteria	Poultry rinse	Fernandez-Pan et al. (2013)
	Soy protein	Oregano oil Thyme oil	*E. coli* O157:H7 *S. aureus* *P. aeruginosa* *Lactobacillus plantarum*	Agar media	Emiroglu et al. (2010)
		Oregano oil Thyme oil	Total viable count *Pseudomonas* spp. Lactic acid bacteria *Staphylococcus* spp. Coliforms	Beef patties	Emiroglu et al. (2010)
	Hake proteins	Citronella oil Coriander oil Tarragon oil Thyme oil	*Brochothrix thermosphacta* *E. coli* *L. innocua* *L. monocytogenes* *Pseudomonas putida* *Salmonella* Typhimurium *Shewanella putrefaciens*	Agar media	Pires et al. (2013)
	Sunflower protein	Clove oil	Total viable bacteria Total mesophiles H_2S-producer microorganisms Luminescent colonies *Pseudomonas* spp. Lactic bacteria *Enterobacteriaceae*	Sardine patties	Salgado et al. (2013)

Essential oils	Chicken feet protein	Marjoram oil Coriander oil Clove bud oil	*L. monocytogenes* *Salmonella* Enteritidis	Cheddar cheese	Lee et al. (2015)
	Perilla seed meal protein	Rosemary oil Lemongrass oil Clove oil	*L. monocytogenes*	Pork sausage	Song et al. (2015)
	Gelatin	Lemongrass oil	Total viable count Psychrophilic bacterial count H_2S-producer microorganisms *Enterobacteriaceae* Lactic acid bacteria	Sea bass slices	Ahmad et al. (2012)
	Gelatin-chitosan	Clove oil	Total viable bacteria H_2S-producer microorganisms *Pseudomonas* ssp. *Enterobacteriaceae* Lactic acid bacteria	Cod fillet	Gomez-Estaca et al. (2010)
	Chitosan	Oregano oil	*L. monocytogenes* *E. coli* O157:H7	Bologna	Zivanovic et al. (2005)
	Carboxymethyl cellulose	*Zataria multiflora* oil	*S. aureus* *B. cereus* *E. coli* *P. aeruginosa* *Salmonella* Typhimurium	Agar media	Dashipour et al. (2015)
	Sodium alginate	Thyme oil Lemongrass oil Sage oil	*E. coli*	Agar media	Acevedo-Fani et al. (2015)
	Pectin	Lime oil	*E. coli* 0157:H7 *Salmonella* Typhimurium *B. cereus* *S. aureus* *L. monocytogenes*	Agar media	Aldana et al. (2015)
	Poly (lactic acid)-cellulose nanocrystals	Oregano oil	*L. monocytogenes*	Mixed vegetables	Salmieri et al. (2014a)

Continued

Table 9.3 Applications of Biopolymer-Based Antimicrobial Packaging—cont'd

	Biopolymer Matrix	Antimicrobial Agent	Microorganisms	Application Medium	References
Living Cells	Acetate cellulose	Bacteriophages	*Salmonella* Typhimurium	Agar media	Gouvea et al. (2015)
	Sodium caseinate	*Lactobacillus sakei*	*L. monocytogenes*	Agar media Beef muscle	Gialamas et al. (2010)
	Sodium caseinate Methylcellulose	*Lactobacillus acidophilus* *Lactobacillus reuteri*	*Listeria innocua*	Agar media	Sánchez-González et al. (2014)
	Sodium caseinate Pea protein Methylcellulose Hydroxypropylmethyl-cellulose	*L. plantarum*	*L. innocua*	Agar media	Sánchez-González et al. (2013)
	Sodium alginate Locust bean gum	*Wickerhamomyces anomalus*	*Penicillium digitatum*	Agar media	Aloui et al. (2015)
	Agar	*Lactobacillus paracasei* L26 *Bifidobacterium lactis* B94	Total viable count *Pseudomonas* spp. Lactic acid bacteria Enterobacteriaceae	Hake fillets	López de Lacey et al. (2014)
Bacteriocins and peptides	Whey protein isolate	E-polylysine	Total viable count *Pseudomonas* spp. Lactic acid bacteria	Beef muscle	Zinoviadou et al. (2010)
	Starch	E-polylysine	*E. coli* *Bacillus subtilis* *Aspergillus niger*	Agar media	Zhang et al. (2015b)
	Whey Protein isolate Alginate Zein Chitosan	E-polylysine	*L. innocua* *S. aureus*	Agar media	Ünalan (2011)
	Poly (lactic acid)-cellulose nanocrystals	Nisin	*L. monocytogenes*	Ham	Salmieri et al. (2014b)
	Chitosan/Poly (lactic acid)	Nisin	*S. aureus*	Agar media	Wang et al. (2015)

	Chitosan	Nisin	*L. monocytogenes* *B. cereus* *S. aureus* *E. coli* *Salmonella* Enteritidis *Clostridium perfringens* *L. acidophilus* *Aspergillus phoenicis*	Agar media	Cé et al. (2012)
Inorganic nanoparticles	Fish protein isolate/fish skin gelatin	Zinc oxide nanoparticles	*L. monocytogenes* *P. aeruginosa*	Agar media	Arfat et al. (2016)
	Gelatin	Zinc oxide nanoparticles	*L. monocytogenes* *E. coli*	Broth media	Shankar et al. (2015)
	Poly (lactic acid)	Zinc oxide nanoparticles	*S. aureus* *E. coli*	Agar media	De Silva et al. (2015)
	Chitosan	Zinc oxide and silver nanoparticles	*B. subtilis* *E. coli* *S. aureus* *Penicillium* *Aspergillus* *Rhizopus* Yeast	Agar media	Li et al. (2010)
	Chitosan-starch	Silver nanoparticles	*E. coli* *S. aureus* *B. cereus*	Agar media	Yoksan et al. (2010)
	Sodium alginate	Silver nanoparticles	*E. coli* *S. aureus*	Agar media	Mohammed Fayaz et al. (2009)
	Agar	Silver nanoparticles	*L. monocytogenes* *E. coli*	Agar media	Rhim et al. (2013b)

connected through a peptide bond between the α-carboxyl and the ε-amino group. E-PL molecules are positively charged and exhibit a broad antimicrobial activity against yeast, fungi, and both gram-positive and gram-negative bacteria (Hiraki, 2000). It has been widely used in Japan as a food preservative for meat or fish, rice, cooked vegetables, and other food products, while it has been approved as GRAS by the US Food and Drug Administration when used up to 50 mg/kg. Consequently, it can be considered as a very promising candidate for the development of innovative active packaging to improve the quality, safety and prolong the shelf life of perishable foods (Zhang et al., 2015a).

9.5.4 FILMS CONTAINING LIVING MICROBIAL CELLS

Biopreservation refers to the extended shelf life and enhanced safety of food products using their natural or controlled microflora (Gialamas et al., 2010). Lactic acid bacteria (LAB) are the main "tool" of biopreservation and enjoy an advantage as they are considered as GRAS. LAB can inhibit the growth of different microorganisms, including bacteria, yeasts, and fungi, and their antagonistic effect against undesirable microorganisms is due to different mechanisms acting synergistically or not, like competition for nutrients, production of organic acids, hydrogen peroxide, enzymes, defective phages, lytic agents and antimicrobial peptides, or bacteriocins (Sánchez-González et al., 2014). In that respect, incorporation of a bacteriocin-producer strain into polymeric matrices could be a low-cost alternative natural preservation method. However, one should keep in mind that significant differences regarding the viability of the cells and the amount of bacteriocins produced have been observed when a variety of biopolymers were tested as film-forming materials. Therefore, the nature of the biopolymer matrix and the knowledge of bacteriocin production kinetics are key to using bioactive films in an effective way (Sanchez-Gonzalez et al., 2013).

9.5.5 INORGANIC NANOPARTICLES

The use of inorganic nanoparticles as antimicrobial agents has also gained popularity recently due to their ability to withstand harsh process conditions that may appear during the packaging fabrication process. Silver nanoparticles have received special attention because of their remarkable and broad spectrum of antimicrobial activity against bacteria, yeasts, fungi, and viruses while being nontoxic to human cells. Currently, a variety of materials ranging from textile-clothing to food-contact materials used in daily life are enriched with silver particles or silver salts (Martínez-Abad et al., 2014). In the United States silver nitrate has been accepted as an additive in bottled water by the US Food and Drug Administration (FDA) (FDA/CFSAN), while the European Food Safety Authority (EFSA, 2011) has provided positive views for silver zeolites, silver zirconium phosphates, and silver-containing glasses with a general restriction of $\leq$0.05 mg/kg food for the whole group. The antimicrobial properties of nanoparticles composed of other inorganic materials than silver have also been investigated. Among others, titanium dioxide (TiO_2) nanoparticles show great potential since they have been found to be effective against a great number of food-borne pathogens such as *Salmonella choleraesuis* subsp., *Vibrio parahaemolyticus*, and *Listeria monocytogenes*. Their antimicrobial activity is photocatalyzed and thus TiO_2-based antimicrobials are only active in the presence of UV light. Regarding food-packaging applications, films incorporating TiO_2 nanoparticles may have the additional benefit of protecting food content from the oxidizing effects of UV irradiation, while maintaining good optical clarity, as TiO_2 nanoparticles are efficient short-wavelength light absorbers with high photostability. However, to our knowledge there is lack of information about its use into biopolymer-based films. The use of zinc oxide (ZnO) nanoparticles is also very attractive due to their high antimicrobial activity, low cost, easy availability, and unique chemical and physical properties. Moreover, ZnO nanoparticles are believed to be nontoxic and are listed as GRAS substances by the FDA (Shankar et al., 2015).

9.6 REGULATIONS AND SAFETY CONCERNS

All these innovative biobased packaging materials, nevertheless, must obey the legitimacy and legislation framework of conventional packaging materials given that the legislative documents talking about bioplastics are difficult to find and there are very few public documents available in this context. Thus, both conventional and biobased materials are treated in exactly the same way in the European food contact material legislation and good manufacturing practice guidelines (Weber, 2000). More specifically, the packaging materials are obligated to meet the various rules and guidelines, such as the European Parliament and Council Directive on Packaging and Packaging Waste (EU Packaging, 1994) and the Framework Regulation (EC) 1935/2004 that gives general requirements for all food-contact materials (Commodities Act, 2014). Furthermore, these are commented on in detail for plastics (guidelines and regulations) in Plastics Regulation 10/2011/EC.

To our knowledge, due to the deliberate interaction of the biobased polymer packaging materials in some cases with the food and/or its environment, these new packaging strategies pose new challenges to the evaluation of their safety as compared to the traditional packaging. Some biobased materials are "old" and well defined, like paper and regenerated cellulose, and legislation on a harmonized European Union (EU) or national level exists. Nevertheless, given that "new" materials have also been developed, the producer is responsible to ensure the safety and suitability for food contact by proving that (1) no hazardous substances can migrate from the packaging into the packed food, (2) the migration of nonhazardous substances should not exceed a certain limit, and (3) the supplier must demonstrate that the packaging complies with the requirements via a Declaration of Compliance (EU Food contact, 2012; Molenveld et al., 2015; Weber, 2000). In addition, EFSA is responsible and it should assure that the new packaging material, which comes in contact with the food, is not harmful. In contrast to the European market, the FDA has responsibility under the Federal Food, Drug, and Cosmetic Act to ensure that no packaging material adulterates food and to conduct a premarket review of certain new food-contact materials and new uses of existing materials (U.S. Environmental Protection Agency, 1999). Although FDA's primary concern in regulating food packaging is to ensure that the packaging is safe for its intended food-contact use, FDA also interprets the National Environmental Policy Act as requiring the agency to evaluate the impact that clearance of a packaging material may have on the environment (Packaging Law, 2002).

When designing food-packaging materials, the migration, obviously, is a significant aspect that needs to be considered. The principal legislation has been laid down in the Framework Directive. However, the existing documentation (regulations and guidelines) on plastics might not be appropriate for innovative biobased plastic-similar materials. For example, biobased materials may contain substances, natural or synthetic, as additives, plasticizers, cross-linking agents, antioxidants, preservatives, etc., which are not common in conventional packaging materials. In the same way, the release and migrational performance of these additives, plus other common additives for food-contact plastics, may well be different in biobased packaging materials compared to conventional plastics (Weber, 2000).

Some expected applications of the biopolymer packaging are the elongation of the shelf life and/or the maintenance or the quality improvement of the packaged food. In order that these requirements to be fulfilled, active substances or active materials and articles that contain these substances can be utilized in the packaging material so as to release or absorb substances into or from the packaged food of the environment surrounding the food. The packaging materials that incorporate these compounds are required to follow the respective rules and guidelines. The Framework Regulation (EC) 1935/2004 allows the introduction of active packaging on the European market, and this regulation provides specific requirements for active materials and articles, including the provisions listed herein:

- They may bring about changes in the composition or organoleptic characteristics of food on the condition that the changes comply with the food legislation.
- Substances deliberately incorporated into active materials and articles to be released into the food or the environment surrounding the food shall be authorized and used in accordance with the relevant EU provisions applicable to food.
- They shall not bring about changes in the composition or organoleptic characteristics of food, for instance, by masking the spoilage of food, which could mislead consumers.
- Active materials and articles already brought into contact with food shall be adequately labeled to allow identification by the consumer of nonedible parts.
- They shall be adequately labeled to indicate that the materials or articles are active.

Until now there is no evidence in the literature that these provisions may create problems during application of the innovative packaging materials.

Furthermore, the Commission Regulation (EC) No 450/2009, is a specific measure under the Framework Regulation that regulates these active materials and articles. This regulation includes additional provisions. More specifically, it declares that the individual substance or group/combination of substances that make up the active component should be safe and comply with the requirements in the Framework Regulation (EC) No 1935/2004 and the Regulation (EC) No 450/2009. Substances should thus undergo a safety assessment by EFSA before they are authorized for use. A union list of substances or group/combination of substances to be used in active components should be drawn up following risk assessment of these substances by EFSA. Substances released from active releasing materials should comply with any restrictions in the existing food law (ie, as authorized food additives) thus, complying with the safety requirement. The overall migration from active releasing materials can exceed the overall migration limits described in EU or national legislation as long as the levels transferred to the food comply with restrictions in the existing food law (eg, as authorized food additives). The transfer of these active substance(s) should not be included in the calculation of the overall migration limit. From the moment they comply with the Framework Regulation the passive parts of the active packaging materials must also comply with the rules applicable to the same materials and articles when they do not contain the active component, such as the Plastics Regulation (EU) No 10/2011.

As far as nanoparticles are concerned, it can be said that although they exhibit high potential for applications in the food-packaging sector as alternative materials with new functional properties, there are still important safety concerns about their incorporation in packaging coming into contact with food products. As for any new technology, nanotechnology carries an ethical responsibility for wise application, recognizing that there are potential unforeseen risks that may come to light with its use in food applications (Sorrentino et al., 2007). Though the available documentation is still very limited about migration of nanoparticles from packaging materials into foods, it is reasonable to assume that migration can occur due to their tiny dimensions. Nano-sized particles do display different properties from those found at the macroscale level since their size would allow them to move through the body (tissues, cell membrane structures) more freely than larger particles. Moreover, their high surface area increases their reactivity and allows a greater contact with cell membranes or cellular organelles, as well as greater capacity for absorption and migration within biological tissues. As a result, their toxicity cannot be extrapolated from studies based on their non-nanosized counterparts, and for this reason further investigations are required before their use in different applications. For the moment, no government has developed any regulatory regime specifically referring to the production and application of nanoparticles. Furthermore, in the context of chemical legislation, particle size does not yet play a role for the registration of new substances. Given that the full understanding of the effect on food safety of the introduction of nanostructures has still to be ascertained, further research is needed on how nanomaterial-based processes and products may interfere with human health before any legislation in this field can be established (Borm et al., 2006; Derfus et al., 2004; Hoet et al., 2004; Oberdorster, 2004; Oberdorster, Oberdorster, and Oberdorster, 2005; Rhim et al., 2013b; Sorrentino et al., 2007). Research and development of bionanocomposite materials for food applications like packaging and other food-contact surfaces is expected to grow in the coming years (Sorrentino et al., 2007). However, in order to achieve the approval for their use, risk assessment studies should be conducted in order to assure their safety.

The primary purpose of food packaging must continue to be maintaining the safety, wholesomeness, and quality of food. As long as these prerequisites are not met, there will always be problems and limitations in the advent of biopolymers as packaging materials. To overcome this problem, sufficient documentation in terms of safety, quality, and performance needs to be available for the legislative authorities before any approval.

9.7 CONCLUSIONS

Biopolymers employed as functional materials in emerging packaging technologies may fulfill the environmental concerns but they show some limitations in terms of performance and stability in contact with food products. Moreover, recent advances in the polymer science field, such as the development of biobased nanoparticles, may lead to new packaging materials with improved physical properties and unique end-use performance. Future research should focus on improving heat sealability and the printability of biodegradable films, in addition to making them cost competitive with plastics. Further studies are also required on microstructure, physicochemical properties, and biodegradability of biobased-packaging materials, as well as more efforts should be directed toward consumer awareness and acceptance of such innovative food-packaging approaches.

INTERNET SITES

www.basf.com
www.paceint.com
www.natureseal.com
www.mantrose.com
www.zeinproducts.com

www.kerry.com
www.novartis.com
www.perfectlyfree.com
www.listerine.com
www.oetker.co.uk
www.loliware.com
www.lavazza.com
www.origamifoods.com
www.vivosfilms.com
www.watson-inc.com
www.tomorrowmachine.se
www.metabolix.com
www.tianan-enmat.com
www.tjgreenbio.com
www.ecoman.com
www.biomer.de
www.novamont.com
www.plantic.com.au
www.biopolymers.com
www.danone.com
www.nuatreworksllc.com
www.stonyfield.com
www.sandros-bio.de
www.polenghigroup.it
www.marketside.com
www.fkur.com
www.beanarella.ch
www.santanna.it

REFERENCES

Acevedo-Fani, A., Salvia-Trujillo, L., Rojas-Grau, M.A., Martin-Belloso, O., 2015. Edible films from essential-oil-loaded nanoemulsions: physicochemical characterization and antimicrobial properties. Food Hydrocolloids 47, 168–177.

Ahmad, M., Benjakul, S., Prodpran, T., Agustini, T.W., 2012. Physico-mechanical and antimicrobial properties of gelatin film from the skin of unicorn leatherjacket incorporated with essential oils. Food Hydrocolloids 28, 189–199.

Alavi, S., Thomas, S., Sandeep, K.P., Kalarikkal, N., Varghese, J., Yaragalla, S., 2015. Polymers for Packaging Applications. Apple Academic Press Inc.

Aldana, D.S., Andrade-Ochoa, S., Aguilar, C.N., Contreras-Esquivel, J.C., Nevarez-Moorillon, G.V., 2015. Antibacterial activity of pectic-based edible films incorporated with Mexican lime essential oil. Food Control 50, 907–912.

Aloui, H., Licciardello, F., Khwaldia, K., Hamdi, M., Restuccia, C., 2015. Physical properties and antifungal activity of bioactive films containing *Wickerhamomyces anomalus* killer yeast and their application for preservation of oranges and control of postharvest green mold caused by *Penicillium digitatum*. International Journal of Food Microbiology 200, 22–30.

Arfat, Y.A., Benjakul, S., Prodpran, T., Sumpavapol, P., Songtipya, P., 2016. Physico-mechanical characterization and antimicrobial properties of fish protein isolate/fish skin gelatin-Zinc oxide (ZnO) nanocomposite films. Food and Bioprocess Technology 9, 101–112.
Armentano, I., Fortunati, E., Burgos, N., Dominici, F., Luzi, F., Fiori, S., Jiménez, A., Yoon, K., Ahn, J., Kang, S., Kenny, J.M., 2015. Bio-based PLA-PHB plasticized blend films: processing and structural characterization. LWT – Food Science and Technology 64, 980–988.
Arrieta, M.P., Samper, M.D., López, J., Jiménez, A., 2014. Combined effect of poly(hydroxybutyrate) and plasticizers on polylactic acid properties for film intended for food packaging. Journal of Polymers and the Environment 22, 460–470.
Azeredo, H., 2009. Nanocomposites for food packaging applications. Food Research International 42, 1240–1253.
Azizi Samir, M.A.S., Alloin, F., Dufresne, A., 2005. Review of recent research into cellulosic whiskers, their properties and their application in nanocomposite field. Biomacromolecules 6, 612–626.
Benetto, E., Jury, C., Igos, E., Carton, J., Hild, P., Vergne, C., Di Martino, J., 2015. Using atmospheric plasma to design multilayer film from polylactic acid and thermoplastic starch: a screening Life Cycle Assessment. Journal of Cleaner Production 87, 953–960.
Biliaderis, C.G., 2009. Structural transitions and related physical properties of startch. In: Whistler, R.L., BeMiller, J.N. (Eds.), Starch: Chemistry and Technology, third ed. Academic Press, pp. 293–371.
Borm, P.J.A., Robbins, D., Haubold, S., Kuhlbusch, T., Fissan, H., Donaldson, K., et al., 2006. The potential risks of nanomaterials: a review carried out for ECETOC. Particle and Fibre Toxicology 3, 11.
Burt, S., 2004. Essential oils: their antibacterial properties and potential applications in foods – a review. International Journal of Food Microbiology 94, 223–253.
Cagri, A., Ustunol, Z., Ryser, E.T., 2004. Antimicrobial edible films and coatings. Journal of Food Protection 67, 833–848.
Campos, C.A., Gerschenson, L.N., Flores, S.K., 2011. Development of edible films and coatings with antimicrobial activity. Food and Bioprocess Technology 4, 849–875.
Cé, N., Noreña, C.P.Z., Brandelli, A., 2012. Antimicrobial activity of chitosan films containing nisin, peptide P34, and natamycin. CYTA – Journal of Food 10, 21–26.
Cerqueira, M.A., Bourbon, A.I., Pinheiro, A.C., Martins, J.T., Souza, B.W.S., Teixeira, J.A., Vicente, A.A., 2011. Galactomannans use in the development of edible films/coatings for food applications. Trends in Food Science & Technology 22, 662–671.
Commission Regulation (EC) No 450/2009. http://eur-lex.europa.eu/LexUriServ/LexUriServ.do?uri=OJ:L:2009:135:0003:0011:EN:PDF.
Commodities Act 2014. http://ec.europa.eu/food/food/chemicalsafety/foodcontact/framework_en.htm.
Cutter, C.N., 2006. Opportunities for bio-based packaging technologies to improve the quality and safety of fresh and further processed muscle foods. Meat Science 74, 131–142.
Dashipour, A., Razavilar, V., Hosseini, H., Shojaee-Aliabadi, S., German, J.B., Ghanati, K., Khakpour, M., Khaksar, R., 2015. Antioxidant and antimicrobial carboxymethyl cellulose films containing *Zataria multiflora* essential oil. International Journal of Biological Macromolecules 72, 606–613.
Derfus, A.M., Chan, W.C., Bhatia, S.N., 2004. Probing the cytotoxicity of semiconductor quantum dots. Nano Letters 4 (1), 11–18.
El-Fawal, G., 2014. Preparation, characterization and antibacterial activity of biodegradable films prepared from carrageenan. Journal of Food Science and Technology-Mysore 51, 2234–2239.
Embuscado, M.E., Huber, K.C., 2009. Edible Films and Coatings for Food Applications. Springer, New York.
Emiroglu, Z.K., Yemis, G.P., Coskun, B.K., Candogan, K., 2010. Antimicrobial activity of soy edible films incorporated with thyme and oregano essential oils on fresh ground beef patties. Meat Science 86, 283–288.
Espitia, P.J.P., Du, W.X., Avena-Bustillos, R.J., Soares, F.F., McHugh, T., 2014. Edible films from pectin: physical-mechanical and antimicrobial properties-a review. Food Hydrocolloids 35, 287–296.
European Food Safety Authority, 2011. Scientific opinion on the safety evaluation of the substance, silver zeolite A, silver content 2–5%, for use if food contact materials. EFSA Journal 9 (2), 1999.

EU Food contact, 2012. http://ec.europa.eu/food/food/chemicalsafety/foodcontact/eu_legisl_en.htm.

EU Packaging, 1994. http://ec.europa.eu/environment/waste/packaging/index_en.htm.

FDA, 2001. FDA/CFSAN/OPA: Agency Response Letter: GRAS Notice No. GRN 000065.

FDA/CFSAN, Food Additives Permitted for Direct Addition to Food for Human Consumption, Silver nitrate-172.167.

Fernandez-Pan, I., Mendoza, M., Mate, J.I., 2013. Whey protein isolate edible films with essential oils incorporated to improve the microbial quality of poultry. Journal of the Science of Food and Agriculture 93, 2986–2994.

Fernandez-Pan, I., Royo, M., Mate, J.I., 2012. Antimicrobial activity of whey protein isolate edible films with essential oils against food spoilers and foodborne pathogens. Journal of Food Science 77, M383–M390.

Gialamas, H., Zinoviadou, K.G., Biliaderis, C.G., Koutsoumanis, K.P., 2010. Development of a novel bioactive packaging based on the incorporation of Lactobacillus sakei into sodium-caseinate films for controlling *Listeria monocytogenes* in foods. Food Research International 43, 2402–2408.

Gomez-Estaca, J., de Lacey, A.L., Lopez-Caballero, M.E., Gomez-Guillen, M.C., Montero, P., 2010. Biodegradable gelatin-chitosan films incorporated with essential oils as antimicrobial agents for fish preservation. Food Microbiology 27, 889–896.

Gouvea, D.M., Santos Mendonca, R.C., Soto, M.L., Cruz, R.S., 2015. Acetate cellulose film with bacteriophages for potential antimicrobial use in food packaging. LWT – Food Science and Technology 63, 85–91.

Han, J.H., 2014. Innovations in Food Packaging, second ed. Elsevier.

Han, L., Qin, Y., Liu, D., Chen, H., Li, H., Yuan, M., 2015. Evaluation of biodegradable film packaging to improve the shelf-life of *Boletus edulis* wild edible mushrooms. Innovative Food Science & Emerging Technologies 29, 288–294.

Hernandez-Izquierdo, V.M., Krochta, J.M., 2008. Thermoplastic processing of proteins for film formation – a review. Journal of Food Science 73, R30–R39.

Hiraki, J., 2000. E-polylysine, its development and utilization. Fine Chemistry 29, 18–25.

Hoet, P.H., Nemmar, A., Nemery, B., 2004. Health impact of nanomaterials? Nature Biotechnology 22 (1), 19.

Huq, T., Salmieri, S., Khan, A., Khan, R.A., Le Tien, C., Riedl, B., Fraschini, C., Bouchard, J., Uribe-Calderon, J., Kamal, M.R., Lacroix, M., 2012. Nanocrystalline cellulose (NCC) reinforced alginate based biodegradable nanocomposite film. Carbohydrate Polymers 90, 1757–1763.

Imre, B., Pukanszky, B., 2013. Compatibilization in bio-based and biodegradable polymer blends. European Polymer Journal 49, 1215–1233.

Jamshidian, M., Tehrany, E.A., Imran, M., Jacquot, M., Desobry, S., 2010. Poly-lactic acid: production, applications, nanocomposites and release studies. Comprehensive Reviews in Food Science and Food Safety 9, 552–571.

Jiang, Z., Neetoo, H., Chen, H., 2011. Control of *Listeria monocytogenes* on cold-smoked salmon using chitosan-based antimicrobial coatings and films. Journal of Food Science 76, M22–M26.

Jimenez, A., Jose Fabra, M., Talens, P., Chiralt, A., 2012. Edible and biodegradable starch films: a review. Food and Bioprocess Technology 5, 2058–2076.

Johansson, C., Bras, J., Mondragon, I., Nechita, P., Plackett, D., Simon, P., Svetec, D.G., Virtanen, S., Baschetti, M.G., Breen, C., Clegg, F., Aucejo, S., 2012. Renewable Fibers and Bio-based Material for Packaging Applications – A Review of Recent Developments. Bioresources 7, 2506–2552.

Khan, A., Khan, R.A., Salmieri, S., Le Tien, C., Riedl, B., Bouchard, J., Chauve, G., Tan, V., Kamal, M.R., Lacroix, M., 2012. Mechanical and barrier properties of nanocrystalline cellulose reinforced chitosan based nanocomposite films. Carbohydrate Polymers 90, 1601–1608.

Kowalczyk, D., Kordowska-Wiater, M., Solowiej, B., Baraniak, B., 2015. Physicochemical and antimicrobial properties of biopolymer-candelilla wax emulsion films containing potassium sorbate – a comparative study. Food and Bioprocess Technology 8, 567–579.

Kristo, E., Biliaderis, C.G., 2007. Physical properties of starch nanocrystal-reinforced pullulan films. Carbohydrate Polymers 68, 146–158.

Kristo, E., Biliaderis, C.G., Zampraka, A., 2007. Water vapour barrier and tensile properties of composite caseinate-pullulan films: biopolymer composition effects and impact of beeswax lamination. Food Chemistry 101, 753–764.

Kristo, E., Koutsoumanis, K.P., Biliaderis, C.G., 2008. Thermal, mechanical and water vapor barrier properties of sodium caseinate films containing antimicrobials and their inhibitory action on *Listeria monocytogenes*. Food Hydrocolloids 22, 373–386.

Lee, J.H., Lee, J., Bin Song, K., 2015. Development of a chicken feet protein film containing essential oils. Food Hydrocolloids 46, 208–215.

Li, X.X., Li, X.Y., Ke, B.L., Shi, X.W., Du, Y.M., 2011. Cooperative performance of chitin whisker and rectorite fillers on chitosan films. Carbohydrate Polymers 85 (4), 747–752.

Li, L.H., Deng, J.C., Deng, H.R., Liu, Z.L., Li, X.L., 2010. Preparation, characterization and antimicrobial activities of chitosan/Ag/ZnO blend films. Chemical Engineering Journal 160, 378–382.

López de Lacey, A.M., López-Caballero, M.E., Montero, P., 2014. Agar films containing green tea extract and probiotic bacteria for extending fish shelf-life. LWT – Food Science and Technology 55, 559–564.

Martínez-Abad, A., Ocio, M.J., Lagaron, J.M., 2014. Morphology, physical properties, silver release, and antimicrobial capacity of ionic silver-loaded poly(L-lactide) films of interest in food-coating applications. Journal of Applied Polymer Science 131.

Mohammed Fayaz, A., Balaji, K., Girilal, M., Kalaichelvan, P.T., Venkatesan, R., 2009. Mycobased synthesis of silver nanoparticles and their incorporation into sodium alginate films for vegetable and fruit preservation. Journal of Agricultural and Food Chemistry 57, 6246–6252.

Molenveld, K., van den Oever, M., Bos, H., 2015. Biobased Packaging Catalogue. Wageningen University.

Oberdörster, E., 2004. Manufactured nanomaterials (fullerenes, C60) induce oxidative stress in the brain of juvenile largemouth bass. Environmental Health Perspectives 112 (10), 1058.

Oberdörster, G., Oberdörster, E., Oberdörster, J., 2005. Nanotoxicology: an emerging discipline evolving from studies of ultrafine particles. Environmental Health Perspectives 823–839.

Plastics Europe Plastics – the Facts, 2013. An Analysis of European Latest Plastics Production, Demand and Waste Data.

Peelman, N., Ragaert, P., De Meulenaer, B., Adons, D., Peeters, R., Cardon, L., Van Impe, F., Devlieghere, F., 2013. Application of bioplastics for food packaging. Trends in Food Science and Technology 32, 128–141.

Pires, C., Ramos, C., Teixeira, B., Batista, I., Nunes, M.L., Marques, A., 2013. Hake proteins edible films incorporated with essential oils: physical, mechanical, antioxidant and antibacterial properties. Food Hydrocolloids 30, 224–231.

Plastics Regulation (EU) No 10/2011. http://eur-lex.europa.eu/LexUriServ/LexUriServ.do?uri=OJ: L:2011:012:0001:0089:en:PDF.

Porta, R., Mariniello, L., Di Pierro, P., Sorrentino, A., Giosafatto, C.V.L., 2011. Transglutaminase crosslinked pectin- and chitosan-based edible films: a review. Critical Reviews in Food Science and Nutrition 51, 223–238.

Packaging Law, 2002. Packaging and Environmental Legislation in the United States: An Overview. A Resource Provided by Keller and Heckman LLP. http://www.packaginglaw.com/special-focus/packaging-and-environmental-legislation-united-states-overview.

Quintavalla, S., Vicini, L., 2002. Antimicrobial food packaging in meat industry. Meat Science 62, 373–380.

da Rocha, M., Loiko, M.R., Tondo, E.C., Prentice, C., 2014. Physical, mechanical and antimicrobial properties of Argentine anchovy (*Engraulis anchoita*) protein films incorporated with organic acids. Food Hydrocolloids 37, 213–220.

Ramos, O.L., Santos, A.C., Leao, M.V., Pereira, J.O., Silva, S.I., Fernandes, J.C., Isabel Franco, M., Pintado, M.E., Xavier Malcata, F., 2012a. Antimicrobial activity of edible coatings prepared from whey protein isolate and formulated with various antimicrobial agents. International Dairy Journal 25, 132–141.

Ramos, O.L., Silva, S.I., Soares, J.C., Fernandes, J.C., Pocas, M.F., Pintado, M.E., Malcata, F.X., 2012b. Features and performance of edible films, obtained from whey protein isolate formulated with antimicrobial compounds. Food Research International 45, 351–361.

Regulation (EC) No 1935/2004. http://eur-lex.europa.eu/LexUriServ/LexUriServ.do?uri=OJ:L:2004:338:0004:0017:en:PDF.

Rhim, J.W., Park, H.M., Ha, C.S., 2013a. Bio-nanocomposites for food packaging applications. Progress in Polymer Science 38, 1629–1652.

Rhim, J.W., Wang, L.F., Hong, S.I., 2013b. Preparation and characterization of agar/silver nanoparticles composite films with antimicrobial activity. Food Hydrocolloids 33, 327–335.

da Silva, J.B., Pereira, F.V., Druzian, J.I., 2012. Cassava starch-based films plasticized with sucrose and inverted sugar and reinforced with cellulose nanocrystals. Journal of Food Science 77, N14–N19.

De Silva, R.T., Pasbakhsh, P., Lee, S.M., Kit, A.Y., 2015. ZnO deposited/encapsulated halloysite-poly (lactic acid) (PLA) nanocomposites for high performance packaging films with improved mechanical and antimicrobial properties. Applied Clay Science 111, 10–20.

Salgado, P.R., Lopez-Caballero, M.E., Gomez-Guillen, M.C., Mauri, A.N., Montero, M.P., 2013. Sunflower protein films incorporated with clove essential oil have potential application for the preservation of fish patties. Food Hydrocolloids 33, 74–84.

Salmieri, S., Islam, F., Khan, R.A., Hossain, F.M., Ibrahim, H.M.M., Miao, C., Hamad, W.Y., Lacroix, M., 2014a. Antimicrobial nanocomposite films made of poly(lactic acid)-cellulose nanocrystals (PLA-CNC) in food applications-part B: effect of oregano essential oil release on the inactivation of *Listeria monocytogenes* in mixed vegetables. Cellulose 21 (6), 4271–4285.

Salmieri, S., Islam, F., Khan, R.A., Hossain, F.M., Ibrahim, H.M.M., Miao, C., Hamad, W.Y., Lacroix, M., 2014b. Antimicrobial nanocomposite films made of poly(lactic acid)-cellulose nanocrystals (PLA-CNC) in food applications: part A-effect of nisin release on the inactivation of *Listeria monocytogenes* in ham. Cellulose 21, 1837–1850.

Sánchez-González, L., Quintero Saavedra, J.I., Chiralt, A., 2013. Physical properties and antilisterial activity of bioactive edible films containing *Lactobacillus plantarum*. Food Hydrocolloids 33, 92–98.

Sánchez-González, L., Quintero Saavedra, J.I., Chiralt, A., 2014. Antilisterial and physical properties of biopolymer films containing lactic acid bacteria. Food Control 35, 200–206.

Sanchez-Gonzalez, L., Saavedra, J.I.Q., Chiralt, A., 2013. Physical properties and antilisterial activity of bioactive edible films containing *Lactobacillus plantarum*. Food Hydrocolloids 33, 92–98.

Sanchez-Gonzalez, L., Vargas, M., Gonzalez-Martinez, C., Chiralt, A., Chafer, M., 2011. Use of essential oils in bioactive edible coatings. Food Engineering Reviews 3, 1–16.

Sangsuwan, J., Rattanapanone, N., Pongsirikul, I., 2015. Development of active chitosan films incorporating potassium sorbate or vanillin to extend the shelf life of butter cake. International Journal of Food Science and Technology 50, 323–330.

Shankar, S., Teng, X., Li, G., Rhim, J.-W., 2015. Preparation, characterization, and antimicrobial activity of gelatin/ZnO nanocomposite films. Food Hydrocolloids 45, 264–271.

Shelef, L.A., 1994. Antimicrobial effects of lactates – a review. Journal of Food Protection 57, 445–450.

Shen, H.L., Wu, J.M., Chen, Y., Zhao, G., 2010. Antimicrobial and physical properties of sweet potato starch films incorporated with potassium sorbate or chitosan. Food Hydrocolloids 24, 285–290.

Song, N.B., Lee, J.H., Song, K.B., 2015. Preparation of perilla seed meal protein composite films containing various essential oils and their application in sausage packaging. Journal of the Korean Society for Applied Biological Chemistry 58, 83–90.

Sorrentino, A., Gorrasi, G., Vittoria, V., 2007. Potential perspectives of bio-nanocomposites for food packaging applications. Trends in Food Science & Technology 18 (2), 84–95.

Steinbuchel, A., 2003. General Aspects and Special Applications. Biopolymers, vol. 10, Wiley.

Ünalan, I.U., Uçar, K.D.A., Arcan, I., Korel, F., Yemenicioğlu, A., 2011. Antimicrobial potential of polylysine in edible films. Food Science and Technology Research 17, 375–380.

U.S. Environmental Protection Agency, 1999. Municipal Solid Waste in the United States: 1999 Facts and Figures. Final Report, 1.

Vieira, M.G.A., da Silva, M.A., dos Santos, L.O., Beppu, M.M., 2011. Natural-based plasticizers and biopolymer films: a review. European Polymer Journal 47, 254–263.

Wang, H., Liu, H., Chu, C., She, Y., Jiang, S., Zhai, L., Li, X., 2015. Diffusion and antibacterial properties of nisin-loaded Chitosan/Poly (L-Lactic Acid) towards development of active food packaging film. Food and Bioprocess Technology 8, 1657–1667.

Weber, C.J., 2000. Biobased Packaging Materials for the Food Industry: Status and Perspectives, a European Concerted Action. KVL.

Yoksan, R., Chirachanchai, S., 2010. Silver nanoparticle-loaded chitosan-starch based films: fabrication and evaluation of tensile, barrier and antimicrobial properties. Materials Science and Engineering C 30, 891–897.

Zhang, L., Li, R., Dong, F., Tian, A., Li, Z., Dai, Y., 2015a. Physical, mechanical and antimicrobial properties of starch films incorporated with ε-poly-L-lysine. Food Chemistry 166, 107–114.

Zhang, Y., Ma, Q., Critzer, F., Davidson, P.M., Zhong, Q., 2015b. Physical and antibacterial properties of alginate films containing cinnamon bark oil and soybean oil. LWT – Food Science and Technology 64, 423–430.

Zinoviadou, K.G., Koutsoumanis, C.P., Biliaderis, C.G., 2009. Physico-chemical properties of whey protein isolate films containing oregano oil and their antimicrobial action against spoilage flora of fresh beef. Meat Science 82, 338–345.

Zinoviadou, K.G., Koutsoumanis, K.P., Biliaderis, C.G., 2010. Physical and thermo-mechanical properties of whey protein isolate films containing antimicrobials, and their effect against spoilage flora of fresh beef. Food Hydrocolloids 24, 49–59.

Zivanovic, S., Chi, S., Draughon, A.F., 2005. Antimicrobial activity of chitosan films enriched with essential oils. Journal of Food Science 70, M45–M51.

CHAPTER

DEVELOPMENT OF FUNCTIONAL FOODS

10

D. Villaño[1,2], A. Gironés-Vilapana[3], C. García-Viguera[1], D.A. Moreno[1]
[1]CEBAS-CSIC, Murcia, Spain; [2]University Catolica San Antonio de Murcia (UCAM), Murcia, Spain; [3]University Miguel Hernández, Orihuela, Spain

10.1 INTRODUCTION

The notion of nutrition has totally evolved over the last century. We have moved from nutrition based on a balanced diet consumed to prevent deficiencies and support normal body functions to the actual stage in which diet is consumed in an optimal way to improve health and to prevent the risk of chronic diseases. Today science supports the idea that food can bring specific components with health benefits beyond those of basic nutrition. The increases in costs of health care, as well as a longer life expectancy, make attractive the idea that the consumption of certain foods may help to improve the state of well-being and reduce the risk of diseases.

It is estimated that the world population will be 9 billion in 2050, parallel to an increase in aging, which will create an economic burden for public health systems in the midst of a global economic crisis. As evidence in the last reports of the Organisation for Economic Cooperation and Development (OECD, 2014), there is a high prevalence of chronic diseases related to lifestyle linked to low physical activity and diet. The future lies in finding sustainable solutions in which food plays a key role in maintaining health.

The concept of functional foods was born in Japan and further developed in the United States, Canada, and Europe. Originally the Japanese Foods for Specified Health Use (FOSHU) included foods expected to have a positive effect on health that must have been scientifically evaluated prior to commercialization.

In Europe, the European Commission funded two projects, Functional Food Science in Europe and Process for the Assessment of Scientific Support for Claims on Foods (PASSCLAIM), which joined European experts on food science and nutrition as well as most relevant food companies with the commitment to explore the scientific evidence supporting the concept of "functional" food/diet/ingredients. The projects were coordinated by the International Life Sciences Institute (ILSI Europe).

A good, nonconsensus definition of functional foods was given by ILSI (Ashwell, 2002), which states that "a food can be regarded as 'functional' if it is satisfactorily demonstrated to affect beneficially one or more target functions in the body, beyond adequate nutritional effects, in a way that is relevant to either an improved state of health and well-being and/or reduction of risk of disease." Moreover, functional foods must remain foods, they are not pills or capsules but they must be consumed as part of an ordinary diet and demonstrate their effects in amounts that can normally be expected to be

Innovation Strategies in the Food Industry. http://dx.doi.org/10.1016/B978-0-12-803751-5.00010-6

consumed in the diet. An extensive review of definitions of functional foods has been provided by Bigliardi and Galati (2013).

First attempts to define a functional food from a practical point of view (Ashwell, 2002) express that it can be a food in which:

- one of its components has been naturally enhanced by growing conditions
- a component has been added to provide benefits
- a component has been removed so that the food has fewer adverse effects or a less negative nutritional profile (ie, salt reduction)
- a component has been modified to improve health
- a component whose bioavailability has been increased to provide greater absorption, or a combination of the previously mentioned situations

The components responsible for a beneficial effect on health are of a diverse nature, such as polyunsaturated fatty acids (PUFA, omega-3), vitamins, polyphenols, probiotics, prebiotics (inulin, beta-glucans), peptides, among others. Examples of these type of functional foods developed in the recent past years are foods with reduced content in salt and fat, or gluten-free, aimed at high-risk populations (diabetic, obese, hypertensive, or celiac patients). Prebiotics are an example of foods added with a beneficial component of carbohydrate nature, not digested in the gastrointestinal tract but available to gut microflora to selectively stimulate the growth of bifidogenic bacteria and avoid the growth of pathogens. Probiotics contain beneficial bacteria in proper concentrations to equilibrate and restore gut microflora conditions when it is unbalanced. Both in Japan and Europe, the market of functional foods has been mainly dominated by probiotics with more than 370 products launched in 2005 (Ouwehand, 2007). However, changes in the legal requirements for the commercialization of these products have affected the market of functional foods, especially in Europe.

The actual framework for the development of functional foods is diverse in the different countries, influencing (and in some cases restricting) greatly the innovation policies of the food companies.

10.2 LEGAL FRAMEWORK FOR FUNCTIONAL FOODS

In this section, the current legislation for functional foods applied in Europe and the United States is briefly described as the most well-known and globally established frameworks.

10.2.1 EUROPE

The claims are a form of communication about functional foods and suggest that the foods possess particular characteristics. In Europe, health claims of food products need to be authorized under Regulation (EC) N° 1924/2006 before being used for labeling and marketing. The assessment to the European Commission on the approval/refusal of the claim is carried out by the European Food Safety Authority (EFSA) by means of its Panel on Dietetic Products, Nutrition, and Allergies, which verifies the scientific substantiation of the health claim. The European Commission (DG Health and Consumers) makes the final decision whether to accept or reject a claim.

The Regulation defines a claim as "any message or representation, which is not mandatory under Community or national Legislation, including pictorial, graphic or symbolic representation, in any form, which states, suggests or implies that a food has a particular characteristic" (EC 2006, Article 2.2.1 Regulation (EC) Nº 1924/2006). The aim is to ensure that any claim made about a food is clear, accurate, truthful, not misleading to the consumer, and based on scientific evidence.

The Regulation classifies three different claims: nutrition claim, health claim, and reduction of disease risk claim.

"Nutrition claim" refers to any claim that states, implies, or suggests that a food has a particular beneficial nutritional property due to:

- the energy it provides, does not provide, or provides at reduced/increased rate and/or,
- the nutrients or other substances it contains, does not contain, or contains at reduced/increased proportions.

"Health claim" refers to any claim that states, implies, or suggests that a relationship exists between a food or one of its constituents and health.

"Reduction of disease risk claim" refers to any claim that states, implies, or suggests that the consumption of a food or one of its constituents reduces a risk factor in the development of a human disease.

Health claims are divided into Article 13 and Article 14 claims. Article 13 includes health claims that describe (1) the role of a nutrient or other substance in the growth, development, and functions of the body, (2) psychological and behavioral functions, (3) slimming or weight control or reduction in the sense of hunger or increase of satiety or reduction in available energy from diet. Article 14 includes health claims specifically referring to the reduction of risk of disease, as well as to children's development and health.

The general conditions for the use of nutrition and health claims are specified in Article 4 of Regulation (EC) N° 1924/2006. It states that functional foods must meet the criteria of an appropriate "nutrient profile" in order to bear claims. It means that the food must comply with particular quantities of specific nutrients and other substances contained so that a claim can be made. The nutrient profile depends on the potential of the food to adversely affect the overall dietary balance of nutrients. In this way it is ensured that the consumer makes healthy choices of food products with good nutritional quality. The establishment of the nutrient profile for a category of foods takes into account the nutritional composition of the foods (main nutrients as carbohydrates, fats, proteins, vitamins, etc.), the dietary recommendations, as well as the dietary habits of the population (EFSA, 2008). Some nutrients for which intakes might exceed the recommended levels include energy density, total fat, saturated fatty acids (SFA), *trans*-fatty acids (TFAs), sugars, and sodium. Nutrients for which intakes might be inadequate in relation to recommended levels include dietary fiber, unsaturated fatty acids, vitamins, and minerals. These nutrient profiles have been set by the Commission. Besides, the nutrient or other substance for which the claim is made must be provided by a quantity of food reasonably expected to be consumed and must be present in the body in quantities sufficient to exert the claimed effect.

Nutrition claims are only permitted if they are listed in the Annex of Regulation (EC) N° 1924/2006. Table 10.1 shows these nutrition claims as well as the conditions of use.

When a food naturally meets the condition(s) laid down in Table 10.1 for the use of a nutritional claim, the term "naturally/natural" may be used as a prefix to the claim.

Concerning health claims, a list of the authorized claims under each article has been provided and updated periodically. The quantity of the food and pattern of consumption required to obtain the claimed beneficial effect should be included in the labeling and reasonably expected to be consumed in the context of a varied and balanced diet.

Table 10.1 Nutrition Claims and Conditions Applying to Them

Nutrient	Nutrition Claims and Conditions of Use								
	Very Low	Low	Free	Reduced	Increased	No Added	Source of	High	Contains
Energy		<40 kcal/100 g; <20 kcal/100 mL	<4 kcal/100 mL	By at least 30%					
Fat		<3 g/100 g; <1.5 g/100 mL	<0.5 g/100 g <0.5 g/100 mL						
Saturated-fat		<1.5 g/100 g; <0.75 g/100 mL (sum of saturated fatty acids and *trans*)	<0.1 g/100 g; <0.1 g/100 mL (sum of saturated fatty acids and *trans*)						
Sugar		<5 g/100 g; <2.5 g/100 mL	<0.5 g/100 g; <0.5 g/100 mL			No added mono- or disaccharides or any other food used for its sweetening properties			
Sodium/salt	<0.04 g/100 g; <0.04 g/100 mL; (except for waters)	<0.12 g/100 g; <0.12 g/100 mL; (different limits for waters)	<0.005 g/100 g						
Fiber							≥3 g/100 g; ≥1.5 g/100 kcal	≥6 g/100 g; ≥3 g/100 kcal	
Protein							≥12% energy value	≥20% energy value	
Vitamins/minerals							15% of the recommended allowance	≥Twice the amount of "Source of"	
Particular nutrient				Reduction ≥30% compared to a similar product, except for micronutrients	Increase ≥30% compared to a similar product and meets the conditions for the claim "source of"				As specified in EC 1924/2006, in particular Art. 5

Adapted from Annex of Regulation (EC) N° 1924/2006 of the European Parliament and of the Council of 20 December 2006. On nutrition and health claims made on foods. Official Journal of the European Union 30.12.2006. p. 9. Adapted and modified from the source http://eur-lex.europa.eu/legal-content/en/ALL/?uri = CELEX:32006R1924.

Health claims other than those referring to the reduction of disease risk and for which scientific substantiated data is available are authorized and included in the list of authorized claims provided for Article 13.1 (published in (EU) N° 432/2012 and updated in (EU) N° 536/2013). An example is Vitamin A. The claim states that "Vitamin A contributes to normal iron metabolism." The condition for use of the claim is that the claim may be used only for food that is at least a source of vitamin A, as referred to in the nutrition claim "Source of (Name of vitamin/s) and/or (Name of mineral/s)" listed in the Annex to the Regulation (EC) N° 1924/2006.

However, Article 13 claims based on newly developed scientific evidence (as stated in Article 13.5) as well as reduction of disease risk claims included in Article 14 must be authorized by EFSA according to a specific procedure that food companies must follow for application and authorization. Each application is evaluated on a case-by-case basis before approval/refusal. Food manufacturers must submit a dossier with full details of the product and substantial scientific evidence that verifies the claim.

Positive EFSA opinions on Article 14 health claims include plant sterols and cholesterol reduction, xylitol and caries reduction, fatty acids and visual development in children, contribution of biotin to normal energy-yielding metabolism, among others (EFSA, 2015; EU Register of nutrition and health claims made on foods). Successful claims are included in the register of nutrition and health claims, which is available for food manufacturers to use throughout the European Union (EU), with the exception of health claims based on proprietary data (Kiely, 2010).

10.2.2 UNITED STATES

The regulation of functional foods in the United States differs somewhat. The US Food and Drug Administration (FDA) is responsible for regulation and supervision of foods and drugs, and was the first of more than 200 laws that constituted one of the world's most comprehensive and effective networks of public health and consumer protection. Today the FDA regulates $1 trillion worth of products a year. It ensures the safety of all food except for meat, poultry, and some egg products; ensures the safety and effectiveness of all drugs, biological products (including blood, vaccines, and tissues for transplantation), medical devices, and animal drugs and feed; and makes sure that cosmetics, medical and consumer products that emit radiation do no harm. There is not a category of functional foods as in Europe, but accordance exists in that the health claims may be based on an authoritative statement of a scientific body (FDA Modernization Act of 1997).

FDA both assesses health claims and makes final decisions about their approval/refusal, differently from Europe, where two entities are involved, EFSA and the EU Commission.

The United States only allows nutrient deficiency health claims, whilst Europe allows reduction of disease claims. The main difference is that the United States may approve health claims based on authoritative statements of a scientific body or institution, and may even allow claims that are new and for which solid scientific evidence does not exist yet (Moors, 2012). The criteria are not as strict as in Europe and health claims are easier to use for marketing purposes.

With respect to regulations of food and flexibility for regulations regarding food claims (Sec. 301), FDA determinations provide for information necessary to:

- enable consumers to develop and maintain healthy dietary practices;
- enable consumers to be informed promptly and effectively of important new knowledge regarding nutritional and health benefits of food;

- ensure that scientifically sound nutritional and health information is provided to consumers as soon as possible;
- enable the FDA Secretary to act promptly to ban or modify a claim.

The FDA provides guidance and regulatory information, with links to Federal Register documents, in their Web page. Guidance documents represent FDA's current thinking on a topic. They do not create or confer any rights for or on any person and do not operate to bind the FDA or the public. Manufacturers can use an alternative approach if the approach satisfies the requirements of the applicable statutes and regulations. The FDA also issues regulations to implement its statutory authority. The regulations can create binding obligations and have the force of law. Links to Federal Register documents (advance notices of proposed rule making, proposed rules, interim final rules, and final rules) are posted in their Web page.

10.2.3 JAPAN

With regard to functional foods in Japan, these are regulated by Food for Specified Health Uses (FOSHU). In their Web page, definition of FOSHU refers to foods containing an ingredient with function(s) for health and officially approved to claim its physiological effects on the human body. FOSHU is also intended to be consumed for the maintenance/promotion of health or special health uses by people who wish to control health conditions, including blood pressure or blood cholesterol. In order to sell a food as FOSHU, the assessment for the safety of the food and effectiveness of the functions for health is required, and the claim must be approved by the Ministry of Health, Labor, and Welfare.

There are some crucial requirements for FOSHU approval, also specified in their Web page:

- effectiveness on the human body is clearly proven;
- absence of any safety issues (animal toxicity tests, confirmation of effects in the cases of excess intake, etc.);
- use of nutritionally appropriate ingredients (eg, no excessive use of salt, etc.);
- guarantee of compatibility with product specifications by the time of consumption;
- established quality control methods, such as specifications of products and ingredients, processes, and methods of analysis.

Table 10.2 shows some FOSHU-approved products.

Some reduction of disease risk claims are also approved such as "calcium and osteoporosis" and "folic acid and neural tube defect."

In addition to "regular" FOSHU, the following types of FOSHU are also available: qualified FOSHU and standardized FOSHU were introduced to facilitate applicants for FOSHU approvals:

- Qualified FOSHU: Food with health function that is not substantiated on scientific evidence that meets the level of FOSHU, or a food with certain effectiveness but without an established mechanism of the effective element for the function will be approved as qualified FOSHU.
- Standardized FOSHU: Standards and specifications are established for foods with sufficient FOSHU approvals and accumulation of scientific evidence. Standardized FOSHU are approved when they meet the standards and specifications.
- Reduction of disease risk FOSHU: Reduction of disease risk claim is permitted when reduction of disease risk is clinically and nutritionally established in an ingredient.

Table 10.2 Approved FOSHU Products

Principal Ingredients Approved (Ingredients Exhibiting Health Functions)	Specified Health Uses
Lactotripeptide, casein dodecapeptide, tochu leaf glycoside (geniposidic acid), sardine peptide, etc.	Foods related to blood pressure
Degraded sodium alginate, dietary fiber from psyllium seed husk, etc.	Cholesterol plus gastrointestinal conditions, triacylglycerol plus cholesterol
Oligosaccharides, lactose, bifidobacteria, lactic acid bacteria, dietary fiber (ingestible dextrin, polydextrol, guar gum, psyllium seed coat, etc.)	Foods to modify gastrointestinal conditions
Chitosan, soybean protein, degraded sodium alginate	Foods related to blood cholesterol level
Indigestible dextrin, wheat albumin, guava tea polyphenol, L-arabinose, etc.	Foods related to blood sugar levels
Calcium citrated malate, casein phosphopeptide, hem iron, fracuto-oligosaccharide, etc.	Foods related to mineral absorption
Soybean isoflavone, milk basic protein, etc.	Foods related to osteogenesis
Middle chain fatty acid, etc.	Foods related to triacylglycerol
Paratinose, maltitiose, erythritol, etc.	Foods related to dental hygiene

Modified from Ministry of Health Labor and Welfare—Japan, http://www.mhlw.go.jp/english/topics/foodsafety/fhc/02.html.

10.3 SCIENTIFIC SUBSTANTIATION OF CLAIMS

Nutrition and health claims must be substantiated by scientific data before a product enters the market. Scientific substantiation is the main aspect (and in some cases the core of obstacles) to be taken into account by food companies for the use of nutrition and health claims. The health claim should be substantiated by human studies. Intervention studies (randomized controlled) are preferred from observational studies (prospective or retrospective cohort studies or case–control studies) in the hierarchy of evidence to substantiate the claim. Data on animal models and in vitro studies can be submitted as supporting evidence, but human studies are mandatory. However, there is a lack of a comprehensive description on how to perform human nutrition intervention studies in general, and particularly for the evaluation and substantiation of health claims on foods. The PASSCLAIM project, which so far is the most conclusive summary, established the criteria for the scientific substantiation of health claims, whereas most regulatory bodies have published guidelines based on or in accordance with PASSCLAIM criteria. Thereby, various collections of advice exist, given fragmentarily in legal regulations or guidance reports from international and national authorities or organizations. For instance, European Food Safety Authority Scientific and Technical Guidance (EFSA, 2011) and the "Application Rules on Health Claims" in Commission Regulation EC 353/2008 (European Commission, 2008) provide only minor information on the conduction of human intervention studies. Within Europe, additional information is available from the Joint Health Claims Initiative (2000), where some criteria on the validity of human studies are listed. The "US Food and Drug Administration Guidance for Industry (US FDA, 2009)" gives a broad description on human intervention studies in a question-and-answer style, which guides through relevant human intervention issues. The varying approaches to health claim evaluation in Europe, the United States, Canada, Australia and New Zealand, China, and Japan have quite recently been reviewed (Jones et al., 2008). Additionally, the "FAO/WHO Codex Alimentarius Commission (2008)" lists basic criteria for health claim evaluation.

Welch et al. (2011) provided "best practice" guidance on how to conduct intervention studies to scientifically substantiate these health benefits of foods, taking into consideration various points:

- Definition of appropriate control products—lack of perfect controls, and the type required, depends on the nature of tested ingredients or foods.
- Responder selection and status, including how to deal with "low," "normal," and "high" responders.
- Criteria for validation of markers—rationale for supporting substantiated and valid markers.
- Design(s) of studies—what is the rationale?

It is imperative to pay attention to some crucial factors when designing, conducting, and reporting human intervention studies to evaluate the health benefits of foods.

10.3.1 DESIGN OF STUDY

- Have a clear hypothesis and appropriate study design.
- Define primary outcomes and methods of measurement, all secondary outcomes and methods of measurement, and all eligibility criteria.
- Have statistical considerations like randomization where possible and double blinding if feasible, single blinding if not.

10.3.2 CONDUCTION OF THE STUDY

- Include an ethical approval and trial registration.
- Define recruitment strategy and process, including settings and dates.
- Select suitable methods to collect and analyze data, define relevant measures, select suitable methods of assessment, and have mechanisms in place to record and respond to adverse events.
- Define acceptable levels of compliance, use appropriate strategies to maximize compliance.
- Select and use rigorous but feasible methods for assessment of compliance.

10.3.3 ANALYSIS AND INTERPRETATION OF RESULTS

- Have a suitable statistical analysis, and devise appropriate analysis methods based on study design and outcome measures.
- Discussion and interpretation, consider study limitations and generalizability of findings.
- Have a clear statement of conclusion.

Guidelines have also been developed by EFSA to help food innovators assist applicants in preparing and presenting their applications for authorization of health claims. The information is presented to assist the applicant in the preparation of a well-structured application. These guidelines outline:

- The information and scientific data that must be included in the application.
- The hierarchy of different types of data and of study designs, reflecting the relative strength of evidence that may be obtained from different types of studies.

- Instructions for presenting summaries of data to highlight the relevant aspects related to the design, results, and quality of the studies.
- The key issues that should be addressed in the application to substantiate the health claim.

It is intended that the guidance remains under review and can be amended and updated as appropriate in the light of experience gained from evaluation of health claims applications. The information provided in the application should follow a common format, and for that the data provided in the application must be organized into five parts:

- Part 1: Administrative and Technical Data.
- Part 2: Food/Constituent Characteristics.
- Part 3: Overall Summary of Scientific Data.
- Part 4: Body of Pertinent Scientific Data Identified.
- Part 5: Appendix to the Application.

This guidance will also be updated as appropriate at a later stage to cover applications for authorization of the health claims that fall under Article 18 (or Article 13(5)) of the Regulation, ie, applications for inclusion of health claims to the Community list of permitted claims provided for in Article 13(3) that are based on newly developed scientific evidence and/or that include a request for the protection of proprietary data.

Finally, applicants have to submit their application to a national competent authority of a member state in the European Union, and if some of the data that are required in this guidance document do not apply to a particular application, reasons/justifications are given for the absence of such data in the application.

In recent years, the application of dietary biomarkers as more objective measures of dietary exposure in nutritional epidemiology has been particularly significant (Jenab et al., 2009). Dietary biomarkers are desirable for their ability to more accurately assess nutritional intake/status versus self-reported methods, validate self-reported intake measures, evaluate intake of dietary items when food-composition databases are inadequate, and to more accurately associate dietary intake with disease risk and nutritional status. These biomarkers of dietary exposure should be valid, reproducible, and able to detect changes in intake over time and be suitable for the general population. Biomarkers have been also used as measures of nutritional status and of exposure to bioactive molecules in foods, as surrogate indicators of food intake, and to validate measures of dietary intake (Potischman and Freudenheim, 2003). Biomarkers are useful when little or no data exist on food composition, as is often the case for bioactive molecules such as glucosinolates or food contaminants such as aflatoxins.

A variety of validated dietary biomarkers identified through the analysis of correlations with dietary intake have been found in epidemiologic studies, being measured in plasma or serum (carotenoids, fatty acids, vitamins, polyphenols, food contaminants, and enzymes), red blood cells (fatty acids, carotenoids, and hemoglobin adducts), and to a lesser extent in urine (polyphenols, vitamins, inorganic compounds, and amino acids). Some of these biomarkers have been used as surrogate biomarkers of food intake, as follows: polyphenols, carotenoids, and vitamin C for fruit and vegetables (Mennen et al., 2006); alkylresorcinols for whole-grain cereals (Ross et al., 2012); isoflavones for soy (Verkasalo et al., 2001); amino acids and fatty acids for meat (Allen et al., 2008); fatty acids for dairy products and fish (Allen et al., 2008); and polyphenols for tea and wine (Mennen et al., 2006). Dietary biomarkers not only include natural food constituents but also certain food additives such as iodine in milk

(Brantsaeter et al., 2009) or food contaminants such as polychlorinated biphenyls in fatty fish (Turunen et al., 2010). Other biomarkers are directly derived from the digestion and gut absorption of food constituents or are endogenous metabolites that have been altered by exposure to specific nutrients. Specificity is another essential characteristic of biomarkers.

Some biomarkers can be highly specific for a particular food or may be common to several foods or characteristic of an entire food group. For instance, proline, betaine, and lycopene are well-established biomarkers for citrus fruits and tomato products, respectively (Lloyd et al., 2011), whereas vitamin C or the sum of carotenoids or flavonoids has been used similarly for fruit and vegetable intake (Mennen et al., 2006). Analytic approaches based on the estimation of combinations of dietary constituents may provide more accurate measurements of dietary exposure. The ratios of two alkylresorcinols characteristic of whole-grain wheat or rye were found to be good indicators of the relative consumption of these cereals (Andersson et al., 2011). However, there are very few such examples in which combinations of biomarkers were used to improve the specificity of dietary exposure measurements.

10.4 FOOD INDUSTRY—FACTORS THAT INFLUENCE PRODUCTION OF AND INNOVATION IN FUNCTIONAL FOODS

The food sector is a stable manufacturing sector in a mature European market (Moors, 2012). The EU food and drink industry had an annual turnover of 1048 billion euro in 2012, being the largest manufacturing sector in the EU. It is a diversified sector, characterized by a wide range of company sizes with small and medium-sized enterprises (SMEs) accounting for a large share of activity (more than 50% of the turnover). With 4.2 million employees, it is the leading employer in the EU (FoodDrinkEurope, 2014). Compared to other industrial sectors, R&D investment in food and drink has been very low in Europe, ie, it has been estimated to be 0.27% of the food and drink industry output.

Innovations applied to functional foods include:

1. the development of a product with new characteristics completely different from those in the market;
2. or a product already in the market but with changes in some ingredients for a better healthy profile;
3. or new optimized production process.

Innovation in functional foods follows the same general pathway of food innovation with additional aspects. The choice of the functional ingredients to add, the current studies on the stability and shelf life of the product, as well as the organoleptic aspects must be considered from the very initial step of the ingredient/food design. Pilot studies at laboratory scale of the preliminary food product must also be performed. A detailed characterization of the food is compulsory for any application to succeed.

Apart from the technological aspect of the food design, other new aspects must be considered in the design of functional foods. Bioavailability studies are necessary to determine the absorption rate and optimal use by the organism, as well as safety and toxicological studies. The recommended dose of the functional ingredient to get a positive effect must be established by human intervention studies, in a clinical and pharmacological perspective. All these ambitious aspects mean high development costs and an important challenge difficult to achieve by a particular food company.

It has been estimated that the cost of bringing a novel food to the market varies globally between 4 and 24 million euro, including R&D costs of safety and efficacy studies and EU-specific regulatory costs (Moors, 2012). Hence, the cost of health claim dossiers is too high for many SMEs and larger manufacturers, while the impact on the end cost of the product is significant. The building of consortia between research institutes, academia, and food industry aims to address these points and serve as a unique opportunity to food innovation, as we will see in the next sections.

The legal framework, in particular the need for scientifically substantiated criteria to validate a claim, as described in the previous section, can be seen as a disincentive to the development, production, and commercialization of functional foods. The large requirements can be an obstacle to food innovation and food companies are reconsidering their role on functional food innovation. The fact that the EU Regulation is more restrictive than the regulations in countries such as United States, and the longer time that it takes for market approval, profoundly affects EU innovation on functional foods.

On the other hand, once approved, the use of the health claim as a marketing tool significantly improves the competitive position of the food company. The EU list of approved claims for Articles 13 and 14 creates a level playing field in which food operators can innovate with legal certainties, since it is obligatory for all new and existing health claims. The list is not extensive, but the products have a stronger market potential because of the high amount of substantiated evidence for the claim.

It must be pointed out that these health claims that are authorized and included in the lists can be used by any food company since they are not protected as patents. Hence, generic products might appear easily in the market, which is considered as a threat to the invented product for the return on investment. The strategy of "product followers" seems more attractive than the risk of searching for a new health claim of the "early movers" as they decrease financial risk and the development costs are lower. In conclusion, a higher number of product followers are expected than early movers who push forward on food innovation.

There are some exceptions stated by the Commission, claims for which their substantiation is based on proprietary data are reserved for exclusive use of the owner of the data for 5 years, unless in the intervening period they are independently substantiated by data from alternative sources (Binns and Howlett, 2009). This means that there is an "early mover advantage" to companies that market the product first, which results in higher sales and market leadership (Moors, 2012).

Other questions arise once the claim is approved, as the return on the investment is not clear, especially in the European young market of functional food consumers. Sales should be high enough to offset the investment.

On the other hand, food companies fear that if the claim is not approved the image of the established brand may be damaged (Binns and Howlett, 2009). They must find an alternative way to market their products without making specific health claims, with the risk of affecting transparency of product information for the consumers.

10.5 OPPORTUNITIES IN FUNCTIONAL FOOD INNOVATION

10.5.1 TOP FOOD INNOVATION TRENDS

Bearing in mind the context discussed previously, some food companies are focusing on a smaller portfolio of food products or on secondary applications. Some attributes related to the food product, to the claim, or to the consumer itself greatly influence consumer tendencies and acceptance of functional foods (Bigliardi and Galati, 2013).

One of the attributes related to the claim is simplicity (labeling must contain simple statements to avoid confusion). Another one is familiarity (the use of a component that is already in the market with health-related arguments seems to create a better perception than a new component). In general, clear and transparent labeling will be the key to upholding the confidence of consumers in food industry (Lahteenmäki et al., 2010; van Trijp and van der Lans, 2007).

Concerning the product, functional foods must meet consumer expectations regarding the traditional food. This means that apart from their added benefits they must retain their organoleptic and nutritional characteristics. This is an opportunity for food innovation. Important attributes are naturalness and taste quality, as the consumer will not compromise on the taste of functional foods for health. With respect to the nutritional value, consumers perceive some nutrients as qualifying (fiber, vitamins, and minerals) or disqualifying (energy, fat, saturated fat, salt, sugar) (Hoefkens et al., 2011). A field of food innovation is to meet the nutrient profiling to obtain healthier products, as reducing sugar, fat, or sodium. Unsaturated and natural fats, present in nuts and butter, are perceived by consumers as healthier fats than *trans*-fats, and hence these two products are becoming a natural alternative to margarines rich in *trans*-fats.

No added sugar, sugar-free, and low-sugar products continue to increase in number. Some big food manufacturers are looking to all areas of health for a more holistic approach. More body, mind, and spiritual connections will be emphasized with particular ingredients (Food Ingredients, 2013). Nature's own functional foods, as fruits and vegetables, will be revisited, especially previously forgotten vegetables as well as alternative and ancient grains, such as chia and quinoa. There are great changes in the demands and profiles of consumers in the 21st century. The consumer takes a leading role, is very demanding, and takes part actively in the selection, looking at very different products: natural, traditional products as well as new products, exotic tendencies—and this tendency represents an enormous potential for innovation in food.

Consumer-related attributes that are worthy of mention are gender, age, and lifestyle. A growing variety of lifestyles are driving the growth of food innovation. Generational differences in eating patterns, flavor preferences, and personal choices are creating new opportunities for the food and beverage companies (Verbeke et al., 2009, 2010). The millennials (born between 1981 and 2000), the children of baby boomers (born between 1946 and 1964), will redefine the food industry in the longer term. They are the most ethnically diverse and the most educated generation to date, who view their food choices as healthier, more expensive, more natural/organic, less processed, and better tasting than their boomer parents (Hartman, 2011). Healthy eating practices are growing in all population groups. Boomers are slightly more interested in the function of the foods they eat than younger segments of the population. They are more likely to look for foods higher in protein, fiber, whole grain, and calcium, and lower in sodium, added sugars, and the type of fat contained (IFIC, 2013). For those older than age 50, the top health concerns for baby boomers and matures (born before 1946) are retaining mental sharpness with aging, cardiovascular disease, bone and joint health/strength, cancer, and eye health (HealthFocus, 2012). Younger shoppers are likely to buy foods/beverages that address allergies, aging, appearance, digestive health, immunity, and mental clarity (Packaged Facts, 2012). Other criteria underlying consumer personalization include cultural mores (halal, kosher, vegan, etc.), or life stage (weaning, infancy, pregnancy, etc.).

All these attributes mentioned herein can be regarded as key opportunities for food innovation.

The last report on food and drink industry trends in Europe (FoodDrinkEurope, 2014) shows that the main drivers of food innovation can be classified into 15 trends, grouped along five axes,

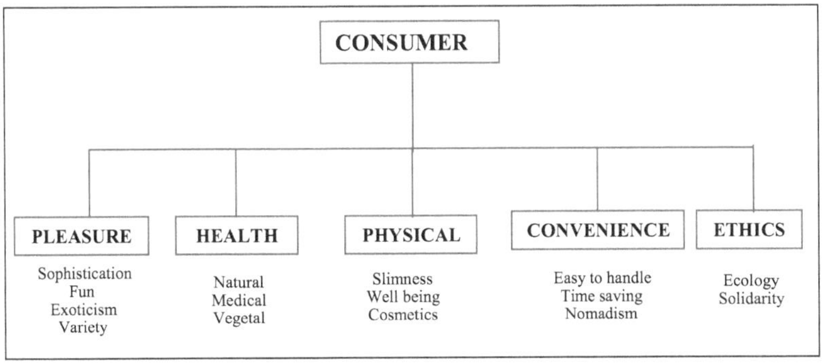

FIGURE 10.1

Drivers of food innovation.

Adapted from FoodDrinkEurope, 2014. Data & Trends of the European Food and Drink Industry. 2013–2014.

corresponding to general consumer expectations: pleasure, health, physical, convenience, and ethics (Fig. 10.1). Pleasure, including a variety of senses, exoticism, fun, and sophistication, was the leading axis with a 57% share in 2013. It was followed by health, including the trends of natural, medical, and vegetal, with a 20% share. Thus, it can be observed that today's pleasures are the leading drivers of food innovation in Europe.

Innovations also affect the food technology process. Functional food development uses both traditional formulation and blending techniques as well as new emerging ones, as nanotechnologies and biomolecular techniques. Examples include preventing food spoilage and deterioration of bioactive compounds, as well as microencapsulation or edible films and coatings.

A major issue in contemporary nutrition science, and hence in food innovation in the longer term, is the personalization factor. Health and disease outcomes are not necessarily predetermined based on the individual's genes. These genetic outcomes are modifiable particularly with respect to diet-related disease. The emerging concept of genetic susceptibility means the ability to change one's health trajectory through the use of a correct diet and lifestyle. Recent technologies have aimed to create personalized nutrition, including nutrigenomics, which considers the interaction between food and an individual's genome and the consequent effect on the phenotype (Grayson, 2010). Dietary intake has been recognized to modulate gene and protein expression and thereby metabolic pathways implicated in correct homeostasis. In addition, genes contribute largely to different responses to diet exposure. In this way, appropriate dietary advice for one individual may be inappropriate for another. The routine practice of consuming personalized foods is still several years away but it will be one of the key consumer trends over the next decade. As personalization of foods will probably be cost prohibitive for industry/consumers, categorization may be the solution. The latest personalized products are targeted to groups or subpopulations of a certain size with certain genetic traits that can be discriminated by other relevant genetic traits.

10.5.2 THE INNOVATION SYSTEM

Functional food innovation is perceived today as a collective effort by industry and academia in a network of institutions with activities focused on nutrition science and on development of innovative food processes. This type of approach is known as an innovation system, being defined as "a network of institutions in the public and private sectors, whose activities and interactions initiate, import, modify and diffuse new technologies" (Freeman, 1987).

The development of national and international research consortia of industry partners, universities, and research centers is a clear opportunity for functional food innovation. Scientists can help to reduce the uncertainty associated with these new functional products by conducting human clinical trials and generating new scientific and technological knowledge.

10.5.3 EUROPEAN AND SPANISH PROJECTS

There are many examples of projects on functional foods in the EU, such as "HEALTHGRAIN," "BIORICE," "LipiDiDiet," "NUTRAHEALTH," among other FP7 projects. HEALTHGRAIN is the acronym for "Exploiting Bioactivity of European Cereal Grains for Improved Nutrition and Health Benefits," an integrated project of the EU's Sixth Framework Program's "Food Quality and Safety" activity. The HEALTHGRAIN project aims to improve the well-being and reduce the risk of metabolic syndrome–related diseases in Europe by increasing the intake of protective compounds in whole grains or their fractions. The aim of the project is to produce health-promoting and safe cereal foods and ingredients of high eating quality. HEALTHGRAIN project carries out an integrated, multidisciplinary effort to establish the variation, process-induced changes and human metabolism of bioactive compounds in the major European bread grains, and to reveal the physiological mechanisms underlying their significance in prevention of metabolic syndrome and related diseases. The target bioactive compounds are vitamins (folate, tocols, choline, etc.), phytochemicals (lignans, sterols, alkylresorcinols, and phenolic acids) and indigestible carbohydrates. In addition, other product characteristics that may add to the metabolic benefits of whole-grain products are promoted.

The BIORICE project brings together six partners distributed in three EU member states and one associated country, and aims to fill the gap of knowledge of the involved SMEs on rice protein by-product pretreatment, peptide isolation, and relative bioactivity and safety testing. In particular, BIORICE research activities will produce added-value bioactive ingredients (semipurified digestates and small-molecular-weight peptides) starting from protein by-products contained in the processing water of the rice starch production stream. The aims of the project are:

- bioconvert into value-added compounds the rice protein by-products present in the starch production processing water;
- establish innovative, not disruptive, industry-driven, and ecosustainable protocols (short-path molecular distillation and/or advanced membrane filtration) to purify and obtain new peptide ingredients;
- establish a wide range of in vitro and cell-based tests to characterize the bioactivity and an innovative protocol to test safety (skin sensitization potency by human reconstructed epidermis containing Langerhans cells) of the new products;
- open the possibility for all the involved SMEs to improve their competitiveness against direct competitors by accessing new markets throughout the introduction of innovative ingredients, obtained via biotechnological and natural processes, having new and/or improved bioactivity, and consequently increase their financial turnover and benefit.

The project results are expected to have a significant impact on the competitiveness of SME. For instance, participants will be able to expand their business by adding new bioactive ingredients and protocols to their product range and enabling new product formulations applicable in food, cosmetic, and nutraceutical sectors.

The European LipiDiDiet project addresses the "Impact of nutritional lipids on neuronal and cognitive performance in aging, alzheimer's disease and vascular dementia." This is based on previous observations that lipids change the risk for dementia. The project is based on two elements. One is applied research documenting the value of nutritional support in persons at risk of getting Alzheimer disease. The other one is basic research generating more knowledge about the possible therapeutic and preventive effects of dietary lipids in model systems of Alzheimer disease and vascular dementia.

Other kinds of European projects in functional foods are FP7 projects. For example, FP7-SME-LUPICARP entitled "Innovative Functional Foods based on Sweet Lupin Protein for Cardiovascular Prevention." aims to assess the health benefits on dyslipidemia prevention of innovative food products based on lupin proteins, starting from some preliminary results provided by two past European collaborative projects (Healthy-Profood and Bioprofibre). The core (Research and Technological Development) RTD activity is the implementation of a multicenter randomized dietary intervention study with LDL-cholesterol as the main end point aimed to compare the hypolipidemic effect of lupin proteins versus animal proteins. All data collected will be exploited at the end of the project by the submission of an application to the European Commission for the approval of a health claim on lupin and cholesterol reduction. The project also included product development and optimization, with particular reference to the nutrition profiles of the foods. To reach these objectives the LUPICARP consortium includes five food-producing SMEs, one lupin ingredient manufacturer, and five RTD performers expert in product innovation, food processing, clinical nutrition, and pharmacology.

Another relevant example of an FP7 European project is the FP7-SME-HYFFI entitled "Hydrocolloids as functional food ingredients for gut health" that targets realizing a commercial opportunity to produce low-molecular-weight polysaccharides from alginate- and agar-bearing seaweeds for applications in food, health, and wellness products by a group of SMEs. The aim was addressed by an integrated work plan comprising one management activity, five interlinked RTD work packages, and a dissemination and exploitation activity.

But in terms of FP7 projects, it is mandatory to emphasize "NUTRAHEALTH." This project is titled as "Improving Research Capacity of TUBITAK MRC Food Institute on Functional Foods, Nutraceuticals, and Natural Health Products" and is funded under FP7-REGPOT. The main objectives of this proposal are (1) to improve the research potential of existing functional foods, nutraceuticals, and NHP research laboratory (herein referred to as NUTRAHEALTH) within the Food Institute, (2) to coordinate networking and cooperative activities in this fostering field through a number of scientific and technical events, and (3) to integrate FP7 projects. Specifically, NUTRAHEALTH intends to improve its research potential in this field. The approach for accomplishing the project objective is multidisciplinary, which includes the following work packages: project management, improving infrastructure, supporting research potential of human resources, organizing scientific events and networking, participating scientific events, communication and dissemination, technical-site visits, and exchange of researchers.

Regarding Spanish projects, a good example of applied research excellence is the PREDIMED project (Prevención con Dieta Mediterranea, Prevention with Mediterranean Diet). The latest is the largest study that has been conducted on nutrition in Spain. It aims to evaluate the effects of Mediterranean diet on primary prevention of chronic diseases. Main Spanish nutrition research centers participate in the project in a big clinical study with three different diet interventions in people at high cardiovascular disease risk. The study evaluates the effect of a Mediterranean diet supplemented with olive oil or nuts on the reduction of the incidence of cardiovascular diseases compared to a low-fat diet. The project involves collaborations with food companies providing extra virgin olive oil and nuts.

Fun-c-Food Project "New Ingredients of Functional Foods to Improve Health" is a 5-year research project (2007–12) to create a network of excellence in research in the area of functional foods, funded by the Consolider Program (CSD 2007-00063). This research program explores the relationship between food and health, focused on functional foods and bioactive food ingredients, with the aim to obtain and characterize new bioactive food ingredients and apply them to develop new functional foods, supported by bioavailability, biological activity, and safety studies (www.functionalfoods.es). Fun-c-Food incorporates 17 research teams from different Spanish universities and public research CSIC centers, with a recognized expertise on complementary areas of knowledge (biotechnology, nutrition, food technology, analysis, genomics, etc.). The aim of this consortium is to cooperate as a unique research group for the successful accomplishment of more ambitious objectives. The implication of active private companies of the food sector is articulated in an "industrial platform" within the Fun-c-Food Consortium. Their contribution ensures the utilization of the knowledge produced in the project.

The type of ingredients and health targets studied are diverse, and different technologies have been developed as result of the research collaborations to face the current challenges in the food arena. An example is the bioactive peptides obtained by optimized conditions of hydrolysis, which are easy to implement in food industry by including them in juices, snack bars, or as nutraceuticals. Some of these peptides have shown to protect the intestinal epithelium and might be useful in intestinal disorders or as adjuvants in pathologies like Crohn disease (Martínez-Maqueda, 2013).

Another functional ingredient that has been developed are the prebiotics derived from lactulose and kojibiose, which have improved digestibility and prebiotic properties for beneficial bacteria at lower doses than the commercially existing ones (Díez-Municio et al., 2014).

A technological innovation developed in this project consists of a hydrolysis process applied to egg white. Egg white is rich in proteins, which, once hydrolyzed, change their rheological properties, particularly exhibiting an increased foaming property. The foaming property is improved compared to conventional foaming products. This characteristic is of interest for food manufacturers as it can be used as a food ingredient to change/improve the texture of both sweet and salted food products, creating opportunities for product innovation (Garcés-Simón et al., 2015).

Another area of research in Fun-c-food is the revalorization of by-products of the food industry. As example, world production of coffee in 2011 was almost 8 million tons; 50% of world coffee production is destined for soluble coffee. This means that huge amounts of low-cost, raw materials are generated. These include coffee silverskin from roasted coffee beans or the by-product remaining after the manufacture of soluble coffee (Martínez-Saez et al., 2014). The by-product is particularly rich in insoluble fiber and can be used in the manufacture of products with the nutritional claim "high in fiber" and hence may have positive effects on intestinal motility in glycemic control in the maintenance of body weight and also may contribute to reducing waste and environmental impact.

The CORNUCOPIA project aims to establish and maintain a cooperative and collaborative network of scientific, technological, and training programs with multidisciplinary and transdisciplinary tasks joining agriculture, food, and nutrition and health areas to gain scientific knowledge and improve competitiveness from the laboratory to the food industry. The network currently has 22 partner groups, comprised of more than 160 people from scientific and technological groups in universities, public research organizations, SMEs, and industries from nine countries: Chile, Colombia, Ecuador,

Guatemala, Mexico, Peru, Portugal, Spain, and Uruguay. The current activity of the networking, training, and cooperation is funded under the CYTED Program (AGL Action 112RT0460, 2012–2015) (http://redcornucopia.org/).There are different activity packages in the collaborative work plan: (1) genotype and environmental effects on the bioactive compounds of Iberian-American fruits; (2) bioavailability, safety, and metabolite profiling of the new ingredients and food matrices; (3) technological and innovative developments; (4) functional evaluation at preclinical level; and (5) validation in studies with human subjects.

CYTED CORNUCOPIA facilitates the knowledge and infrastructure sharing within the network and generates impacts in training, publications, and innovations in the region. Examples are new ingredients, foods, and beverages developed by incorporating the Iberian-American fruits' sources of bioactive compounds (ie, isotonic drinks with maqui and lemon, gluten-free flours, or novel cucumber lines, etc.). In the context of the networking, these products are being tested for quality and composition and evaluated for their ability to improve and maintain the physical and cognitive performance in adult health. The bioactivity models for these tests already have been optimized, shared, and made available to the public in this synergistic networking.

There is great value in thematic networks involving consolidated and emerging groups, focused on obtaining nutritive and healthy foods with added value (ie, functional foods, beverages, nutraceuticals) using sustainable practices and systems in Iberian-American countries.

These examples highlight the importance of these types of projects to obtain and characterize new functional ingredients, which can be applied in the development of functional foods by the food industry. Some of these products have already been patented by the research centers. Patents can be key drivers of food innovation, and the companies that have acquired these technologies have access to products that make a difference and increase their competitiveness in the sector. Besides, the development of new ingredients in the research institutions contributes to the reduction of risks, costs, time of production, and time to market as the product has already been tested somewhat. In most cases, collaborations between academia and industry continue after the licensing agreement, with R&D contracts to further develop the technology and to share knowledge and expertise in the area.

10.5.4 FOOD TECHNOLOGY PLATFORMS

Technology platforms are other examples of the innovation system concept. Cooperation between food companies is becoming increasingly important in functional food innovation, especially in Europe, for sharing risks and costs of the research needed to submit health claim dossiers. The organizations involved in these platforms can operate at different stages in the functional food value chain: production of ingredients, producers of functional foods, market distributors, and subcontracting companies. Food industry associations serve as a forum for discussion and updates about EU regulations and the European Confederation of Food and Drink Industries, FoodDrinkEurope, is an important stakeholder in the representation of many functional food companies in their discussions with the EC and EFSA.

The challenging opportunities for improving welfare and well-being in Europe through research and innovation in the European agro-food industry, together with the size, nature, and regional importance of this industry sector justified the creation of the European food technology platform "Food for Life" in 2005. It aims to be a platform for quicker and more effective, consumer-oriented food innovation and critical mass. It is a forum for ensuring an effective multidisciplinary/integrating approach with the aim of delivering innovative, novel, and improved food products for, and to, national, regional, and global markets in line with consumer needs and expectations. Working groups of the platform include Food and Health, Food Quality and Manufacturing, Food and Consumer, Food Safety, Sustainable Food Production and Food Chain Management, Communication, Training and Technology Transfer. This is followed by a dialog with policy makers at European and national level to ensure that the interests of consumers and other stakeholders are fully addressed.

10.6 CONCLUSIONS

Innovation policies of food companies are greatly influenced by the legal framework concerning functional foods. The high costs of development and the uncertainty of the investment return are obstacles for functional food innovation. Food companies are diversifying the strategies on food innovation; one of the ways is to meet the nutrient profiling recommended by authorities, to obtain healthier products. Another way is to search for natural sources of functional products as ancient grains or forgotten vegetables traditionally consumed. In these and other cases, food industry should always follow the tendencies of consumer personal preferences and lifestyle patterns, keeping in mind that these are constantly moving in the current society of the 21st century. In the actual context of demographic, economic, and social challenges for the food industry, the building of consortia between research institutes, academia, and the food industry serves as a unique opportunity for food innovation.

ACKNOWLEDGMENTS

The authors express their gratitude to the Spanish Ministry of Economy and Competitiveness for the funding through the CICYT project AGL2013-46247-P and the CONSOLIDER-INGENIO 2010 Research Project FUN-C-FOOD (CSD2007-00063) as well as CYTED Program (Ref. 112RT0460) CORNUCOPIA Thematic Network (URL: red-cornucopia.org). AGV also thanks the CSIC and the European Social Funds for the JAE predoctoral grant.

REFERENCES

Allen, N.E., Grace, P.B., Ginn, A., Travis, R.C., Roddam, A.W., Appleby, P.N., Key, T., 2008. Phytanic acid: measurement of plasma concentrations by gas-liquid chromatography-mass spectrometry analysis and associations with diet and other plasma fatty acids. The British Journal of Nutrition 99, 653–659.

Andersson, A., Marklund, M., Diana, M., Landberg, R., 2011. Plasma alkylresorcinol concentrations correlate with whole grain wheat and rye intake and show moderate reproducibility over a 2-to 3-month period in free-living Swedish adults. Journal of Nutrition 141, 1712–1718.

Ashwell, M., 2002. Concepts of functional foods. ILSI Europe Concise Monograph Series 1–40.

Bigliardi, B., Galati, F., 2013. Innovation trends in the food industry: the case of functional foods. Trends in Food Science & Technology 31, 118–129.

Binns, N., Howlett, J., 2009. Functional foods in Europe: international developments in science and health claims. European Journal of Nutrition 48, S3–S13.

Brantsaeter, A.L., Haugen, M., Julshamn, K., Alexander, J., Meltzer, H.M., 2009. Evaluation of urinary iodine excretion as a biomarker for intake of milk and dairy products in pregnant women in the Norwegian Mother and Child Cohort Study (MoBa). European Journal of Clinical Nutrition 63, 347–354.

Codex Alimentarius Commission, 2008. Proposed Draft Annex to the Codex Guidelines for the Use of Nutrition and Health Claims: Recommendations on the Scientific Substantiation of Health Claims. CRD 17. November 2008 http://www.ccnfsdu.de/fileadmin/user_upload/PDF/2008/CRD_17.pdf (accessed August 2015).

Commission Regulation (EU) No 432/2012 of 16 May 2012. Establishing a list of permitted health claims made on foods, other than those referring to the reduction of disease risk and to children's development and health. Official Journal of the European Union 25.05.2012. p. 1.

Commission Regulation (EU) No 536/2013 of 11 June 2013 Amending Regulation (EU) No 432/2012 establishing a list of permitted health claims made on foods other than those referring to the reduction of disease risk and to children's development and health. Official Journal of the European Union 12.06.2013. p. 4.

Commission Regulation (EC) No 353/2008 Establishing Implementing Rules for Applications for Authorisation of Health Claims as provided for in Article 15 of Regulation (EC) No 1924/2006 of the European Parliament and of the Council. OJ L109, 11–16.

Díez-Municio, M., Montilla, A., Moreno, F.J., Herrero, M., 2014. A sustainable biotechnological process for the efficient synthesis of kojibiose. Green Chemistry 16, 2219–2226.

European Food Safety Authority (EFSA), 2008. The setting of nutrient profiles for foods bearing nutrition and health claims pursuant to Article 4 of the Regulation (EC) N° 1924/2006. The EFSA Journal 644, 1–44.

EFSA NDA Panel (EFSA Panel on Dietetic Products, Nutrition and Allergies), 2011. Scientific and technical guidance for the preparation and presentation of an application for authorisation of a health claim (revision 1). EFSA Journal 9 (5), 2170. http://dx.doi.org/10.2903/j.efsa.2011.2170 36 pp.

EFSA NDA Panel (EFSA Panel on Dietetic Products, Nutrition and Allergies), 2015. Scientific opinion on biotin and contribution to normal energy-yielding metabolism: evaluation of a health claim pursuant to Article 14 of Regulation (EC) No 1924/2006. EFSA Journal 13 (7), 4181. 9 pp.

EU Register of nutrition and health claims made on foods http://ec.europa.eu/nuhclaims/.

Food and Drug Administration. 105th "H.R.1411: Food and Drug Administration Regulatory Modernization Act of 1997". U.S. House of Representative Bill Summary & Status. Library of Congress THOMAS. US Congress (April 23, 1997).

FoodDrinkEurope, 2014. Data & Trends of the European Food and Drink Industry. 2013–2014.

Food Ingredients, December 2013. The Top 10 Trends 2014.

Freeman, C., 1987. Technology Policy and Economic Performance. Pinter, London.

Garcés-Simón, M., Sandoval, M., Molina, E., López-Fandiño, R., Miguel, M., 2015. Egg protein hydrolysates: new culinary textures. International Journal of Gastronomy and Food Science. http://dx.doi.org/10.1016/j.ijgfs.2015.04.001.

Grayson, M., 2010. Nutrigenomics. Nature 468, S1.

Hartman, 2011. Culture of Millennials.

HealthFocus, 2012. U.S. Trend Study. HealthFocus Intl, St. Petersburg, Fla. www.healthfocus.com.

Hoefkens, C., Verbeke, W., Van Camp, J., 2011. European consumers' perceived importance of qualifying and disqualifying nutrients in food choices. Food Quality and Preference 22 (6), 550–558.

IFIC, 2013. Food & Health Survey. The International Food Information Council Foundation, Washington, DC. www.foodinsights.com.

Jenab, M., Slimani, N., Bictash, M., Ferrari, P., Bingham, S.A., 2009. Biomarkers in nutritional epidemiology: applications, needs and new horizons. Human Genetics 125, 507–525.

Joint Health Claims Initiative, 2000. Code of Practice on Claims on Foods. http://www.jhci.org.uk/info/code.pdf (accessed August 2015).

Jones, P.J.H., Asp, N.-G., Silva, P., 2008. Evidence for health claims on foods: how much is enough? Introduction and general remarks. Journal of Nutrition 138, S1192–S1227.

Kiely, M., Black, L.J., Plumb, J., Kroon, P.A., Hollman, P.C., Larsen, J.C., Speijer, G.J., Kapsokefalou, M., Sheehan, D., Gry, J., Finglas, P., EuroFIR Consortium, 2010. EuroFIR eBASIS: application for health claims submissions and evaluations. European Journal of Clinical Nutrition 64, S101–S107.

Lahteenmäki, L., Lampila, P., Grunert, K., Boztug, Y., Ueland, Ø., Aström, A., 2010. Impact of health-related claims on the perception of other product attributes. Food Policy 35 (3), 230e239.

Lloyd, A.J., Beckmann, M., Fave, G., Mathers, J.C., Draper, J., 2011. Proline betaine and its biotransformation products in fasting urine samples are potential biomarkers of habitual citrus fruit consumption. The British Journal of Nutrition 106, 812–824.

Martínez-Maqueda, D., Miralles, B., Cruz-Huerta, E., Recio, I., 2013. Casein hydrolate and derived peptides stimulate mucin secretion and gene expression in human intestinal cells. International Dairy Journal 32, 13–19.

Martínez-Saez, N., Ullate, M., Martin-Cabrejas, M.A., Martorell, P., Genovés, S., Ramon, D., del Castillo, M.D., 2014. A novel antioxidant beverage for body weight control based on coffee silverskin. Food Chemistry 1, 227–234.

Mennen, L.I., Sapinho, D., Ito, H., Galan, P., Hercberg, S., Scalbert, A., 2006. Urinary flavonoids and phenolic acids as biomarkers of intake for polyphenol-rich foods. The British Journal of Nutrition 96, 191–198.

Moors, E.H.M., 2012. Functional foods: regulation and innovations in the EU. Innovation: The European Journal of Social Science Research 25, 424–440.

OECD, 2014. Health at a Glance. Dic.

Ouwehand, A.C., 2007. Antiallergic effects of probiotics. Journal of Nutrition 137, 794S–797S.

Packaged Facts, February 2012. Targeted Health & Wellness Foods and Beverages.

Potischman, N., Freudenheim, J.L., 2003. Biomarkers of nutritional exposure and nutritional status: an overview. Journal of Nutrition 133 (Suppl.), 873S–874S.

Regulation (EC) N° 1924/2006 of the European Parliament and of the Council of 20 December 2006. On nutrition and health claims made on foods. Official Journal of the European Union 30.12.2006. p. 9.

Ross, A.B., Bourgeois, A., Macharia, H.N., Kochhar, S., Jebb, S.A., Brownlee, I.A., Seal, C.J., 2012. Plasma alkylresorcinols as a biomarker of whole-grain food consumption in a large population: results from the WHOLE heart Intervention Study. The American Journal of Clinical Nutrition 95, 204–211.

Turunen, A.W., Mannisto, S., Kiviranta, H., Marniemi, J., Jula, A., Tiittanen, P., Suominen-Taipale, L., Vartiainen, T., Verkasalo, P.K., 2010. Dioxins, polychlorinated biphenyls, methyl mercury and omega-3 polyunsaturated fatty acids as biomarkers of fish consumption. European Journal of Clinical Nutrition 64, 313–323.

US FDA, 2009. Guidance for Industry: Evidence-based Review System for the Scientific Evaluation of Health Claims. http://www.fda.gov/Food/GuidanceComplianceRegulatoryInformation/GuidanceDocuments/FoodLabelingNutrition/ucm073332.htm (accessed August 2015).

van Trijp, H.C.M., van der Lans, I.A., 2007. Consumer perceptions of nutrition and health claims. Appetite 48 (3), 305e324.

Verbeke, W., 2010. Consumer reactions to foods with nutrition and health claims. Agro Food Industry Hi-Tech 21 (6), 5–8.

Verbeke, W., Scholderer, J., Lahteenmaki, L., 2009. Consumer appeal of nutrition and health claims in three existing product concepts. Appetite 52 (3), 684–692.

Verkasalo, P.K., Appleby, P.N., Allen, N.E., Davey, G., Adlercreutz, H., Key, T.J., 2001. Soya intake and plasma concentrations of daidzein and genistein: validity of dietary assessment among eighty British women (Oxford arm of the European Prospective Investigation into Cancer and Nutrition). The British Journal of Nutrition 86, 415–421.

Welch, R.W., Antoine, J.M., Berta, J.L., Bub, A., De Vries, J., Guarner, F., et al., 2011. Guidelines for the design, conduct and reporting of human intervention studies to evaluate the health benefits of foods. British Journal of Nutrition 106, S3–S15.

CHAPTER

11 FOOD USE FOR SOCIAL INNOVATION BY OPTIMIZING FOOD WASTE RECOVERY STRATEGIES

C.M. Galanakis[1], J. Cvejic[2], V. Verardo[3], A. Segura-Carretero[4]

[1]Galanakis Laboratories, Chania, Greece; [2]University of Novi Sad, Novi Sad, Serbia; [3]University of Almería, Almería, Spain; [4]University of Granada, Granada, Spain

11.1 INTRODUCTION

Until the end of the 20th century, disposal of food waste was not considered as a matter of concern. Particularly, increase of food production without improving the efficiency of the food systems was the prevalent policy. This consideration increased generation of wasted food along supply chains. In the 21st century, escalating demands for processed foods have required identification of concrete directions to minimize energy demands and economic costs as well as reduce food losses and waste. On the other hand, degradation signs of natural resources (eg, decline in land, water, and biodiversity) create critical concerns about meeting the future demands at the global level. In the next 50 years, not only the population increase but also increasing urbanization and rising incomes will bring a sharp growth in food processing industries and alter food supply chains worldwide. Pressures in natural resources to feed world population and waste disposal costs are global priorities, whereas the recovery of resources will play a vital role for the management strategies in the years to come. Today the challenge is keeping a balance between future demands and sustainable supply (Otles et al., 2015). The latter can only be met by introducing innovative science and technology, reducing food loss, improving food system governance, and establishing sustainable diets (Foresight, 2011; FAO, 2012, 2014; HLPE, 2014).

Food processing industries generate a huge amount of liquid and solid wastes. According to FAO (2014), a management strategy for resource optimization via waste reduction at source is producing the greatest benefit to society. Besides, environmental concerns and increasing attention toward sustainability have forced renewed and stringent worldwide changes in the environmental legislative frameworks and requirements for waste disposal. For example, the US Environmental Protection Agency (EPA) describes the concept of the "US Food Waste Challenge" and the 3R (reduce-recover-recycle) approach: "reduce" for food loss and waste, "recover" for wholesome food for human consumption, and "recycle" for other uses including animal feed, composting, and energy generation (EPA, 2011). The 3R application helps to minimize the amount of waste to disposal, to get a more effective waste management, and finally minimize associated health and environmental risks (Murugan and Ramasamy, 2013). On the other hand, European Union strategy focus on waste prevention (more efficient production technologies), internal recycling of production waste, source-oriented improvement of waste

Innovation Strategies in the Food Industry. http://dx.doi.org/10.1016/B978-0-12-803751-5.00011-8

quality, reuse of products, and more recently circular bioeconomy and biobased products (Otles et al., 2015). Other international directives and regulations are enforcing handling and treatment of wastes. The main methodologies for waste minimization and valorization include general methods such as incineration, anaerobic fermentation, biofuel conversion methods, composting and vermicomposting, landfill, or agricultural applications like animal feed and fertilizers. Indeed, the potential of food wastes to create new markets has been underestimated until very recent years. Today consumers' consciousness about environmental issues and legislative pressures increases the requirements of new methods for the recovery of food waste rather than its disposal.

In any case, dealing with food wastes is difficult in many aspects. Inadequate biological stability and existence of pathogens causes an increase in microbial activity. High water content (specifically for meat and vegetable wastes) has an important effect on transport costs. Besides, food wastes with high fat content are susceptive to oxidation, causing deterioration due to the continuous enzymatic activity (Russ and Meyer-Pittroff, 2004). Nevertheless, food processing by-products (the organic residues of processed raw materials) are of specific interest because of their existence in concentrated locations as well as their low degree of deterioration. Skins, husks, hulls, vegetable and fruit peels, seeds, animal meat, bones, or eggshells may be considered as wastes, but on the other hand, they contain considerable amounts of high added-value reusable ingredients (Chandrasekaran, 2013). The latter could be converted to products with market value (Galanakis, 2013; Galanakis and Schieber, 2014; Rahmanian et al., 2014), providing an economically attractive option by implementing feasible strategies, supported at national and international levels. Food processing by-products that can be used as high added-value components possess specific characteristics that allow them to improve targeted aspects of the final product quality. Added value can be related to the shelf life or color of the final product, health benefits, formulation, or other product feature. For nutraceutical applications or conversion of by-products into chemicals, valuable components should be extracted from the matrix in an appropriate way. The focus of scientific research in this area is mainly on the recovery of functional compounds derived from agricultural and food processing by-products. These sources are abundant, easily reachable, and more stable compared to the wastes produced at the end of food supply chain (Galanakis, 2012). Due to regulatory and technical reasons, such as traceability and health-and-safety issues, by-products with a high potential for valorization are fruit- and vegetable-derived wastes (Lin et al., 2013). For instance, vegetable and plant residues may be used as a source to recover dietary fibers, polyphenols, flavonoids, carotenoids, glucosinolates, protein concentrates as well as pectin in order to produce food supplements, additives, and flavors (Patsioura et al., 2011; Tsakona et al., 2012; Galanakis, 2012; Galanakis et al., 2013a,b, 2015b; Galanakis, 2011, 2015a,b; Heng et al., 2015; Roselló-Soto et al., 2015; Deng et al., 2015). On the other hand, dairy residues and meat by-products contain proteins, peptides, salts, fatty substances, and lactose, which could also be used as food additives and functional hydrolyzates (Galanakis et al., 2014).

A regular generous intake of dietary fiber reduces risk for developing coronary heart disease, stroke, hypertension, diabetes, obesity, and certain gastrointestinal diseases. Increasing the intake of food with high fiber content or fiber supplements lowers blood pressure, aids weight loss, and appears to improve immune function (Brown et al., 1999; Montonen et al., 2003; Steffen et al., 2003; Whelton et al., 2005). In the observed protective effect of dietary fiber intake, associated bioactive compounds such as minerals, vitamins, and antioxidants may have important complementary beneficial effects. The role of phenolic compounds (eg, anthocyanins, catechins, proanthocyanidins, flavanols, stilbenes), as components partly responsible for the protective effects of a fruit-and-vegetable-rich diet is highly important for their use as nutraceuticals and functional food components. Although these compounds are not essential for short-term well-being, their low long-term intake is related to beneficial effects on human health and disease prevention, mostly related to so-called "modern diseases" such as cancers, cardiovascular and neurodegenerative diseases (Riboli and Norat, 2003; Letenneur et al., 2007; Crowe et al., 2011; Spencer and Crozier, 2012). As high added-value ingredients, carotenoids are used for improving food color, and also, due to their biological activities, for promoting beneficial health properties of final products. A positive link is suggested between higher dietary intake, tissue concentrations of carotenoids, and lower risk of chronic diseases. Due to their biological activities, carotenoids act in prevention of cardiovascular diseases and specific cancers, whereas they are important dietary sources of vitamin A (Rao and Rao, 2007). Commercial pectins are used as a stabilizer, thickener, gelling agent, emulsifier, and drug vehicle in the food and pharmaceutical industries (Wicker et al., 2014). Pectins have also been shown to possess a variety of pharmacological activities among which are immunostimulating, antimetastasis, and antiulcer activities. Antitumor applications of these compounds have been suggested, too (Cobs-Rosas et al., 2015; Leclere et al., 2013). Regarding mentioned properties, pectins have a broad use as added-value components.

11.2 FOOD WASTE RECOVERY FOR SUSTAINABLE FOOD SYSTEMS

A food system is defined as the sum of all the diverse elements and activities that lead to the production and consumption of food. In particular, it gathers all the elements (environment, people, inputs, processes, infrastructures, institutions, etc.), activities related to the production, processing, distribution, preparation, and consumption of food, and their socioeconomic and environmental outcomes (HLPE, 2014). In addition, it interfaces further with a wide range of other systems (energy, transport, etc.) and faces various constraints. In a food system, sustainability could be illustrated through the product stewardship concept. Product stewardship can be defined as the shared responsibilities that all the participants in a product's life cycle have for minimizing its environmental and health impacts (Product Stewardship Institute, 2011).

Nowadays, organizations and specifically food production systems can no longer ignore the need to act in a sustainable way. Food products must be politically, economically, environmentally, socially, and technologically sustainable (Beer and Lemmer, 2011). Thereby, the relative organizations have specific responsibilities to society that go beyond their economic and legal obligations. In simpler words, they should not only care about the profits but also deal with their impacts on all of the involved stakeholders, ie, shareholders, employees, customers, suppliers, etc. (McGuire, 1963). By balancing the social and environmental elements of sustainability, long-term profitability could be achieved. In order to maximize the creation of shared value, the organizations should adopt a long-term approach to sustainability such as exploring opportunities for developing innovative products that will enable societal well-being (European Commission, 2010).

Following these considerations, effective food waste management will benefit all supply chain members. Reductions in energy and raw material usage can reduce costs significantly and simultaneously increase the environmental performance of the food system. This will be achieved through the efficient use of the materials and energy used for production. Efficient use of materials in the food waste recovery process has two meanings. Firstly, utilizing the material that otherwise will have been thrown away, and secondly using/processing that material in an efficient way. Subsequently, utilization of processing food wastes by recovering high added-value ingredients could significantly reduce their amounts, develop innovative products, and finally create new opportunities and benefits for everyone related to a food production system (Otles et al., 2015). Using the appropriate food waste valorization methods means that high added-value compounds are recovered and utilized to develop new food products or even extending their shelf life to be consumed in a longer period of time (Oreopoulou and Tzia, 2007). The creation of new food products could steadily increase food availability. For instance, polyphenols and carotenoids from fruit by-products could be used as natural food or beverage preservatives as they extent the shelf life of the product and increase antioxidant activity (Galanakis et al., 2015a). This progress could increase food availability by delaying a product's deterioration, having the product available to the shelf for more days, and improve people's livelihoods as a whole (Otles et al., 2015). Through this process, fruit by-products intended to be thrown away are transformed into a new material with an economic value. On the other hand, pronounced antioxidant properties as well as other mentioned biological effects of these compounds enable their extensive application in the production of dietary supplements and health-promoting functional food. Potential health benefits of dietary phenolics on cardiovascular health, such as reduction of hypertension, beneficial lipid profile, and inhibition of platelet activation have been suggested. Besides, phenolic compounds could be effective in reversing neurodegenerative pathology, while the consumption of foods and extracts rich in flavonoids may induce improvements in cognitive performance (Del Rio et al., 2013; Spencer, 2008, 2010a,b).

In another example, olive mill wastewater could be used as a source of polyphenols and pectin (Galanakis et al., 2010a,b,c,e). If those materials are not produced, subsequently there would be a need to purchase new ones. Likewise, recovering high added-value compounds from food wastes enables compliance with different types of food regulations. The latter are used to check regulatory compliance of other food products. Finally, the recovery of valuable components from food wastes could also promote the viability and diversity of rural and urban economies. Existing and new methods of successful food waste recovery could create new job opportunities. This is a very new area and there is a lot of potential for creating innovative and sustainable solutions (Otles et al., 2015).

11.3 UNIVERSAL RECOVERY STRATEGY

The "universal recovery strategy" (Fig. 11.1) was recently introduced by Galanakis (2015a,b) as a holistic approach for the management of all the important issues concerning the recovery of any target compound from any food waste source. The ultimate goal is to ensure the development of economically feasible and sustainable recovery methodologies, as well as safe final products. The strategy contains also the concept of a 5-Stages Universal Recovery Process (Fig. 11.2), which was earlier introduced for the proper management of the selected stages, as well as conventional and emerging technologies applied in the particular topic (Galanakis, 2012).

When someone tries to recover valuable compounds from a food waste material, the first issue to deal with is waste minimization prior to recovery processing. The second problem is the generation of the initial source in different forms and compositions following the regional, seasonal, and processing characteristics in each case. In addition, food processing by-products that are of particular interest contain lower concentrations of valuable compounds compared to the initial sources (eg, fruits or vegetables). This fact leads to lower recovery yield, higher processing cost, and finally lower revenues. Thereby, the different forms and compositions of by-products for a particular source should be

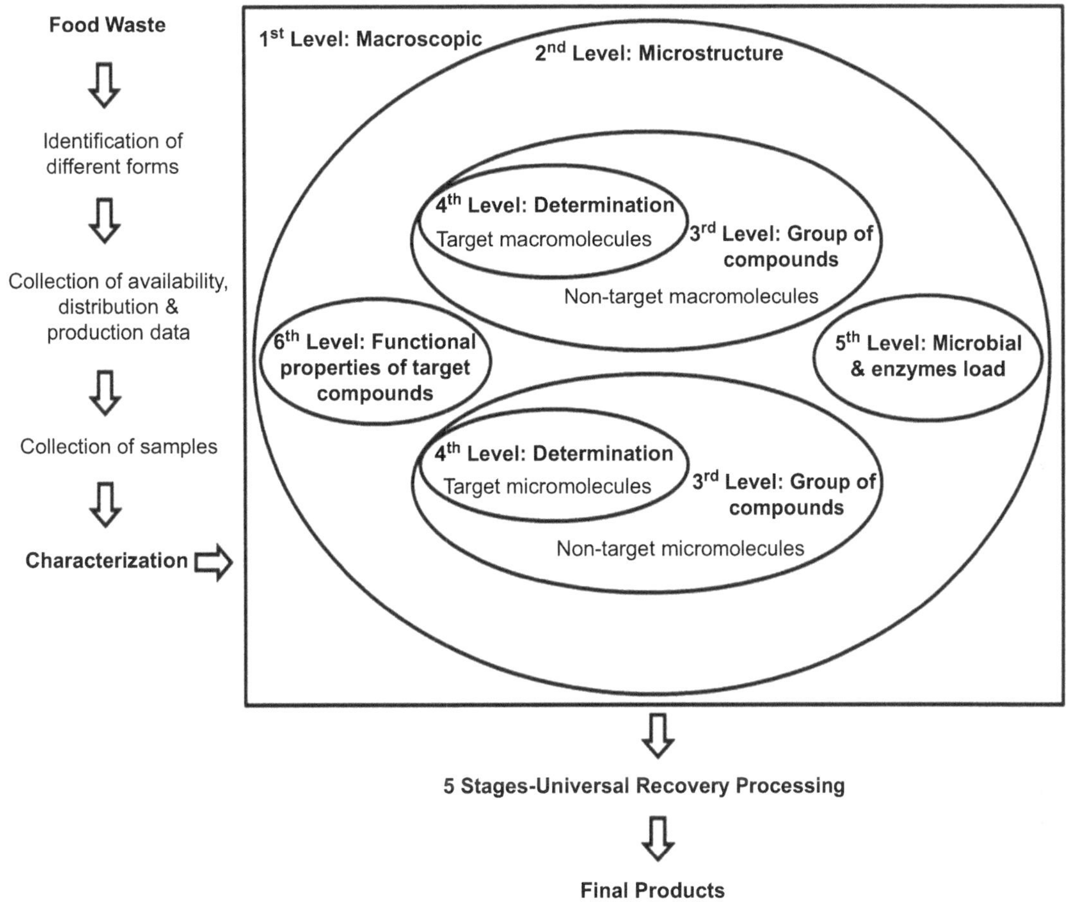

FIGURE 11.1

The Universal recovery strategy.

Adapted from Galanakis, C.M., 2015b. The universal recovery. In: Galanakis, C.M. (Ed.), Food Waste Recovery: Processing Technologies and Industrial Techniques (Chapter 3).

identified. Another problem is that food wastes and by-products require both preservation and fast treatment since they are susceptible to microbial growth due to their already processed nature. This means that collection from the source is also very important, whereas extended transportation should be avoided. Therefore, gathering all the necessary information concerning food waste availability (eg, abundance and distribution in different industries or locations, production frequency, seasonal generation, produced quantities) is the next step. This information should be used for the proper collection and mixing of food wastes at the source, in order to minimize variations in the content of their components and subsequently avoid variations in the final product. In addition, these data are necessary in business plans that evaluate the overall potentiality and economic growth of the methodology under development.

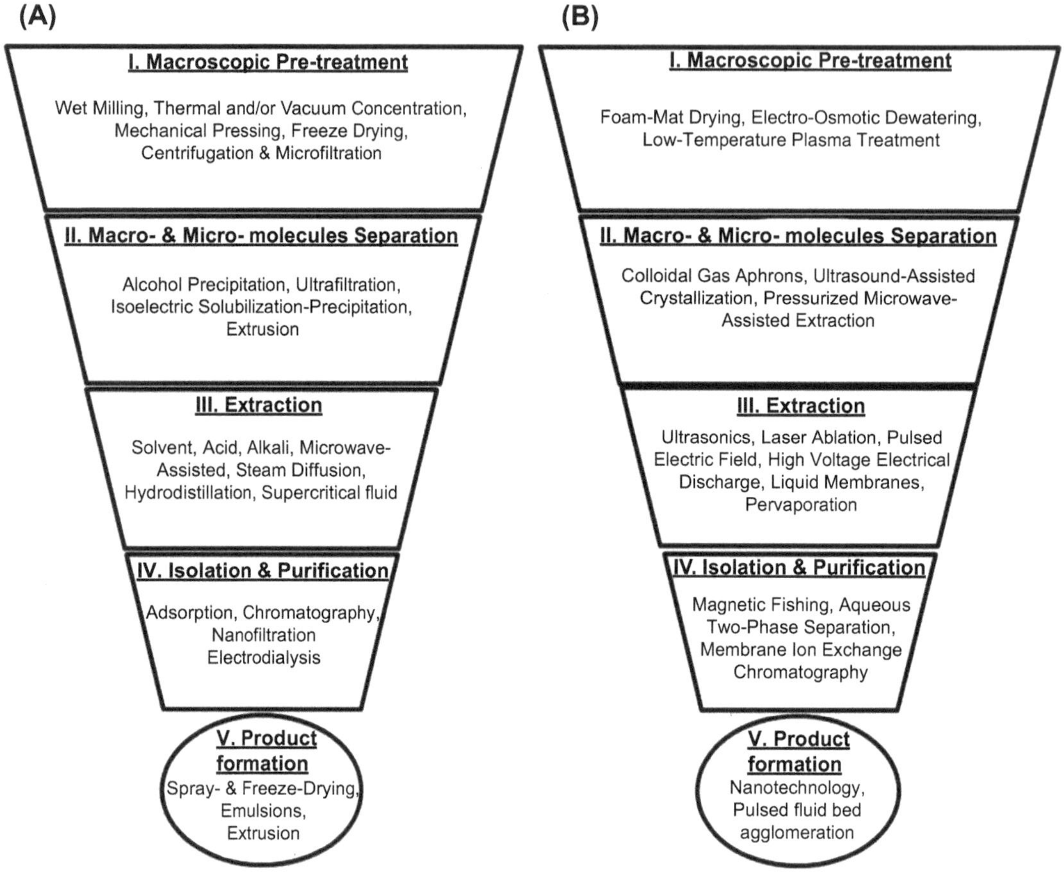

FIGURE 11.2

The 5-stages universal recovery process. (A) Conventional techniques, (B) Emerging technologies.

Adapted from Galanakis, C.M., 2012. Recovery of high added-value components from food wastes: conventional, emerging technologies and commercialized applications. Trends in Food Science & Technology 26, 68–87.

Collection and characterization of food waste samples follows the collection of the aforementioned information. The six-level characterization (macroscopic, microstructure, target and nontarget macro- and micromolecules, microbial and enzymes load, functional properties) of the samples (Fig. 11.1) is of particular interest for the selection of the appropriate technologies in the 5-Stages Universal Recovery Process (Fig. 11.2). For instance, the content of the different macroscopic waste phases (eg, water, oils, and solids) will help the adjustment of the initial substrate prior to macro- and micromolecule separation. The second level of characterization (microstructure) will provide an overview of waste matrix prior to designing the important extraction stage. Besides, gathering compounds in groups (eg, total phenols, carotenoids, dietary fibers, and proteins) provides a clear view and a fast screening of the main macro- and micromolecule content prior to designing their separation. Microbial and enzyme load accelerates substrate deterioration and diminishes functional properties of target compounds. For instance, polyphenoloxidases oxidize polyphenols, whereas pectin methyl esterase and pectic lyase diminish pectin's gelling properties by changing its structure.

The 5-Stage Universal Recovery Process (Fig. 11.2) follows the principles of advanced analytical chemistry such as substrate preparation, extraction, and purification of the target compounds. In addition, separation progresses from the macroscopic to the macromolecular and then micromolecular level. The ultimate goal in this case is to develop a methodology that ensures the highest recovery yields of several compounds by discharging minimum quantities of by-products in the environment. Besides, it is important to provide a nondestructive separation of valuable compounds in different streams. The utilization of food grade materials, processes, and green solvents is desirable in order to ensure the edibility of the final product, while the purification of target compounds from co-extracted impurities and toxic substances is in most cases necessary. Other important issues include the adaption to the demands of industrial processing, as well as the avoidance of compounds' autoxidation and loss of their functional properties. The latter could be conducted by encapsulating target compounds during product formation (fifth stage). In general, recovery downstream processing could be accomplished in the five recovery stages, although depending on the case, one or two steps can be removed (eg, if the target compound is a protein) or the order changed. The 5-Stage Universal Recovery Process could be utilized for the simultaneous recovery of different ingredients in two or more streams, eg, polyphenols in an ethanolic extract and pectin in the corresponding ethanol insoluble residue. The implementation of conventional (Fig. 11.2A) or emerging (Fig. 11.2B) technologies depends on several parameters such as convenience, energy consumption, and processing costs. Other important parameters include the preservation of compounds' functional properties as well as the stability, safety, and sensory characteristics of the final product. Emerging technologies promise to overcome problems associated with overheating, loss of functionality, and instability, but their positive effect on general cost and safety issues are still under debate (Galanakis, 2015a,b). Finally, it is important to produce safe and healthy products in response to market demands, ensuring that all consumers have access to nutritious food and accurate information about the respective food products.

11.4 IMPLEMENTATION OF THE STRATEGY FOR THE DEVELOPMENT OF COMMERCIALLY VIABLE PRODUCTS

A collection of commercial products recovered from food by-products is presented in Table 11.1. Patent application name was, herein, used to match market existing products with patented processes. Nevertheless, this matching may not always be correct as companies typically keep their production line secret, whereas respective data cannot be found in literature studies (Galanakis, 2012). For instance, one of the first by-products that has been valorized for recovery purposes is citrus peel. Its industrial commercialization concerns the solvent extraction of a "sugar syrup" containing essential oils, flavonoids, sugar, and pectin (Bonnell, 1983). The latter product is used as a sweetener and flavor substance in juices, replacing artificial compounds such as saccharine or aspartame. Due to the high concentrations of biologically active secondary metabolites remaining in citrus peel molasses, this raw material, as well as citrus peel, is also suitable as a source for functional food and/or food supplement ingredients (Manthey and Grohmann, 1996; Nafisi-Movaghar et al., 2013; Hanley Clark et al., 2015). The main flavonoids present include hesperidin, diosmin, and naringin. These compounds have been reported to have beneficial health properties, such as anti-inflammatory and anticancer potential activities (Manthey et al., 2001). Although present in lesser amount, hydroxycinnamates have been related to purported health actions, connected to their roles as antioxidants, anticancer and antimicrobial agents (Bocco et al., 1998; Kaul and Khanduja, 1998).

Table 11.1 Patented Methodologies Leading to Commercial Products Recovered From Food By-Products

Source	Patents Application Number	Applicant/ Company	Title/Treatment Steps	Products/Brand Names	Potential/ Commercialized Applications	Inventors/ References
Citrus peel	AU1983/0011308D	Tropicana Products Inc. (Florida, USA)	Treatment of citrus fruit peel	Sugar syrup	Food natural sweetener	Bonnell (1983)
Citrus peel	US2013/0064947	Del Monte Corporation (California, USA)	Treatment of citrus fruit peel	Juice, fiber, naringin, and oil	Food & feed supplement	Nafisi-Movaghar et al. (2013)
Citrus waste	US2015/0065698 A1	University of York (York, UK)	Treatment of citrus fruit peel	Pectin, D-limonene and flavonoids	Food supplement	Hanley Clark et al. (2015)
Apple peel	US2005/0147723A1	Leahy Orchards Inc. (Quebec, Canada)	Process for the production and use of apple peel powder	AppleActiv GHX, AppleActiv DAPP	Food supplement	Liu (2005)
Apple peel	US2013/0165396A1	National Research Council Of Canada, Dalhousie University	Application of products of apple peel for pharmaceutical use	Apple peel extract	Food supplement and for medical uses	Rupasinghe et al. (2013)
Pomegranate seeds	US7943185B1	Pom Wonderful, Llc (Los Angeles, CA, USA)	Process for the extraction of pomegranate seed oil	Pomegranate seed oil	Food supplement and for medical uses	Anderson et al. (2011)
Fruit and vegetable residues, unsellable fruits with defects	US2001/6296888	Provalor (Hoofddorp, The Netherlands)	Squeezing/decantation/centrigufation	Juices	Health drinks	Nell (2001)
Tomato pomace	PCT/EP2007/061923	Biolyco SRL (Lecce, Italy)	Process for the extraction of lycopene	Lycopene	Food antioxidant & supplement	Lavecchia and Zuorro (2008)
Coffee Silverskin	WO 2013/004873	Consejo Superior de Investigaciones Cientificas/CIAL (Madrid, Spain)	Application of products of coffee silverskin antiaging cosmetics and functional food	Bioactive silverskin extract	Cosmetics, Nutrition and Health	del Castillo et al. (2013)

Spent coffee grounds	WO 2006/036208	Ajinomoto General Foods, Inc. (Tokyo, Japan), Kraft Foods Global Brands LLC (Northfield, Illinois, United States)	Mannooligosaccharide composition for body fat reduction	Mannooligosaccharides	Functional Food Ingredient	Asano et al. (2006)
Palm agro-waste	WO 2014/042509 A1	Sime Darby Malaysia Berhad (Malaysia)	Process for the extraction of lecithin	Lecithin	Food supplement and for medical uses	Mee et al. (2014)
Chicken skin	US2011/0160471 A1	Umeda Jimusho Ltd., Marudai Food Co., Ltd., Boocs Medical Corporation	Process for the extraction of phospholipids	Phospholipids	Cosmetics, Nutrition and Health	Nadachi and Mawatari (2011)
Pig brains	WO1996/000077A1	Institut de Recherche Biologique (France)	Application of product as therapeutic and dietetic uses	Phospholipids	Food supplement and for medical uses	Ponroy (1996)
Olive mill waste	PCT/US2001/027132	CreAgri, Inc. (Hayard, USA)	Method of obtaining a hydroxytyrosol-rich composition from vegetation water	Hydroxytyrosol/Hidrox	Food supplements & cosmetics	Crea (2002)
Olive mill wastewater	PCT/SE2007/001177	Phenoliv AB (Lund, Sweden)	Olive waste recovery	Olive phenols & dietary fibers containing powders	Natural antioxidants in foodstuff & fat replacement in meatballs, respectively	Tornberg and Galanakis (2008)
Olive skin	WO2012/017108 A1	University of Granada (Spain)	Process for the extraction of maslinic and oleanolic acid	Maslinic and oleanolic acid	Food supplements & cosmetics	Mateo-Hernandez et al. (2012)
Cheese whey	PCT/SE1993/000378	Alfa-Laval Food Engineering AB (Lund, Sweden)	Method for obtaining high-quality protein products from whey	α-lactoalbumin & β-lactoglobulin containing product	Food supplements & additives	Jensen and Larsen (1993)

Continued

Table 11.1 Patented Methodologies Leading to Commercial Products Recovered From Food By-Products—cont'd

Source	Patents Application Number	Applicant/ Company	Title/Treatment Steps	Products/Brand Names	Potential/ Commercialized Applications	Inventors/ References
Cheese whey	PCT/US2002/010485	Davisco International Foods Inc. (Le Sueur, USA)	Isolation of glycoproteins from bovine milk	Whey protein Isolate/Bipro	Food supplements	Davis et al. (2002)
Bovine colostrum	US2005/0092684A1	Advanced Protein Systems (Phoenix, AZ, USA)	Process for the fractionation of colostrum	Immulox	Food supplement and for medical uses	Keech and Jimenez-Flores (2005)
Shrimp & crab shell	CN1994/1001978	Qingdao Zhengzhongjiahe Export & Import Co., Ltd. (Shandong, China)	Preparation of chitosan derivative fruit & vegetable antistaling agent	Chitosan (≥85%) food grade	Food thickener & fruit antistaling agent	Shenghui (1995)
Salmon viscera, heads, skin, frames, and trimmings	-	Aquaprotein, Chile	Hydrolysis, mechanical extraction and spray drying	Protein hydrolyzates and salmon oil	Petfood, pig weaning, animal breeding, injuries recuperation in animals	http://www.aquaprotein.com/
Animal lungs	US5024995A	Chiesi Farmaceutici SPA	Process for the extraction of pulmonary surfactant	Natural pulmonary surfactant	Medical uses	Robertson and Curstedt (1991)
Food waste	US2006/0088921A1	Yu J. and University of Hawaii	Process for the production of biodegradable thermoplastic materials	Polyhydroxyalkanoates	Food packaging	Yu (2006)
Shrimp waste	US2011/0282042A1	Coyote Foods Biopolymer and Biotechnology, S.R.L. (Saltillo, Mexico)	Process for the production of chitin	Chitin	Food packaging	Contreras-Esquivel et al. (2011)

Adapted from Galanakis, C.M., Martínez-Saez, N., del Castillo, M.D., Barba, F.J., Mitropoulou, V.S., 2015b. Patented and commercialized applications. In: Galanakis, C.M. (Ed.), Food Waste Recovery: Processing Technologies and Industrial Techniques. Elsevier-Academic Press (Chapter 15).

Due to the presence of various associated bioactive compounds (eg, polyphenols and carotenes), citrus fiber as well as apple fiber are in general of better quality than other dietary fibers (Fernández-Ginés et al., 2003; Wolfe and Liu, 2003). Furthermore, the appropriate characteristics such as volume replacement, thickening, or texturizing found in the fiber concentrates from apple and citrus fruit residues suggest various potential applications in the development of foods reduced in calories and rich in dietary fiber (Figuerola et al., 2005).

Nafisi-Movaghar et al. (2013) recently proposed a complete use of citrus peels for the recovery of juice (from juice sacs attached to the peel after hand or mechanical peeling), dietary fiber, naringin, and oil. The processes involve pressing peels to release juice from vesicles and then grinding peels in water to create slurry. Peel color (naringenin) and oil are removed by a flotation technique. Decolorized citrus peel particles or pulp are debittered with water in steps of boiling and washing. Dietary fiber obtained after debittering is dried. Researchers at the University of York (Hanley Clark et al., 2015) described a method of isolating pectin, D-limonene, flavonoids, and cellulose from citrus material wherein said method comprises the microwave-assisted hydrothermal low-temperature treatment of citrus material. The proposed method presented a number of important advantages such as mobility and flexibility, it reduces CO_2 burden, it is rapid, it can be continuous, and it has high energy efficiency.

Liu (2005) proposed a new method for the conversion of apple peel to powder by subjecting the apple peel to a phytochemical preservation treatment prior to drying and grinding. The method has been developed at the Cornell Research Foundation, Inc., however Leahy Orchards (a parent company to AppleActives) is the exclusive licensee for the patent (US2005/0147723A1) concerning the production method and the uses of apple peel powder. Well-known powerful antioxidant activity of apple peel is due to phenolic compounds mainly localized in that part of a fruit. Rich in phenolics, apples are also among the main sources of proanthocyanidins in the diet (Santos-Buelga and Scalbert, 2000). Leahy Orchards formulated two commercial products, namely AppleActiv GHX and AppleActiv DAPP (extracted from raw apple peels and containing a rich polyphenol blend with high levels of antioxidants, flavonoids, quercetin, triterpenoids, ursolic acid, vitamins, and minerals). Recently, Rupasinghe et al. (2013) patented the pharmaceutical and nutraceutical applications of apple skin extracts, which can reduce cholesterol levels and inhibit low-density lipoprotein (LDL) oxidation. Furthermore, flavonoid-rich apple peel extract has potential antihypertensive properties as well as anti-inflammatory effect on *Helicobacter pylori*-associated gastritis (Pastene et al., 2010; Balasuriya and Rupasinghe, 2012). Demonstrated protective effect against gastrointestinal damage supports the concept that apple peel polyphenols may also be useful in development of products for prevention and/or treatment of nonsteroidal anti-inflammatory drug-associated side effects (Carrasco-Pozo et al., 2011).

In another effort, Anderson et al. (2011) provided a novel method for the preparation of pomegranate seeds and the subsequent extraction, stabilization, and deodorization of pomegranate seed oil. Pomegranate seed oil contains approximately 65% punicic acid, which is a conjugated linolenic acid (9cis, 11trans, 13cis-CLNA). This acid has been reported to possess medicinal properties including anticarcinogenic activity against breast cancer and skin cancer. Antioxidant activity of a pomegranate seed preparation and cold-pressed pomegranate seed oil have also been demonstrated, whereas the use of these materials as potential natural food preservatives and/or health protective agents has been suggested. Furthermore, pomegranate seed oil may have a potential role as a cardioprotective and anti-inflammatory agent (Schubert et al., 1999). With the new method for the treatment of pomegranate seeds the inventors obtained stable and deodorized oil contrary to cold-pressed pomegranate seed oil.

The recovery of water-insoluble carotenoids and specifically lycopene from tomato by-products is under industrial development, too (Lavecchia and Zuorro, 2008). Lycopene is one of the most popular natural pigments (red), whereas the US Food and Drug Administration (FDA) approved its use to color processed meats as an alternative to carmine. Besides, the Food Safety and Inspection Service has established that extracts and concentrates (GRN 000156) of tomato lycopene of ≤50 and ≤100 mg/kg, respectively, could be utilized as coloring agent in ready-to-eat meat, poultry, and egg products (USDA, 2014). Aside from their coloring properties, carotenoids are also well known for their biological activities. Carotene and lycopene have been shown to be inversely related to the risk of cardiovascular diseases and certain cancers (Johnson, 2002; Agarwal and Rao, 2000). Unlike some other carotenoids, lycopene does not have provitamin A properties. On the other hand, the unsaturated nature of lycopene produces potent antioxidant properties of this compound (Rao and Rao, 2007). Coffee silverskin also contains bioactive compounds (ie, prebiotic carbohydrates, dietary fiber, and antioxidants) (Borrelli et al., 2004; Pourfarzad et al., 2013; Ballesteros et al., 2014). The recovery of bioactives from coffee silverskin has been proposed using subcritical water at moderate temperature (50°C or higher) and high pressure (1500 psi) without prior milling (del Castillo et al., 2013). The antioxidant properties of this patented extract survive the in vitro gastrointestinal digestion process and remaining antioxidants bioaccessible to exert their function in the body (Galanakis et al., 2015b). It has been proposed as an antioxidant additive in food and cosmetics manufacture, with potential excipient (preservative, flavoring) functions as well as antiaging and anticellulite activities. Besides, it has a potential to reduce body fat accumulation (Martinez-Saez et al., 2014). It was recently suggested that apart from its potential as a food supplement, coffee silverskin may represent an innovative functional ingredient exploitable to increase the general antioxidant capacity of a wide range of food products (Bresciani et al., 2014).

Today the new trend in the field of food waste recovery is the recovery of phenolic compounds from olive mill waste (Galanakis, 2011; Galanakis et al., 2010d,e; Rahmanian et al., 2014; Roselló-Soto et al., 2015). For instance, a popular methodology includes the acid treatment of olive mill wastewater, prior to an incubation process that converts oleuropein to hydroxytyrosol. The next step could include a supercritical fluid extraction and a column operating in the countercurrent mode, where a nonselective porous membrane is the barrier interface between the hydroxytyrosol-containing fluid and the dense gas. Product formation is conducted using freeze- or spray-drying (Crea, 2002). Hydroxytyrosol could be used as functional supplement, food preservative in bakery products, or life-prolonging agents (Liu et al., 2008). It has been demonstrated that olive mill waste polyphenols and (especially hydroxytyrosol) are efficient in inhibiting hyperglycemia and oxidative stress induced by diabetes. Based on obtained data, administration of hydroxytyrosol was suggested in the prevention of diabetic complications associated with oxidative stress (Hamdena et al., 2009). Furthermore, another scientific study supports the hypothesis that hydroxytyrosol may exert a protective activity against cancer, suggesting that hydroxytyrosol, an important component of virgin olive oil, may have a major role in its anticancer activity (Fabiani et al., 2002). Hidrox is a commercially available hydroxytyrosol product from CreAgri (Hayward, USA), granting a generally recognized as safe (GRAS) status. According to several scientific studies, Hidrox possesses several beneficial (ie, anti-inflammatory and antimicrobial) properties.

Fig. 11.3 illustrates an example for the recovery of phenols and dietary fibers from olive mill wastewater, using the aforementioned 5-Stages Universal Recovery Process. In this case, the pretreatment step includes two processes (centrifugation and skimming) in order to remove remaining fats from three-phase olive mill wastewater. Then a vacuum concentration process removes a part of the

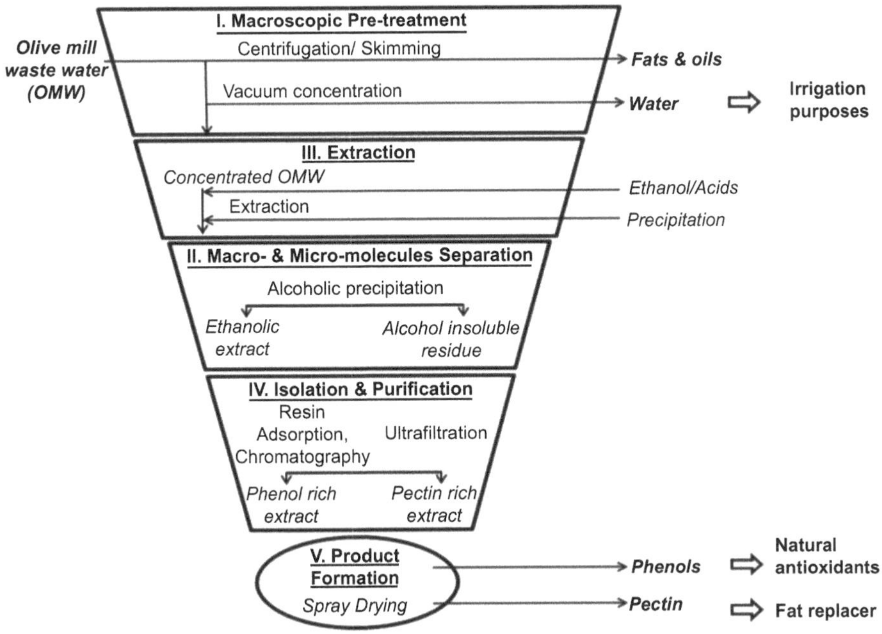

FIGURE 11.3

Recovery of valuable compounds from olive mill wastewater using the *5-stages universal recovery process.*

Adapted from Galanakis, C.M., Martínez-Saez, N., del Castillo, M.D., Barba, F.J., Mitropoulou, V.S., 2015. Patented and commercialized applications. In: Galanakis, C.M. (Eds.), Food Waste Recovery: Processing Technologies and Industrial Techniques. Elsevier-Academic Press (Chapter 15).

contained water. The next step includes the treatment of the substrate with acids and ethanol, generating two streams: an alcohol insoluble residue rich in dietary fibers and an ethanolic extract rich in phenols. Purification of both streams can be conducted using ultrafiltration (Galanakis, 2015a,b). The extracts obtained from the olive mill wastewater can be considered as an important source of antioxidant fiber with associated phenolic compounds. Regarding that these antioxidant compounds are thought to reach the colon intact and beneficially act on human health, this property is considered to be of nutritional significance (Saura-Calixto, 2011; Rubio-Senent et al., 2015). Pectin derived from olive mill wastewater has been proved to restrict oil uptake of low-fat meatballs during deep-fat frying (Galanakis et al., 2010c), whereas the recovered phenols could be utilized as antioxidants in vegetable oils, bakery and meat products. Besides, olive pulp extracts have been approved by the FDA with GRAS status (GRN No. 459) for being used as an antioxidant in baked goods, beverages, cereals, sauces and dressings, snacks, and functional foods at a level up to 3000 mg per kilogram in the final food (FDA, 2014). In another effort, olive skin has been successfully used to extract maslinic acid and oleanolic acid (Mateo-Hernández et al., 2012). Maslinic acid has the potential to provide significant natural defense against colon cancer, and the anticancer activity observed for olive fruit extracts seems to partly originate from this compound. Maslinic acid is therefore a promising new constituent in preparations intended to be used for the chemoprevention of colon cancers (Reyes-Zurita et al., 2009). The present invention relates to a method of maslinic acid and oleanolic acid comprising three stages: separation of the skin from olive pit, two consecutive extractions of the skin, and purification of the extracts. The recovery of maslinic and oleanolic acid is developed using two consecutive Soxhlet extractions with different solvents.

Wine production waste is another important source of natural antioxidants that are considered completely safe in comparison with synthetic antioxidants. Grape pomace is an industrial waste from the wine process, consisting basically of grape seeds, skins, and stems. It represents a rich source of different products including tartrates and malates, citric acid, grape seed oil, hydrocolloids, dietary fiber, as well as phenolic compounds. The latest, due to their poor extraction during winemaking, are present in considerable amount in by-products. Grape skin present in the pomace is the main source of phenolics, followed by the seed and leaf (Xia et al., 2010). Anthocyanins, procyanidins, flavanols, phenolic acids, alcohols, and stilbenes such as the well-known compound resveratrol, are some of the phenolics present in grape pomace. As a result of the considerable amount of bioactive phenolic compounds, grape pomace could be suitably used as dietary supplements or as ingredients in functional foods and cosmetics (Kammerer et al., 2004). The extraction of bioactive compounds from wine production waste has been proposed using several methods such as fractionation of grape seed or acidolysis of selected grape pomace fraction in the presence of cysteamine or electrodialysis (Shrikhande, 2000; Alonso et al., 2002; Gonzales-Paramas et al., 2004; Andres et al., 1997). More recently developed, the biological extraction process, using hydrolytic enzymes, has several advantages over chemical processes, eg, by considering the preservation active ingredients' stability (low temperatures and pressures are applied) as well as regarding its environmental impact (enzymatic method of extraction excludes the use of xenobiotics or toxic reagents). In addition, the obtained extract shows pronounced antioxidant properties, pointing out the potential usage in functional food and dietary supplements production. Furthermore, it has been suggested that compared to other grape pomace extracts, an enzymatically obtained extract has better availability, stability, as well as biological activity (Rodríguez-Rodríguez et al., 2012).

Cheese whey is generated in different forms and compositions depending on the characteristics of cheese manufacture, and comprises one of the most known food waste sources used for recovery purposes. The main target compounds include protein concentrates and lactose that are used in nutritional supplements and confectionery, respectively. Separation of these compounds is typically conducted using electrodialysis, ultrafiltration, and nanofiltration (Davis et al., 2002; Galanakis et al., 2014). For instance, a patented two-step microfiltration process for the sequential concentration of α-lactoalbumin and β-lactoglobulin has been invented by Jensen and Larsen (1993). Hydrolyzed whey proteins are known for their ability to reduce total and LDL-cholesterol levels in mammals (Davis et al., 2003) and thus could be used for the production of functional foods. Bovine colostrum is another by-product of dairy industry that has been investigated for food and therapeutic usage. Keech and Jiménez-Flores (2005) established a method and a system to isolate target peptides and proteins from bovine colostrum. Briefly, the colostrum is first passing at least one ion-exchange column. The latest includes an anionic and/or a cationic resin. The ion-exchange column is selected to remove large particles at a maximum fluid pressure of 30 pounds per square inch (psi). The large particles can be released from the ion-exchange column with a rinse solution having the same pH value selected to release the large particles. The pH value of the fluid can then be adjusted to <5.0, and then fluid can be filtered with a microfilter or ultrafilter. This process has been patented by APS BioGroup to produce several commercial products such as colostrum and Immulox powders, with immune system balancing effects and high level of proline-rich peptides.

Finally, the meat and fishery industries generate a large amount of by-products that are a good source of nutrients and can be used as food ingredients and additives. Among the different meat by-products, air floatation skimming sludge has typically had an important nutritive value. However, it is lost due to microbial degradation. On the other hand, a practical precipitation process of food waste sludge from dissolved air floatation units and sugar by-products has been patented (Lee, 2002). Fish oils contain omega-3 and omega-6 fatty acids that could be used in the food and pharmaceutical industries. Other commercial applications are based on the recovery of fish protein hydrolyzates and fish oil for feed from fish guts, skins, heads, and bones (Galanakis et al., 2015b). Nadachi and Mawatari (2011) used chicken skin that contains a great amount of human-form sphingomyelin and plasmalogen-form glycerophospholipid to produce highly purified products. The production process includes drying of total lipids extracted from the skins followed by an extraction treatment with a specific mixture solution to separate an insoluble and a soluble portion. The insoluble portion is used to obtain crude sphingomyelin and the soluble one to generate plasmalogen-form glycerophospholipid. Ponroy (1996) noticed the therapeutic or dietetic use of a preparation based on phospholipids extracted from pig brains for treating aging disorders consisting of memory, attention, intellectual efficacy, humor, affectivity, and thinking difficulty disorders in a human. Animal lungs are increasingly used as source of natural surfactants. Chiesi Farmaceutici SPA patented a method of preparation and pharmaceutical compositions of natural pulmonary surfactant for medical uses. The surfactant (obtained through filtration, centrifugation, and extraction by gel chromatography) induces advanced therapeutic properties in the treatment of infant and adult respiratory distress syndromes.

Finally, several applications about the use of food wastes as source of polymer for sustainable packaging have been reported. Yu (2006) proposed a system and method for converting organic wastes to biodegradable thermoplastic materials including polyhydroxyalkanoates (PHAs). The method consists of treating the organic wastes with an acidogenic microbial population to form fermentative organic acids, and polymerization of the organic acids by PHA-producing microbial species to form PHAs. Chitin, a long-chain biopolymer of an N-acetylglucosamine, has been used for edible films or coating production. In recent years, several extraction methods of chitin from food and food wastes have been proposed. Contreras-Esquivel et al. (2011) developed a new production method of chitin through employment of a microwave process under pressure and/or autoclave with organic acids. This new method eliminates salts and proteins in a single stage and reduces contamination levels.

11.5 MANAGEMENT OF INTELLECTUAL PROPERTY

The recovery of valuable compounds from food by-products involves the development of new methods, industrial processes, and new applications of recovered compounds in food products. This new knowledge comprises intellectual property, which provides competitive advantages (eg, applications as a marketing edge) for the researchers, organizations (ie, universities, institutes), as well as involved companies (ie, start-up, spin-off, spin-out). Intellectual property has a commercial value that increases business value since it could be implemented as a potential revenue stream through licensing and attraction of investments. Subsequently, it should be protected and exploited following proper scientific and commercial strategies, ie, filing patent applications where appropriate, registering trademarks and copyrights, as well as taking appropriate steps to protect trade secrets. A patent forbids other individuals to produce, sell, use, import, and possess the invention.

Since intellectual property law is complex, the inventors should be aware of the basic rules (ie, procedures, filing rules, costs, expected revenues, rights, etc.). All issues relating to rights over the invention between the organization, its employees, and any other partner (who may have participated either financially or technically in developing the invention) should be clarified. In addition, the decision to file an application for a patent should be carefully taken since legal systems vary between countries. In particular, national patents are used to protect inventions within the home country market and as a basis to provide "priority" and extend protection to other countries. On the other hand, international patent applications are used to extend the prepublication period up to 30 months and protect inventions in different countries. Besides, the inventors must decide a specific and robust patent/IP rights strategy as early as possible. The decision should be taken when a solution is invented, or even earlier, when a need comes around and must be fulfilled. For instance, start-up companies typically build their credibility by protecting intellectual property during business formation. This process provides safety to investors and creates a solid foundation that can be capitalized later. It is also important to approach the idea as a financial asset, conduct a cost–benefit analysis, and check alternatives such as secrecy and utility models (Galanakis et al., 2015b).

The case of polyphenols and specifically hydroxytyrosol recovery from olive mill wastewater, using the 5-Stages Universal Recovery Process (Section 11.4), could be used to exemplify the process and the required decisions for an effective selection of intellectual property strategy. For instance, the first choice could be to apply for a national patent in the country of origin of the invention (eg, in Greece). The patent could include both the innovative recovery process as well as the products of high marketing potential (eg, functional foods). Following this scenario, the obtained patent gives 1-year priority and at the same time protects commercialization possibility with a low and early investment. After filing a national patent, application could be extended to a regional patent like the European Patent Office, if the anticipated products could be marketed in all European countries. An alternative approach is to find an investor or licensee who invests in the early stage of marketing products, providing the patent holder finance to develop new products. The selection of the target market is very critical, too. For instance, consumers in Mediterranean countries are well aware of olive oil taste and health benefits, and thus prefer consuming this product instead of another food containing additives of olive polyphenols. In Northern European countries, consumers are not as used to the olive oil taste and thus could find it interesting and "exotic" to approach a food product containing olive polyphenols. Besides, some countries (ie, the United States) provide specific marketing advantages to olive polyphenols. Today commercial products containing hydroxytyrosol are already marketed in the United States, where the FDA granted these products with GRAS status and health claims. On the other hand, the European Food Safety Authority (EFSA) handles market release and health claims of hydroxytyrosol-containing products in a more conservative manner. For instance, hydroxytyrosol has received European market release acceptance only for low-cholesterol spreads and health claims for olive oil. This importantly different approach of market release and health claims between the FDA and EFSA influences the success of the developed products. These factors should be taken into account when selecting patent filing strategy (Galanakis et al., 2015b).

11.6 PROBLEMS

In general innovations in the field are driven by spin-off or start-up companies, which are typically created by researchers. These kinds of companies are considered as small and medium enterprises (SMEs), although in most cases they are even smaller (microenterprises). Another common practice is licensing of patents and methodologies developed from researchers to SMEs. In this case, a typical problem is that academics need to publish and face problems managing intellectual property rights. Besides, SMEs are struggling to implement innovations in the field due to limitations and numerous barriers, such as the lack of adequate collaborations, the relative low capability of managing perceived challenges, the minimal external experience (competition), and the limited financial resources for internal research and development. Other problems include the inadequate human resources and competencies, the absence of production facilities, restrictions in securing intellectual properties, the narrow business portfolio, the low market power, and finally the lower "status" as an innovation partner. Besides, in most cases, internal university funds are insufficient to develop links with industry (Saguy and Sirotinskaya, 2014). Finally, developments in the chain's environment, like the availability of new technologies or restrictions imposed by governmental and supragovernmental legislation, can also trigger innovation efforts (Sarkar and Costa, 2008).

To overcome these difficulties, the companies should adapt alternative strategies, such as the open innovation model. Open innovation is "the use of purposive inflows and outflows of knowledge to accelerate internal innovation, and expand the markets for external use of innovation, respectively" (Chesbrough et al., 2006). This strategy is today essential to survive and gain a competitive advantage in most business environments, where companies must use both external and internal ideas, open channels for knowledge access, employ external technology and solutions, and purchase or license inventions (Saguy and Sirotinskaya, 2014). This way, progress in the field is leveraging. However, there is always the risk of "outsourcing" and losing progressively internal capabilities to develop one's own innovations (Aron and Singh, 2005). On the other hand, large companies active in the field typically try to select technologies as mature and proven as possible in order to reduce implementation and technological risks. However, this process allows intervention with the inventors only in the final stages, and this way the chances to customize and produce radical innovations are reduced (Huston and Sakkab, 2006). This filter certainly limits the potential of the open innovation model and generates problems, as the internal research and development capabilities of the company are less utilized.

Commercialization of compound recovery from food wastes includes several steps like laboratory research, pilot plan set-up, development of full-scale production, etc. These parameters are necessary in order to ensure the sustainability of the process, the economic benefit for the involved food industry, and the perpetual establishment of the derived products in the market. Besides, a working strategy, focused absolutely on the extraction technologies and not on the investigation of tailor-made applications, is doomed to fail (Galanakis, 2012, 2013). To this prospect, it is very important to provide clean label ingredients for processed food products (eg, natural preservatives, functional compounds), without impacting flavor or texture and while maintaining minimum shelf life requirements. In addition, the development of the production line should be designed near, but not inside, the food industries in order to ensure minimum transportation and at the same time meet their hazard analysis and critical control points requirements (Galanakis, 2015a,b). Scale-up process should be conducted without diminishing the functional properties (eg, antioxidant, viscoelastic) of the target compounds and simultaneously to develop a product that meets consumers' high-quality organoleptic standards (Galanakis and Schieber, 2014). This is difficult since compound recovery development confronts the typical scale-up problems (eg, mixing and heating time) as any food manufacture procedure. More specifically, transition of batch to continuous processes is accompanied with extended mixing and heating time, heavier handling, increased air incorporation, and higher degree of scrutiny. All of these parameters generate numerous interactions and loss of product functionality. Subsequently, process cost is increased, as industrially recovered compounds are used in food formulations in higher concentrations compared to laboratory-recovered compounds. The broad variation of target and nontarget compounds from source to source could also affect the functionality and the organoleptic character of the final product. For this reason, it is important to mix selectively different by-products at the beginning of the process, taking into account their content in bioactives and impurities, as well as other basic characteristics. Another complicated issue is waste collection from the source. This process often requires additional transportation cost and control of microbial growth. Thereby, a proper management of collection process is required, such as cooling, freezing of the material, or the addition of preservatives (Galanakis, 2012).

Common problems often arise from market needs for healthier products. Authorities around the world (especially in Europe via EFSA) have tightened up the way in which companies can advertise health benefits. Indeed, health claims have been approved for a small number of compounds (ie, hydroxytyrosol in olive oil) and products (ie, cholesterol-reducing yogurts and butters). This policy is driven by the need for protecting consumers from dubious claims. Nevertheless, demonstration of proven health benefits is very costly for the companies activate in the field, as identified previously. This fact creates implications for stifling innovation in the field, as the obtainment of the required data is not affordable to many and most companies (typically start-ups with low funding) cannot afford them. Besides, the risk of claims rejection by the corresponding authorities is too high (Galanakis et al., 2015b).

The legislation challenges regulating health-beneficial dietary products are tied to the specificity of the products, which have the characteristics of both food and biologically active ingredients. Nevertheless, the compendial testing is not required for the marketing authorization application of dietary supplements or food containing added active ingredients. Therefore, it would be advisable to clearly define the manufacturing and quality control criteria related to composition and content range of active substances as well as manufacturing development of product. However, as previously mentioned, product can be satisfactorily used by consumers only if there is a reasonable balance between quality requirements, manufacturing qualifications and market/economics rules are taken into consideration. Currently the manufacturer's label usually provides only limited information about the origin and composition of extract used in final product formulation. The increased consumption of functional foods, dietary supplements, and related products, as well as availability of a wide range of items, makes their tracking and quality control a difficult task, which has to be fulfilled in order to ensure positive effects for the user. Safety assessment and market release permission of purified active compounds is usually more demanding. This procedure includes long and sophisticated tests on different species of laboratory animals, similar to synthetic antioxidants. On the other hand, if the product is an enriched natural extract, the market release criteria are not so strict. This is happening because natural extracts are considered to exist inherently in foods and thus safety concerns are limited (Galanakis, 2012). The main objective for better control of the dietary products manufacturing and increased surveillance in this area should certainly be the safety and health care of consumers. A clearer label of the dietary supplements, functional foods, and products containing recovered added-value components would provide much better insight into the quality and composition of products, and would enable nutritionists, pharmacists, and doctors to be more confident when recommending these products.

11.7 SOLUTIONS

Today the growing complexity of globalized food chains, including health risks, food safety scares, and habitat depletion requires integrated perspectives for the implementation of innovations in the field (Lowe et al., 2008). For instance, to prevent stifling of innovation from the mounting safety concerns and strict regulations, and simultaneously address emerging wellness issues, a new direction for developing, acquiring, and implementing the vast potential of scientific breakthroughs is needed (Saguy and Sirotinskaya, 2014). Perhaps the establishment of a new label (similar to organic foods) or reduced taxes to relevant products could reveal the potential of recovering high added-value ingredients from food by-products and reutilizing them in foods. In general, the key point for commercialization is to develop a recovery strategy that allows flexibility and provides alternative scenarios for each stage of processing (Galanakis, 2012).

What is strongly recommended is the implementation of nonthermal technologies, addition of green solvents, and safer materials (possessing GRAS status). The development of tailor-made applications for the recovered products (crude or highly purified) is necessary, too, as target compounds may not be as beneficial as proposed theoretically, and more importantly, it is difficult to survive competition in the market of functional foods. Integral investigations that include recovery protocols and preservation assays are also necessary. The latter parameters would ensure sustainability of the final product and industrial exploitation (Galanakis et al., 2015b). With regard to the recovery stages, cheaper methodologies are those containing fewer recovery steps. On the other hand, these methodologies generate crude products containing lower concentrations of the compounds of interest. Subsequently, the functional properties of the developed products are diminished since some of the target compounds are replaced with coextracted ingredients. Product formation (the fifth one from the 5-Stages Universal Recovery Process) is the most essential step, as encapsulation enhances functionality and extents the shelf life of the products. Researchers in the field will soon deal with the prospect of applying emerging technologies and particularly nanotechniques with an ultimate goal to optimize overall efficiency of suggested methodologies (Galanakis, 2012). What is important to state is that the final product should be precisely designed for a well-defined market. For instance, natural antioxidants to be used in fresh, mom-approved, and kid-friendly foodstuffs with clean ingredient lines and minimum 90-day refrigerated shelf life. Finally, unlike the need to delight the consumer and minimize environmental impact, developers should also ensure that the product process and specifications are ready for commercial production (Thota, 2012).

It is also of particular interest that the companies activate in the field and academia move on to a mutual commitment of resources and time by entering into a collaborative and long-term engagement. Selection of research topics should be, as far as possible, guided not only by scientific interest but also by manufacturers' and markets' specific needs and requests, thus providing data and information useful for selection and implementation of appropriate solutions regarding development of the targeted product. This approach would positively influence application of obtained research results, use of scientific breakthroughs' potential, and consequently, valorization of enterprise–academia collaboration. An initial small-scale collaboration could offer unique opportunities (eg, agile decision-making, flexibility of entrepreneurial culture, business specialization, and niche products). Flexibility and expertise give SMEs an advantage in building and nurturing relationships for win–win collaborations. In any case, researchers face difficulties in finding industrial partners. A common way of solving this problem is to search for partners in dedicated industrial platforms and networks. For instance, "European Institute for Food Processing", "HighTech Europe," and "Food for Life" are three interdisciplinary and interactive European platforms that link industrial needs and available research-and-development results, implement innovations into SMEs, and promote the participation of SMEs in specific innovation management, respectively. All these networks can also help develop applications in the field of food waste recovery. The success of industry–academia collaborations depends on finding common goals, negotiating plans that fit mutual needs, and obtaining financial and intellectual payoffs for both parties.

To make a significant change, a new mindset is required, in which academia is not only an intellectual center and a generator of knowledge but also a contributor of innovation.

Despite the aforementioned problems (Section 11.6), becoming an "organic" member of an industrial open innovation effort is of high priority and should be considered. In order to succeed, all participants of the open innovation ecosystem need to take a proactive role in what is known as the "fourth helix": university, industry, government, and private sector. Banks, private and venture capital, other private and government funds, and angels have a paramount role in making open innovation a reality for SMEs and therefore should be included (Saguy, 2013). It is also proposed that SMEs improve new product development speed and quality of delivery of new products. This could be conducted by setting bold and strategic goals for growth as well as core values that sustain the entire organization and build integrity, ethics, and trust. Besides, research has shown that innovation can only occur if the business structure encourages collaboration between the employees and empowers them to access and share internal knowledge (Panduwawala et al., 2009). Moreover, cocreation is probably the most powerful way of innovation, benefiting both the participants and the ecosystem as a whole.

11.8 MEETING MARKETS' AND CONSUMERS' NEEDS

The ultimate goal of innovations in the field is to provide benefits to consumers. The extent to which consumers perceive these benefits determines the acceptance or rejection of the developed innovation. Food marketing literature indicates that few innovations are widely accepted by consumers. For instance, it has been noted that people increasingly display a preference for natural entities that have been produced without human intervention. Nowadays people are more aware of risks confronting them, even ones that are distant in time (threats to future generations) or space (ones that reach beyond their locality) (Lowe et al., 2008).

Following these trends, modern new product development should aim at the fulfillment of consumer needs and the realization of consumer value rather than at the development of products or enabling technologies per se (Costa and Jongen, 2006). This is important to state due to the recently opened debate concerning the safety of products recovered from food wastes and the impact (beneficial or not) of recycling them inside the food chain (Galanakis, 2012). Today the food industry faces a large number of challenges comprised of instantaneously changing and evolving consumer needs, shortened product life cycles, the competitive time-to-market race, and cluttered retail shelf space (Bellairs, 2010). Likewise, consumers are more heterogeneous and whimsical than ever, which makes their food choices harder to understand and predict. They require more and more food products closely tailored to their individual needs and preferences and aiming at their health. The growing importance of values such as quality of life, well-being, or protecting the planet's environment are also exerting influences on the way consumers perceive and evaluate foods and production systems, thereby increasingly determining choice (Costa and Jongen, 2006). These aspects support the proliferation of marketing schemes, such as "organic," "fair trade," and "protected designation of origin," which valorize particular production methods and establish the basis for market differentiation, as well as new niche markets for local producers and rural territories (Lowe et al., 2008). Indeed, within the era of open innovation, consumers have gained a more critical role, in which they are defined as cocreators. They discuss their food-product experiences, and evaluate their appeal, value, acceptance, and effectiveness by sharing needs, ideas, feelings, emotions, and experiences and providing feedback (Saguy and Sirotinskaya, 2014).

With more technology-based reductions of processes' and ingredients' costs being frequently marketed as new, enticing benefits to the consumer, it becomes increasingly harder for buyers to perceive the added-value of "slightly new" food products and thus reward incremental innovation (Costa and Jongen, 2006). This problem could be solved following the evolutionary idea noted in the Section 11.7: the establishment of a new label (similar to organic foods) to increase the potentiality of commercializing valuable compounds from food by-products. Besides, the food chain has become one of the primary sites of political resistance to globalization. To reinforce policy reforms and regulatory changes, they have also sought to encourage consumers to buy "environmentally friendly" products. This in turn has stimulated "green" marketing and environmental quality assurance schemes (Lowe et al., 2008). Finally, a culture of experimentation requires a change to a more flexible motto: to fail fast, to fail early, and to ask "what if" questions to create unique customer value. The key is to address problems early in the design phase and develop products based on customer needs (Thota, 2012).

REFERENCES

Agarwal, S., Rao, A.V., 2000. Carotenoids and chronic diseases. Drug Metabolism Drug Interaction 17 (1–4), 189–210.

Alonso, A., Guillean, D., Barroso, C., Puertas, B., Garcia, A., 2002. Determination of antioxidant activity of wine byproducts and its correlation with polyphenolic content. Journal of Agricultural and Food Chemistry 50, 5832–5836.

Anderson, S., Dreher, M., Green, R., 2011. Method and Composition for Producing a Stable and Deodorized Form of Pomegranate Seed Oil. US7943185B1.

Andres, L., Riera, F., Alvarez, R., 1997. Recovery and concentration by electrodialysis of tartaric acid from fruit juice industries waste waters. Journal of Chemical Technology & Biotechnology 70, 247–252.

Aron, R., Singh, J., 2005. Getting offshoring right. Harvard Business Review 83 (12), 135–143.

Asano, I., Fujii, S., Mutoh, K., Takao, I., Ozaki, K., Nakamuro, K., Matsushima, T., 2006. Mannooligosaccharide Composition for Body Fat Reduction. Patent WO 2006/036208.

Balasuriya, N., Rupasinghe, V., 2012. Antihypertensive properties of flavonoid-rich apple peel extract. Food Chemistry 135, 2320–2325.

Ballesteros, L.F., Teixeira, J.A., Mussatto, S.L., 2014. Chemical, functional, and structural properties of spent coffee grounds and coffee silverskin. Food and Bioprocess Technology 7 (12), 3493–3503.

Beer, S., Lemmer, C., 2011. A critical review of "Green" procurement: life cycle analysis of food products within the supply chain. Worldwide Hospitality and Tourism Themes 3 (3), 229–244.

Bellairs, J., 2010. Open innovation gaining momentum in the food industry. Cereal Foods World 55 (1), 4–6.

Bocco, A., Cuvelier, M.E., Richard, H., Berset, C., 1998. Antioxidant activity and phenolic composition of citrus peel and seed extracts. Journal of Agricultural and Food Chemistry 46, 2123–2129.

Bonnell, J.M., 1983. Treatment of Citrus Fruit Peel. Australian Government – IP, Australia. 1130883 (A).

Borrelli, R.C., Esposito, F., Napolitano, A., Ritieni, A., Fogliano, V., 2004. Characterization of a new potential functional ingredient: coffee silverskin. Journal of Agricultural and Food Chemistry 52, 1338–1343.

Bresciani, L., Calani, L., Bruni, R., Brighenti, F., Del Rio, D., 2014. Phenolic composition, caffeine content and antioxidant capacity of coffee silverskin. Food Research International 61, 196–201.

Brown, L., Rosner, B., Willett, W.W., Sacks, F.M., 1999. Cholesterol lowering effects of dietary fiber: a meta-analysis. American Journal of Clinical Nutrition 69, 30–42.

del Castillo, M.D., Ibáñez, E., Amigo, M., Herrero, M., Plaza, M., Ullate, M., 2013. Application of Products of Coffee Silverskin in Anti-Aging Cosmetics and Functional Food. WO 2013/004873.

Carrasco-Pozo, C., Speisky, H., Brunser, O., Pastene, E., Gotteland, M., 2011. Apple peel polyphenols protect against gastrointestinal mucosa alterations induced by indomethacin in rats. Journal of Agricultural and Food Chemistry 59, 6459–6466.

Chandrasekaran, M., 2013. Need for valorization of food processing by-products and wastes. In: Chandrasekaran, M. (Ed.), Valorization of Food Processing By-Products. Taylor & Francis Group, Florida, pp. 91–108.

Chesbrough, H., Vanhaverbeke, W., West, J., 2006. Open Innovation: Researching a New Paradigm. Oxford University Press, Oxford, UK.

Cobs-Rosas, M., Concha-Olmos, J., Weinstein-Oppenheimer, C., Zuniga-Hansen, M.E., 2015. Assessment of antiproliferative activity of pectic substances obtained by different extraction methods from rapeseed cake on cancer cell lines. Carbohydrate Polymers 117, 923–932.

Contreras Esquivel, J.C., Balvantin Garcia, C., Valdez Peña, A.U., Flores Davila, C.P., 2011. Obtainment of Chitin from Shrimp Waste by Means of Microwave and/or Autoclaving in Combination with Organic Acids in a Single Stage. US2011/0282042A1.

Costa, A.I.A., Jongen, W.M.F., 2006. New insights into consumer-led food product development. Trends in Food Science & Technology 17, 457–465.

Crea, R., 2002. Method of Obtaining a Hydroxytyrosol-Rich Composition from Vegetation Water. World Intellectual Property Organization. WO/2002/0218310.

Crowe, F.L., Roddam, A.W., Key, T.J., Appleby, P.N., Overvad, K., Jakobsen, M.U., Tjonneland, A., Hansen, L., Boeing, H., Weikert, C., Linseisen, J., Kaaks, R., Trichopoulou, A., Misirli, G., Lagiou, P., Sacerdote, C., Pala, V., Palli, D., Tumino, R., Panico, S., Buenode-Mesquita, H.B., Boer, J., van Gils, C.H., Beulens, J.W., Barricarte, A., Rodriguez, L., Larranaga, N., Sanchez, M.J., Tormo, M.J., Buckland, G., Lund, E., Hedblad, B., Melander, O., Jansson, J.H., Wennberg, P., Wareham, N.J., Slimani, N., Romieu, I., Jenab, M., Danesh, J., Gallo, V., Norat, T., Riboli, E., 2011. Fruit and vegetable intake and mortality from ischaemic heart disease: results from the European prospective investigation into cancer and nutrition (EPIC)-heart study. European Heart Journal 32, 1235–1243.

Davis, M., Su, S., Ming, F., Yang, M., Ichinomiya, A., Melachouris, N., 2002. Isolation of Glycoproteins from Bovine Milk. World Intellectual Property Organization. WO/2002/080961.

Davis, M.E., Nelson, L.A., Keenan, J.M., Pins, J.J., 2003. Reducing Cholesterol with Hydrolyzed Whey Protein. World Intellectual Property Organization. WO/2003/063778.

Del Rio, D., Rodriguez-Mateos, A., Spencer, J.P.E., Tognolini, M., Borges, G., Crozier, A., 2013. Dietary (Poly) phenolics in human health: structures, bioavailability, and evidence of protective effects against chronic diseases. Antioxidants & Redox Signaling 18 (14), 1818–1892.

Deng, Q., Zinoviadou, K.G., Galanakis, C.M., Orlien, V., Grimi, N., Vorobiev, E., Lebovka, N., Barba, F.J., 2015. The effects of conventional and non-conventional processing on glucosinolates and its derived forms, isothiocyanates: extraction, degradation and applications. Food Engineering Reviews 7 (3), 357–381.

EPA, 2011. Generators of Food Waste. http://www.epa.gov/osw/conserve/materials/organics/food/fd-gener.htm#food-hier (accessed 18.02.12.).

European Commission, 2010. Preparatory Study on Food Waste Across EU 27-Technical Report. .

Fabiani, R., De Bartolomeo, A., Rosignoli, P., Servili, M., Montedoro, G.F., Morozzi, G., 2002. Cancer chemoprevention by hydroxytyrosol isolated from virgin olive oil through G1 cell cycle arrest and apoptosis. European Journal of Cancer Prevention 11 (4), 351–358.

FAO, 2012. Sustainable Diets and Biodiversity. Directions and Solutions for Policy, Research and Action. Rome http://www.fao.org/docrep/016/i3004e/i3004e.pdf.

FAO, 2014. Definitional Framework of Food Loss. http://www.fao.org/fileadmin/user_upload/save-food/PDF/FLW_Definition_and_Scope_2014.pdf.

Food and Drug Administration (FDA), 2014. http://www.accessdata.fda.gov/scripts/fdcc/?set=GRASNotices&id=459.

Fernández-Ginés, J.M., Fernández-López, J., Sayas-Barbera, E., Pérez-Álvarez, J.A., 2003. Effects of storage conditions on quality characteristics of bologna sausages made with citrus fiber. Journal of Food Science 68 (2), 710–715.

Figuerola, F., Hurtado, M.L., Estevez, A.M., Chiffelle, I., Asenjo, F., 2005. Fibre concentrates from apple pomace and citrus peel as potential fibre sources for food enrichment. Food Chemistry 91, 395–401.

Foresight, 2011. Foresight Project on Global Food and Farming Futures Synthesis Report C7: Reducing Waste. The Government Office for Science, London.

Galanakis, C.M., Schieber, A., 2014. Editorial. Special issue on recovery and utilization of valuable compounds from food processing by-products. Food Research International 65, 299–484.

Galanakis, C.M., Tornberg, E., Gekas, V., 2010a. A study of the recovery of the dietary fibres from olive mill wastewater and the gelling ability of the soluble fibre fraction. LWT-Food Science & Technology 43, 1009–1017.

Galanakis, C.M., Tornberg, E., Gekas, V., 2010b. Clarification of high-added value products from olive mill wastewater. Journal of Food Engineering 99, 190–197.

Galanakis, C.M., Tornberg, E., Gekas, V., 2010c. Dietary fiber suspensions from olive mill wastewater as potential fat replacements in meatballs. LWT-Food Science & Technology 43, 1018–1025.

Galanakis, C.M., Tornberg, E., Gekas, V., 2010d. Recovery and preservation of phenols from olive waste in ethanolic extracts. Journal of Chemical Technology & Biotechnology 85, 1148–1155.

Galanakis, C.M., Tornberg, E., Gekas, V., 2010e. The effect of heat processing on the functional properties of pectin contained in olive mill wastewater. LWT-Food Science & Technology 43, 1001–1008.

Galanakis, C.M., Goulas, V., Tsakona, S., Manganaris, G.A., Gekas, V., 2013a. A knowledge base for the recovery of natural phenols with different solvents. International Journal of Food Properties 16, 382–396.

Galanakis, C.M., Markouli, E., Gekas, V., 2013b. Fractionation and recovery of different phenolic classes from winery sludge via membrane filtration. Separation and Purification Technology 107, 245–251.

Galanakis, C.M., Chasiotis, S., Botsaris, G., Gekas, V., 2014. Separation and recovery of proteins and sugars from Halloumi cheese whey. Food Research International 65, 477–483.

Galanakis, C.M., Kotanidis, A., Dianellou, M., Gekas, V., 2015a. Phenolic content and antioxidant capacity of Cypriot wines. Czech Journal of Food Sciences 33 (2).

Galanakis, C.M., Martínez-Saez, N., del Castillo, M.D., Barba, F.J., Mitropoulou, V.S., 2015b. Patented and commercialized applications. In: Galanakis, C.M. (Ed.), Food Waste Recovery: Processing Technologies and Industrial Techniques. Elsevier-Academic Press (Chapter 15).

Galanakis, C.M., 2011. Olive fruit and dietary fibers: components, recovery and applications. Trends in Food Science and Technology 22, 175–184.

Galanakis, C.M., 2012. Recovery of high added-value components from food wastes: conventional, emerging technologies and commercialized applications. Trends in Food Science & Technology 26, 68–87.

Galanakis, C.M., 2013. Emerging technologies for the production of nutraceuticals from agricultural by-products: a viewpoint of opportunities and challenges. Food and Bioproducts Processing 91, 575–579.

Galanakis, C.M., 2015a. Separation of functional macromolecules and micromolecules: from ultrafiltration to the border of nanofiltration. Trends in Food Science & Technology 42 (1), 44–63.

Galanakis, C.M., 2015b. The universal recovery. In: Galanakis, C.M. (Ed.), Food Waste Recovery: Processing Technologies and Industrial Techniques (Chapter 3).

González-Paramas, A., Esteban-Ruano, S., Santos-Buelga, C., Pascual-Teresa, S., Rivas-Gonzalo, J., 2004. Flavanol content and antioxidant activity in winery byproducts. Journal of Agricultural and Food Chemistry 52, 234–238.

Hamdena, K., Alloucheb, N., Damakb, M., Elfekia, A., 2009. Hypoglycemic and antioxidant effects of phenolic extracts and purified hydroxytyrosol from olive mill waste in vitro and in rats. Chemico-Biological Interactions 180 (3), 421–432.

Hanley Clark, J., Pfaltzgraff, L.A., Lvovich Budarin, V., De Bruyn, M., 2015. Microwave Assisted Citrus Waste Biorefinery. US 2015/0065698 A1.

Heng, W.W., Xiong, L.W., Ramanan, R.N., Hong, T.L., Kong, K.W., Galanakis, C.M., Prasad, K.N., 2015. Two level factorial design for the optimization of phenolics and flavonoids recovery from palm kernel by-product. Industrial Crops & Products 63, 238–248.

HLPE, 2014. Food Losses and Waste in the Context of Sustainable Food Systems. A report by the High Level Panel of Experts on Food Security and Nutrition of the Committee on World Food Security.

Huston, L., Sakkab, N., 2006. Connect and develop –inside P&G's new model for innovation (cover story) Harvard Business Review 84 (3), 58–66.

Jensen, J., Larsen, P.H., 1993. Method for Obtaining High-quality Protein Products from Whey. World Intellectual Property Organization. WO/1993/021781.

Johnson, E.J., 2002. The role of carotenoids in human health. Nutrition in Clinical Care 5 (2), 47–49.

Kammerer, D., Claus, A., Carle, R., Schieber, A., 2004. Polyphenol screening of pomace from red and white grape varieties (*Vitis vinifera* L.) by HPLC-DAD-MS/MS. Journal of Agricultural and Food Chemistry 52, 4360–4367.

Kaul, A., Khanduja, K.L., 1998. Polyphenols inhibit promotional phase of tumorigenesis: relevance of superoxide radicals. Nutrition and Cancer 32, 81–85.

Keech, A., Jimenez-Flores, R., 2005. Novel Colostral Fractionation Process. US2005/0092684A1.

Lavecchia, R., Zuorro, A., 2008. Process for the Extraction of Lycopene. World Intellectual Property Organization. WO/2008/055894. Production, Composition, and Application of Coffee and Its Industrial Residues.

Leclere, L., Cutsem, P.V., Michiels, C., 2013. Anti-cancer activities of pH- or heat-modified pectin. Frontiers in Pharmacology 4 (128), 1–8.

Lee, J.H., 2002. Precipitation Recovery Process for Food Waste Sludge. US Patent 6368657 B1.

Letenneur, L., Proust-Lima, C., Le, G.A., Dartigues, J.F., Barberger-Gateau, P., 2007. Flavonoid intake and cognitive decline over a 10-year period. American Journal Epidemiology 165, 1364–1371.

Lin, C.S.K., Pfaltzgraff, L.A., Herrero-Davila, L., Mubofu, E.B., Abderrahim, S., Clark, J.H., Koutinas, A.A., Kopsahelis, N., Stamatelatou, K., Dickson, F., Thankappan, S., Mohamed, Z., Brocklesby, R., Luque, R., 2013. Food waste as a valuable resource for the production of chemicals, materials and fuels. Current situation and global perspective. Energy Environmental Science 6, 426–464.

Liu, J., Schalch, W., Wang-Schmidt, Y., Wertz, K., 2008. Novel Use of Hydroxytyrosol and Olive Extracts/Concentrates Containing it. WO Patent App. WO/2008/128552.

Liu, R.H., 2005. Apple Peel Powder, Methods of Making, and Uses Thereof. US2005/0147723A1.

Lowe, P., Phillipson, J., Lee, R.P., 2008. Socio-technical innovation for sustainable food chains: roles for social science. Trends in Food Science & Technology 19, 226–233.

Manthey, J.A., Grohmann, K., 1996. Concentrations of hesperidin and other orange peel flavonoids in citrus processing byproducts. Journal of Agricultural and Food Chemistry 44, 811–814.

Manthey, J.A., Guthrie, N., Grohmann, K., 2001. Biological properties of citrus flavonoids pertaining to cancer and inflammation. Current Medicinal Chemistry 8, 135–153.

Martínez-Saez, N., Ullate, M., Martin-Cabrejas, M.A., Martorell, P., Genovés, S., Ramon, D., 2014. A novel antioxidant beverage for body weight control based on coffee silverskin. Food Chemistry 150, 227–234.

Mateo-Hernández, F., Muñoz-Ortega, M., González-Santoyo, F., 2012. Method for Obtaining Maslinic Acid and Oleanolic Acid. WO Patent App. WO/2012/017108 A1.

McGuire, J.W., 1963. Business and Society. McGraw-Hill, New York, USA.

Mee, T.Y., Mohd, Z.B.M.Z., Appleton, D.R., Ahmad, J.A., Harikrishna, A.L.K., 2014. Extracting Lecithin from Palm Agro-waste. WO2014/042509 A1.

Montonen, J., Knekt, P., Jarvinen, R., Aromaa, A., Reunanen, A., 2003. Whole-grain and fiber intake and the incidence of type 2 diabetes. American Journal of Clinical Nutrition 77, 622–629.

Murugan, K., Ramasamy, K., 2013. Environmental concerns and sustainable development. In: Chandrasekaran, M. (Ed.), Valorization of Food Processing By-Products. Taylor & Francis Group, Florida, pp. 739–756.

Nadachi, Y., Mawatari, S., 2011. Method for Production of Highly Pure Phospholipid, and Highly Pure Sphingomyelin and Plasmalogen-type Glycerophospholipid Produced by the Method. US2011/0160471 A1.

Nafisi-Movaghar, K., Druz, L.L., Pepper Victoria, C., 2013. Process for Conversion of Citrus Peels into Fiber, Juice, Naringin, and Oil. US2013/0064947.

Nell, P., 2001. Method for the Extraction of Vegetable Juices from Vegetable Residue and/or from Vegetable Remnants Residue. US Patent 6296888.

Oreopoulou, V., Tzia, C., 2007. Utilization of plant by-products for the recovery of proteins, dietary fibers, antioxidants, and colorants. In: Oreopoulou, V., Russ, W. (Eds.), Utilization of By-Products and Treatment of Waste in the Food Industry. Springer Science+Business Media, New York, pp. 209–232.

Otles, S., Despoudi, S., Bucatariu, C., Kartal, C., 2015. Food waste management, valorisation & sustainability in the food industry. In: Galanakis, C.M. (Ed.), Food Waste Recovery: Processing Technologies and Industrial Techniques. Elsevier-Academic Press (Chapter 1).

Panduwawala, L., Venkatesh, S., Parraquez, P., Zhang, X., 2009. Connect and develop: P&G's big stake in open innovation. Retrieved from: http://www.openinnovate.co.uk/pgs-connect-anddevelop/.

Pastene, E., Speisky, H.N., García, A., Moreno, J., Troncoso, M., Figueroa, G., 2010. In vitro and in vivo effects of apple peel polyphenols against *Helicobacter pylori*. Journal of Agricultural and Food Chemistry 58, 7172–7179.

Patsioura, A., Galanakis, C.M., Gekas, V., 2011. Ultrafiltration optimization for the recovery of β-glucan from oat mill waste. Journal of Membrane Science 373, 53–63.

Ponroy, Y., 1996. Therapeutic and Dietetic Uses of a Brain Phospholipid-Based Complex. WO1996/000077A1.

Pourfarzad, A., Mahdavian-Mehr, H., Sedaghat, N., 2013. Coffee silverskin as a source of dietary fiber in bread-making: optimization of chemical treatment using response surface methodology. LWT - Food Science and Technology, 50, 599–606.

Product Stewardship Institute, 2011. http://productstewardship.us/displaycommon.cfm?an=1&subarticlenbr=55 (accessed 17.04.11.).

Rahmanian, N., Jafari, S.M., Galanakis, C.M., 2014. Recovery and removal of phenolic compounds from olive mill wastewater. Journal of the American Oil Chemists' Society 91, 1–18.

Rao, A.V., Rao, L.G., 2007. Carotenoids and human health. Pharmacological Research 55, 207–216.

Reyes-Zurita, F.J., Rufino-Palomares, E.E., Lupiáñez, J.A., Cascante, M., 2009. Maslinic acid, a natural triterpene from *Olea europaea* L., induces apoptosis in HT29 human colon-cancer cells via the mitochondrial apoptotic pathway. Cancer Letters 273 (1), 44–54.

Riboli, E., Norat, T., 2003. Epidemiologic evidence of the protective effect of fruit and vegetables on cancer risk. American Journal of Clinical Nutrition 78, 559S–569S.

Robertson, B., Curstedt, T., 1991. Natural Pulmonary Surfactant, Method of Preparation and Pharmaceutical Compositions. US5024995A.

Rodríguez-Rodríguez, R., Justo, M.L., Claro, C.M., Vila, E., Parrado, J., Herrera, M.D., Sotomayor, M.A., 2012. Endothelium-dependent vasodilator and antioxidant properties of a novel enzymatic extract of grape pomace from wine industrial waste. Food Chemistry 135, 1044–1051.

Roselló-Soto, E., Barba, F.J., Parniakov, O., Galanakis, C.M., Grimi, N., Lebovka, N., Vorobiev, E., 2015. High voltage electrical discharges, pulsed electric field and ultrasounds assisted extraction of protein and phenolic compounds from olive kernel. Food & Bioprocess Technology 8 (4), 885–894.

Rubio-Senent, F., Rodríguez-Gutiérrez, G., Lama-Munoz, A., García, A., Fernández-Bolanos, J., 2015. Novel pectin present in new olive mill wastewater with similar emulsifying and better biological properties than citrus pectin. Food Hydrocolloids 50, 237–246.

Rupasinghe, H.P.W., Wang, Y., Thilakarathna, S.K.P.H., 2013. Apple Skin Extracts for Treating Cardiovascular Disease. US2013/0165396A1.

Russ, W., Meyer-Pittroff, R., 2004. Utilizing waste products from the food production and processing industries. Critical Reviews in Food Science and Nutrition 44, 57–62.

Saguy, I.S., Sirotinskaya, V., 2014. Challenges in exploiting open innovation's full potential in the food industry with a focus on small and medium enterprises (SMEs). Trends in Food Science & Technology 38, 136–148.

Saguy, I.S., 2013. Academia-industry innovation interaction: paradigm shifts and avenues for the future. In: Yanniotis, S., Taoukis, P., Stoforos, N.G., Karathanos, V.T. (Eds.), Advances in Food Process Engineering Research and Applications. Springer, New York, pp. 645–656 (Chapter 32).

Santos-Buelga, C., Scalbert, A., 2000. Proanthocyanidins and tannin-like compounds – nature, occurrence, dietary intake and effects on nutrition and health. Journal Scientific Food and Agriculture 80, 1094–1117.

Sarkar, S., Costa, A.I.A., 2008. Dynamics of open innovation in the food industry. Trends in Food Science & Technology 19 (11), 574–580.

Saura-Calixto, F., 2011. Dietary fiber as a carrier of dietary antioxidants: an essential physiological function. Journal of Agricultural and Food Chemistry 59, 43–49.

Schubert, S.Y., Lansky, E.P., Neeman, I., 1999. Antioxidant and eicosanoid enzyme inhibition properties of pomegranate seed oil and fermented juice flavonoids. Journal of Ethnopharmacology 66, 11–17.

Shenghui, Z., 1995. Preparation of Chitosan Derivative Fruit and Vegetable Antistaling Agent. State Intellectual Property Office of the People's Republic of China. CN 1106999.

Shrikhande, A., 2000. Wine byproducts with health benefits. Food Research International 33, 469–474.

Spencer, J.P.E., Crozier, A., 2012. In: Packer, L., Cadenas, H. (Eds.), Flavonoids and Related Compounds: Bioavailability and Function. Oxidative Stress and Disease, vol. 30. CRC Press, Boca Raton, FL.

Spencer, J.P., 2008. Flavonoids: modulators of brain function? British Journal of Nutrition 99 (E1), ES60–ES77.

Spencer, J.P., 2010a. Beyond antioxidants: the cellular and molecular interactions of flavonoids and how these underpin their actions on the brain. Proceedings of the Nutrition Society 69, 244–260.

Spencer, J.P., 2010b. The impact of fruit flavonoids on memory and cognition. British Journal of Nutrition 104 (3), S40–S47.

Steffen, L.M., Jacobs, D.R., Stevens, J., Shahar, E., Carithers, T., Folsom, A.R., 2003. Associations of whole-grain, refined grain, and fruit and vegetable consumption with risks of all-cause mortality and incident coronary artery disease and ischemic stroke: the atherosclerosis risk in communities (ARIC) study. American Journal of Clinical Nutrition 78, 383–390.

Thota, H., 2012. Kraft and Unilever Pathways to Innovate Leadership. Innova, pp. 26–30.

Tornberg, E., Galanakis, C.M., 2008. Olive Waste Recovery. World Intellectual Property Organization. WO/2008/082343.

Tsakona, S., Galanakis, C.M., Gekas, V., 2012. Hydro-ethanolic mixtures for the recovery of phenols from Mediterranean plant materials. Food & Bioprocess Technology 5, 1384–1393.

USDA, 2014. http://www.fsis.usda.gov/wps/portal/fsis/topics/regulatory-compliance/new-technologies/new-technology-information-table.

Whelton, S.P., Hyre, A.D., Pedersen, B., Yi, Y., Whelton, P.K., He, J., 2005. Effect of dietary fiber intake on blood pressure: a metaanalysis of randomized, controlled clinical trials. Journal of Hypertension 23, 475–481.

Wicker, L., Kim, Y., Kim, M.J., Thirkield, B., Lin, Z., Jung, J., 2014. Pectin as a bioactive polysaccharide – extracting tailored function from less. Food Hydrocolloids 42 (2), 251–259.

Wolfe, K.E., Liu, R.H., 2003. Apples peels as value-added food ingredient. Journal of Agricultural and Food Chemistry 51, 1676–1683.

Xia, E.Q., Deng, G.F., Guo, Y.J., Li, H.B., 2010. Biological activities of polyphenols from grapes. International Journal of Molecular Sciences 11, 622–646.

Yu, J., 2006. Production of Biodegradable Thermoplastic Materials from Organic Wastes. US2006/0088921A1.

CHAPTER

ADOPTION OF ICT INNOVATIONS IN THE AGRI-FOOD SECTOR: AN ANALYSIS OF FRENCH AND SPANISH INDUSTRIES

12

V. Martinez-Gomez, J. Domenech, F. Mas-Verdú

Universitat Politècnica de València, Valencia, Spain

12.1 INTRODUCTION

The development of information and communications technologies (ICTs) over the last two decades has considerably reduced the costs of processing and transferring information. Seemingly this has diminished the relevance of the firm's physical location and the importance of the distance from other entities with which relationships may form. According to the theory of rational expectations, firms adopt innovations such as ICTs if they expect positive returns from these ICTs (Au and Kauffman, 2003). Given that the distance to highly populated areas is one of the disadvantages of being in a rural location, firms located in rural areas will benefit more from adopting ICTs. Thus, ICTs may act as a driver of productive activities in rural areas. From the perspective of territorial innovation models, however, the adoption of innovations such as ICTs may occur more quickly in urban areas because of the lack of infrastructure and qualified workers in rural areas (Moulaert and Sekia, 2003).

The agri-food industry is a common industrial activity in rural areas because of the proximity to raw materials, among other reasons. In Spain the agri-food industry plays a prominent role in the economy. The agri-food industry is a very relevant sector in the countries studied. It accounts for 17% of Spain's industrial gross domestic product (GDP), more than 20% of industrial employment, and 2.5% of overall employment (FIAB, 2013). In France, the figures are of a similar scale: the agri-food industry accounts for around 15% of turnover from French industry and more than 13% of manufacturing jobs (IGR/CGAAER, 2012). These figures are in line with those presented at the European Union (EU) level, where the food and drink industry is the largest manufacturing sector (12.8% of value added in manufacturing), the leading employer in the EU with 15% of employment in manufacturing, and contributes to 1.8% of the EU's gross value added (data from Food Drink Europe, 2015).

The predominant view in the literature is that innovation intensity is lower in the agri-food sector than in other sectors (Connor and Schiek, 1997; Garcia-Martinez and Briz, 2000; Lopez et al., 2003; Capitanio et al., 2009). Results from business innovation surveys conducted by the French and Spanish national statistical services seem to confirm this trend. In the French agri-food industry, R&D spending as a proportion of income is approximately one-third of that of other French firms, according to the DIRDE survey (Zarka and Laroche, 2015).

Innovation Strategies in the Food Industry. http://dx.doi.org/10.1016/B978-0-12-803751-5.00012-X

Hence, this research addresses a sector that is highly relevant in socioeconomic and territorial terms yet is less innovation intensive than other industrial sectors are. Notably, however, the food sector is more technology intensive than the primary agriculture sector is estimated to be and is indeed more technology intensive than some other industrial sectors (Fearne et al., 2013).

One of the reasons used to explain such behavior within these sectors is the small size of the firms, which sometimes outsource their R&D activities to research centers (Roeland and den Hertog, 1999). Authors have also claimed that the acquisition of inputs from other sectors usually represents the greatest source of incorporation and use of innovations in the agri-food system (Hauknes and Knell, 2009). Garcia-Martinez and Burns (1999) made similar claims regarding the Spanish agri-food industry, citing the incorporation of innovation through the acquisition of machinery and equipment.

Nevertheless, the availability of new technologies and the high degree of internationalization of agri-food businesses shows that this sector is growing in terms of technological intensity and is increasing its R&D spending (Filippaios et al., 2009; Alarcón and Sánchez, 2013). The latter authors also reported that agri-food firms perform internal research, while Garcia et al. (2012) and López-García et al. (2013) provided empirical evidence that innovation intensity in the agri-food sector is higher than it is in other sectors, which a priori are considered more innovative.

The European agri-food industry has recently taken steps to adopt ICTs, which has allowed the industry to improve its innovation capacity (Alarcón et al., 2013). The rate of firms with domain names is similar to that of other industrial sectors, although differences by size remain. More than 90% of large firms have domain names compared to 60–80% of small and medium enterprises (SMEs). In recent years, however, SMEs have endeavored to acquire domain names. Around 25% of SMEs purchase from suppliers online, whereas more than 40% of large firms do the same. The use of the Internet to sell products is less common, although the gap between SMEs and large firms is smaller. For example, in 2008, 11.5% of food SMEs sold their products online to consumers compared to 15.9% of large food firms.

Against this backdrop, this chapter concerns a study that uses data on Spanish and French agri-food firms to analyze the relationship between adoption of ICTs and business location. Following Fink (1998), adoption of ICTs can contribute to efficiency gains in the production process, to increased management effectiveness, and to improved business performance. The evolution of the technologies available has not modified substantially the previous gains due to the adoption of ICTs; López et al. (2014) state that there are two critical processes for which the adoption of ICTs is crucial to agri-food firms, providing them with competitive advantages: the production processes and the exchange of information among companies. However, some barriers hinder ICTs' adoption in the sector. They are a certain lack of skilled personnel to fully benefit from those advantages, the weak organization level provoking a low degree of planning, and the lack of proactivity in the adoption. Thereafter, several elements from the literature on the adoption of ICTs are explored, first describing the sample, the variables, and the model used in the research. Finally, the key results of the empirical analysis are reported and discussed.

12.2 THEORETICAL FRAMEWORK

The analysis of the literature reveals two main categories of variables that affect the adoption of innovations—in this case, the adoption of ICTs by firms. The first category comprises characteristics of the firm, whereas the second consists of characteristics of the business environment. We now examine these two categories.

12.2.1 CHARACTERISTICS OF THE FIRM

According to the sources used in this research, one firm characteristic that influences the adoption of innovations is size, although the evidence in the literature is ambiguous. Larger firms tend to have access to greater financial and R&D resources. Moreover, these firms have better internal organization (Mowery et al., 1996). Thus, larger firms are more likely to innovate. Nevertheless, authors such as Liao et al. (2003) point out that smaller firms are more agile and flexible and are less bureaucratic than larger firms. These traits let smaller firms create, introduce, and adopt innovations more easily than larger firms can. This ambiguity in the literature may owe to techniques used to measure the variables. We have already mentioned that small size in the agri-food industry is considered something of an obstacle to performing in-house research (Roeland and den Hertog, 1999). In addition, the most recent literature shows the existence of a positive relationship between the size of the firm and innovation (Rogers, 2004; Idris and Tey, 2011).

Particularly with reference to the adoption of ICTs, empirical evidence shows that smaller firms use fewer ICTs than larger ones. As an example, in Spain the penetration of the Internet is over 97% in small, medium, and large firms, and about 66% for microfirms (ONTSI, 2012); while in general, the literature reports evidence about a lagged adoption by smaller firms in several countries and contexts (see, for example, Bhagwat and Sharma (2007) or Harindranath et al. (2008)).

Therefore, firm size seems to be a key factor according to the evidence. Particular features of smaller firms hindering access have been proposed and sought by a number of researchers. Ritchie and Brindley (2005) analyze the different barriers and diffusion agents to adoption of ICTs by SMEs, including strategic, technological, organizational, and behavioral. More specifically, Sin Tan et al. (2010) list seven categories of possible barriers. They are the unsuitability of ICTs' adoption for businesses, the unavailability of ICT personnel and network infrastructure, the high ICTs' cost, the expensive software, the unbalance between the investment required and its costs, the uncertainty related to ICT regulations, and the lack of confidence in ICT security. In their empirical validation on Malaysian SMEs, only security and unsuitability for businesses appear negatively related to ICTs adoption. To our knowledge, the most comprehensive list of factors affecting the use and adoption of ICTs by SMEs is provided by Wymer and Regan (2005), who identify 26 factors. They classify them as technology-related or business-related, with expertise, financial, and external elements within the latter factors. Their empirical results for US companies highlight the consistent role of the financial/cost as the key factor. Harindranath et al. (2008) also identify cost as the most relevant factor to the investment in ICTs in their UK sample.

To conclude then with the reasons for the lag of ICT adoption by SMEs, literature has identified a huge set of barriers, whose relevance differs substantially across countries, sectors, and organizational types. In other words, as Panayiotou and Katimertzoglou (2015, p. 509) state, "...Internet adoption of the smaller businesses is not clearly understood."

Several studies highlight the positive correlation between innovation and expansion to foreign markets (Castellani and Zanfei, 2007; Gorodnichenko et al., 2010). These two phenomena frequently form a virtuous circle (Kafouros et al., 2008) whereby exporting and innovation interact, leading to complementary effects (Hughes, 1986; Kleinknecht and Oostendorp, 2002).

Because of the high levels of efficiency required to operate in international markets, exporting firms are usually more innovation intensive than those that do not export (Bernard et al., 2005). Greater competitive pressure in the international markets usually makes innovation a key element for exporting (Filipescu et al., 2009). Thus, firms that export acquire knowledge and experience from their international activities, which increases their absorptive capacity for innovation (Cohen and Levinthal, 1990).

In the agri-food industry, recent evidence reveals the connections between innovation and internationalization. The agri-food firms with the greatest in-house and external R&D expenditures are the firms that most consistently sell their products abroad (Alarcón and Sánchez, 2014).

Many conceptual frameworks link internationalization strategies with firms' innovation processes. For instance, from an innovation-related perspective (Andersen, 1993), internationalization decisions go hand in hand with other processes such as the development of technological capabilities of the firm. The network approach (Johanson and Mattsson, 1988), which complements the innovation-related perspective, shows that a large amount of a firm's potential for internationalization depends on the firm's relationships with clients, suppliers, competitors, and institutions. Therefore, the adoption of ICTs as a means of innovation may be both a requirement and a way of strengthening these relationships.

The training and qualification level of the workers is also among the determinants of the adoption and use of innovations. Several studies (Minvaeva et al., 2003; Vinding, 2006) report the existence of a positive relationship between the qualifications of the firm's human capital and the firm's absorptive capacity. Giunta and Trivieri (2007), Hollenstein (2004), and Martins and Oliveira (2008) have shown the same relationship for the special case of the adoption of ICTs.

The literature suggests that in rural environments workers' qualifications tend to be less conducive to the capacity to adopt and use advanced technologies (OECD, 2009). This finding implies the existence of interesting relationships among the firm's location, technology adoption, and workers' qualifications.

12.2.2 THE BUSINESS ENVIRONMENT

The environment of the firm can strengthen or undermine innovation processes (Audretsch, 2002). A firm's innovation potential is conditioned not only by internal resources (technology, human, etc.) but also by the firm's capacity to combine its resources and capabilities with those available in the territorial environment.

The territory is the framework within which the firm undertakes its economic activity. The characteristics of this territory determine many factors that condition the strategies and performance of the firm, including the availability and accessibility of resources in the firm's local environment. Other aspects are the infrastructures to transport goods and people, general infrastructures (health, education, and leisure), which can help to retain local inhabitants or attract new residents, as well as proximity and ease of access to knowledge and training opportunities, which can affect the qualifications of the inhabitants. The role of institutions is another determining factor. Specific support policies or the existence of a regional or national innovation system affect the degree of development of territories. As Doloreux and Parto (2004) point out, regions capable of developing are those with the necessary institutions, the right cooperative structures, and an adequate level of knowledge and skills.

Numerous studies (Funke and Niebuhr, 2005; Alecke et al., 2006) have shown the positive role of location and geographic proximity in boosting firms' innovation potential. Location encourages interaction among different agents in the innovation process. Thus, the location of firms is becoming a key part of the innovation process. Location is especially relevant in the agri-food industry because agri-food activity is closely linked to the territory. Fort et al. (2005) point out that in the agri-food industry the effect of territorial links may be less applicable, especially if product innovation is the main type of innovation.

In this vein, some recent empirical studies seem to conclude that location does not play a crucial role in the innovation activities carried out by agri-food firms, confirming hence the weak effect of the territorial links mentioned previously. In this line, a recent empirical study of agri-food firms in a Spanish region by Fearne et al. (2013) found that firms located in rural areas are as innovative as those located in urban areas. Hence, the urban/rural location of the firm seems irrelevant in terms of innovation. Likewise, García Álvarez-Coque et al. (2013) analyzed innovative behavior in the agri-food industry, finding that "rurality does not seem to be per se a handicap for innovative firms" (García Álvarez-Coque et al., 2013, p. 9).

In contrast, organizational innovation, such as the type of innovation analyzed in this chapter, might actually have strong ties with the territory. The first studies (Premkumar and Roberts, 1999) that specifically analyzed the relationship between location and adoption of ICTs showed that firms located in rural environments could become trapped in a vicious circle—the lack of telecommunications infrastructures reduces the demand for this type of service, which in turn limits future investment in such technologies. Over time, however, telecommunications infrastructures in rural areas have improved in developed countries, and this argument has become less relevant. In fact, Forman et al. (2005) suggested that adopting ICTs is a way of overcoming the relative isolation of firms in rural areas, making distance less of an issue.

Forman et al. (2005) presented evidence of greater Internet adoption by US firms located in rural areas than by US firms located in urban areas. More recently, Domenech et al. (2014) found that agri-food businesses in rural areas are more likely to adopt ICT innovations. Galliano and Roux (2008), however, reported no such evidence in their study on French industry. Conversely, in fact, their results imply that location does not affect the adoption of the Internet by French firms, and, moreover, the determinants of adoption are relatively similar in urban and rural firms.

The literature also reports a relationship between the degree of competition in a sector and the amount of innovation, although the empirical evidence is contradictory. Thus, Hashmi (2011) denoted a positive relationship between competition and innovation in the United States, whereas, also in the United States, Prasad (2008) found different results by industry. In Sweden, Tingvall and Poldahl (2006) reported results that vary depending on how the competition is measured. In Finland, Kilponen and Santavirta (2007) found evidence of an inverted U curve between innovation and competition within the sector. Likewise, Aghion et al. (2005) also reported a nonlinear, inverted U curve that reflects two possible effects. The first effect is a Schumpeterian effect: competition negatively affects innovation because greater competition means that the future income from incorporating innovations will be lower. The second effect is an "avoid the competition" effect (Arrow, 1962), which implies a positive relationship between innovation and competition. According to this effect, greater competition means greater incentives to innovate if doing so reduces competitive pressure.

12.3 METHOD

12.3.1 SAMPLE, VARIABLES, AND MODEL

Data from 1327 agri-food firms (NACE Rev. 2, codes 10 and 11) were gathered, of which 587 were French and 740 were Spanish. We randomly selected the sample and took the data from two similar databases belonging to Bureau Van Dijk. For the Spanish firms, we used Sistema de Análisis de

Balances Ibéricos (SABI). For the French firms, we used ORBIS. The variables captured the following data for each firm for 2011 (variable names in parentheses):

- Size of the firm (SIZE): continuous variable measured using the logarithm of the number of employees in the firm.
- Mean salary (WAGE): continuous variable measured as the mean salary per employee. This variable acted as a proxy for the average level of qualifications among employees.
- The firm's commitment to exports (EXPORT): dichotomous variable that took the value 1 if the firm was engaged in export activity and 0 otherwise.

This group of variables was supplemented by data from other sources regarding the location of the firm, the firm's adoption of ICTs, and the competition within the firm's subsector (variable names in parentheses):

- Urban or rural location (RURAL): dummy variable that took the value 1 if the firm's headquarters was in a rural community and 0 if the firm's headquarters was in an urban community. To define urban and rural communities, we used the definition provided by the OECD (1994). Hence, a community was defined as rural if the population density was less than 150 inhabitants per square kilometer.
- Adoption of the Internet (WEB): dummy variable that took the value 1 if the firm had a working corporate Website and 0 if the firm lacked such a Website. The variable also took the value 0 if we were unable to find a Website for the firm because the Website was in construction or was not working or because the firm did not have its own domain name. This would occur if the Website appeared within another Website (eg, on the Website of an association of firms or the group the firm belonged to). We constructed the variable by searching on Google for the official name of the firm and its tax identification number. Spanish law requires firms to include this information on their Websites.
- Market concentration (CONC): the level of competition within a sector can be measured by the market concentration within that sector. The lower the market concentration within a sector, the closer that sector is to having perfect competition. We used the CR_{10} concentration ratio for each subsector of the agri-food industry. The firms in the sample were grouped into three concentration levels for each country.

Using these variables, we estimated the following probit model:

Adoption of ICTs = *f*(characteristics of the firm, characteristics of the business environment)

López et al. (2014) indicate that Websites are becoming a display window for agri-food firms and quite often constitutes the first stage in the adoption of ICTs. Consequently, the adoption of ICTs was measured using the variable WEB. The characteristics of the firm were size of the firm (SIZE), mean salary (WAGE), and export activity (EXPORT). The characteristics of the business environment were location (RURAL) and competition within the subsector (CONC).

12.3.2 DESCRIPTIVE ANALYSIS

Table 12.1 shows the main descriptive statistics for the sample. The most prevalent firms were nonexporting firms located in urban environments. Only 34% of the firms had working corporate Websites.

The adoption of the Internet varied across different environments. Figs. 12.1–12.3 show the adoption of Internet technology over time depending on the location of the firm for the two countries and for the whole sample, respectively. In both countries, adoption of the Internet was substantially

Table 12.1 Descriptive Statistics

	Media	S.D.	(1)	(2)	(3)	(4)	(5)
1. SIZE	2.124	1.261					
2. WAGE	34.963	40.636	−0.171				
3. EXPORT	0.200	0.400	0.437	0.065			
4. RURAL	0.313	0.464	0.000	0.013	0.080		
5. CONC	1.180	0.432	0.051	0.068	0.144	0.038	
6. WEB	0.341	0.474	0.392	−0.023	0.366	0.082	0.138

Compiled by the authors.

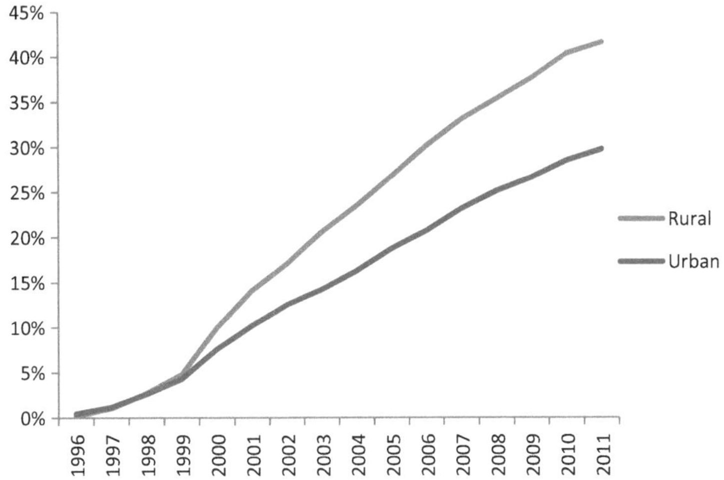

FIGURE 12.1

Annual adoption of a Website in the agri-food industry in urban and rural environments in Spain (Accumulated %).

Compiled by the authors.

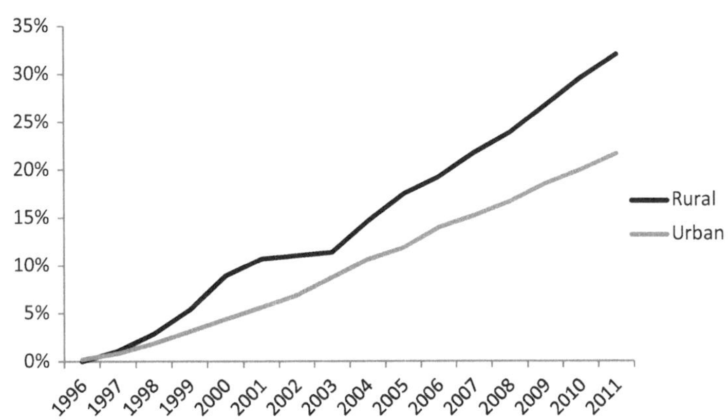

FIGURE 12.2

Annual adoption of a Website in the agri-food industry in urban and rural environments in France (Accumulated %).

Compiled by the authors.

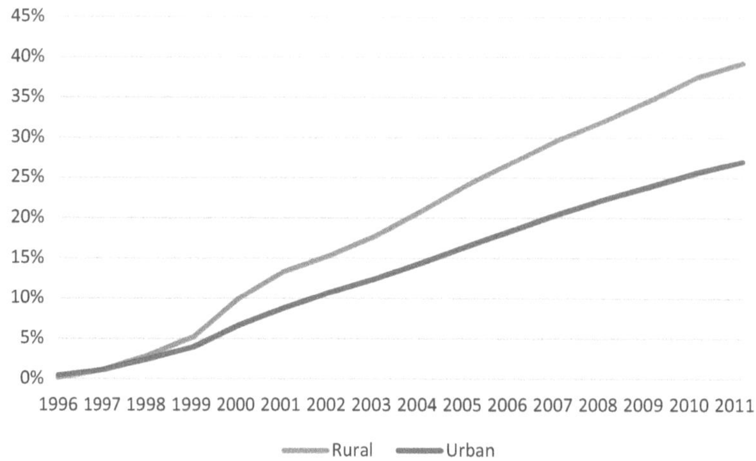

FIGURE 12.3

Annual adoption of a Website in the agri-food industry in urban and rural environments. Total sample (Accumulated %).

Compiled by the authors.

greater and occurred more quickly in firms located in rural areas than in those located in urban areas. Thus, it seems that location could be a key determinant in the adoption of this type of innovation.

In Spain (Fig. 12.1), the differences became more significant after 2000, whereas, in France (Fig. 12.2), the differences emerged earlier (around 1998). Overall, the mean level of adoption in each type of environment was higher in Spain.

12.4 RESULTS

Table 12.2 shows the results of the estimates. Column one presents the results of the regression for all firms in the sample. The variables with the strongest relationships with ICTs are size of the firm and export activity. In the two cases, the sign found is positive, as expected according to the literature. On the other hand, none of the business environment variables, including location, seem to significantly affect the adoption of ICTs. This finding seems to contradict the earlier descriptive analysis, which implied that location was a relevant variable. Because of this apparent contradiction, we refined our analysis by performing separate analyses for France and Spain.

Columns two and three present the separate estimates for French and Spanish firms. Comparing results for the two countries reveals that the variables that are positively related to the adoption of ICTs are the same as in the first analysis with the same sign and similar scale. Therefore, for the sample, the country effect makes no difference to the ICT-adoption behavior of the agri-food industry. The analysis shows the positive effects of exporting and size of the firm (both characteristics of the firm) on the dependent variable.

Of the environmental variables, firms in rural environments seem to have a greater tendency to adopt ICTs. The significance of the effect is low, however, so evidence is inconclusive regarding this

Table 12.2 Determinants of the Adoption of ICTs

	(1) Sample	(2) French Firms	(3) Spanish Firms
SIZE	0.604[c]	0.478[c]	0.725[c]
	(0.063)	(0.096)	(0.087)
WAGE	0.000	0.001	0.010
	(0.002)	(0.001)	(0.007)
EXPORT	1.146[c]	0.983[c]	1.266[c]
	(0.171)	(0.259)	(0.247)
RURAL	0.312[a]	0.454[a]	0.435[a]
	(0.140)	(0.212)	(0.197)
CONC2	0.815[c]	0.730[b]	1.115[c]
	(0.183)	(0.266)	(0.261)
CONC3	0.048	1.255	−0.848
	(0.475)	(0.755)	(0.687)
Intercept	−2.493[c]	−2.715[c]	−2.787[c]
	(0.185)	(0.289)	(0.293)
N	1327	587	740
Pseudo-R^2	0.2790	0.2419	0.3383

Standard error in parentheses.
[a] $p < 0.1$.
[b] $p < 0.05$.
[c] $p < 0.01$.
Compiled by the authors.

finding. The market concentration of the subsector is significant if firms operate in subsectors with moderate market concentration but not if firms operate in subsectors with other levels of market concentration.

12.5 CONCLUSIONS

The adoption of ICTs is a key innovation for firms in terms of both internal organizational systems and relationships with other entities. The literature is sometimes ambiguous regarding the way that characteristics of the firm and business environment affect the adoption of ICTs.

Therefore, this chapter presents an empirical study of the determinants of ICT adoption by agri-food firms from two EU countries. The study analyzes several variables for a sample of 1327 firms from the agri-food sector. The variables cover the adoption of the Internet, the location of the firm, and some characteristics of the firm (size, export activity, etc.).

Our results show that characteristics of the firm such as size and export activity are among the main determinants of the adoption of innovations such as ICTs. Large firms and firms that export are more likely to adopt ICTs. In these two variables, our results are consistent with the main findings reported in the literature on innovation adoption and, specifically, the adoption of ICTs. In contrast, our results fail to confirm other findings from the literature regarding employees' qualifications, since one of the barriers identified was the lack of qualifications of the firms' personnel. For the agri-food industry, we found no relationship between the adoption of ICTs and employees' mean salary, which we used as a proxy for employees' qualification level.

In terms of environmental variables, descriptive data suggest that firms in rural environments adopt ICTs earlier than firms in urban environments, and our empirical study indicates that rural firms are more likely to adopt ICTs, although this result is less significant than the results for the other determinants. These findings may indicate that the effect of urban/rural location loses explanatory capacity as online transactions and online contact with suppliers and customers become more common because this new online business activity replaces interactions in situ. Thus, our results regarding the effect of this variable did not fully clarify the inconclusive findings in the literature. Regarding market concentration, results show that a moderate level of market concentration is positively related to a tendency to adopt ICTs. This variable is as determinant as the characteristics of the firm we highlighted previously. In addition, this result confirms the inverted U relationship reported in the literature, whereby two opposing effects act simultaneously.

As a final remark, the findings are similar in French and Spanish firms, with scarcely any differences arising in the factors related to the tendency to adopt ICTs. Given the relative similarity between the French and Spanish agri-food industries and, by extension, a large part of the EU, the present chapter identifies common trends in the adoption of ICTs. Nevertheless, one question remains unresolved. Do the agri-food industries in other countries with other socioeconomic determinants display similar innovative behavior in terms of adoption of ICTs?

ACKNOWLEDGMENTS

This research was partially funded by the Generalitat Valenciana, project GV/2015/073.

The authors also thank the Spanish Ministry of Science and Innovation, project TIN2013-43913-R, for its financial support.

REFERENCES

Aghion, P., Bloom, N., Blundell, R., Griffith, R., Howitt, P., 2005. Competition and innovation: an inverted-U relationship. Quarterly Journal of Economics 701–728.

Alarcón, S., Sánchez, M., 2014. Cómo innovan y qué resultados de innovación consiguen las empresas agrarias y alimentarias españolas. Cuadernos de Estudios Agroalimentarios 6, 63–82.

Alarcón, S., Sánchez, M., 2013. External and internal R&D. Capital investment and business performance in the Spanish agri-food industry. Journal of Agricultural Economics 64 (3), 654–675.

Alarcón, S., González, L., Sánchez, M., 2013. Strategies for the development of new products in the Spanish agri-food industry. In: Zacharoula, A., Smarthrakis, V., Louca, S., Vlachopoulou, M. (Eds.), E-Innovation for Sustainable Development of Rural Resources During Global Economic Crisis. IGI Global, pp. 181–198.

Alecke, B., Alsleben, F., Scharr, F., Untiedt, G., 2006. Are there really high-tech clusters? the geographic concentration of German manufacturing industries and its determinants. Annals of Regional Science 40 (1), 19–42.

Andersen, O., 1993. On the internationalization process of firms: a critical analysis. Journal of International Business Studies 24 (2), 209–231.

Arrow, K., 1962. Economic welfare and the allocation of resources for invention. In: Nelson, R. (Ed.), The Rate and Direction of Invention Activity: Economic and Social Factors. Princeton University Press, Princeton, New Jersey.

Au, Y.A., Kauffman, R.J., 2003. What do you know? Rational expectations in information technology adoption and investment. Journal of Management Information Systems 20, 49–76.

Audretsch, D., 2002. The innovative advantage of US cities. European Planning Studies 10, 165–176.

Bhagwat, R., Sharma, M.K., 2007. Information system architecture: a framework for a cluster of small- and medium-sized enterprises (SMEs). Production Planning & Control 18 (4), 283–296.

Bernard, A.B., Bradford, J., Schott, K.P., 2005. Firms in international trade. Journal of Economic Perspectives 21 (3), 105–130.

Capitanio, F., Coppola, A., Pascucci, S., 2009. Indications for drivers of innovation in the food sector. British Food Journal 111 (8), 820–838.

Castellani, D., Zanfei, A., 2007. Internationalisation, innovation and productivity: how do firms differ in Italy? The World Economy 30 (1), 156–176.

Cohen, W.M., Levinthal, D.A., 1990. Absorptive capacity: a new perspective on learning and innovation. Administrative Science Quarterly 35 (1), 128–152.

Connor, J.M., Schiek, W.A., 1997. Food Processing: An Industrial Powerhouse in Transition, second ed. John Wiley & Sons, New York.

Doloreux, D., Parto, S., 2004. Regional Innovation Systems: A Critical Synthesis. Institute for New Technologies. United Nations University, Maastricht.

Domenech, J., Martinez-Gomez, V., Mas-Verdú, F., 2014. Location and adoption of ICT innovations in the agri-food industry. Applied Economics Letters 21 (6), 421–424.

Fearne, A., Álvarez-Coque, J.M., López-García Usach, T., Sánchez, M., 2013. Innovative firms and the urban/rural divide: the case of agro-food system. Management Decision 51 (6), 1293–1310.

FIAB, 2013. Informe Económico 2012.

Filipescu, D.A., Rialp, A., Rialp, J., 2009. Internationalisation and technological innovation: empirical evidence on their mutual relationship. Advances in International Marketing 20, 125–154.

Filippaios, F., Papanastassiou, M., Pearce, R., Rama, R., 2009. New forms of organisation and R&D internationalisation among the world's 100 largest food and beverages multinationals. Research Policy 38 (6), 1032–1043.

Fink, D., 1998. Guidelines for the successful adoption of information technology in small and medium enterprises. International Journal of Information Management 18 (4), 243–253.

Food Drink Europe, 2015. Data & Trends. European Food and Drink Industry. 2014–2015.

Forman, C., Goldfarb, A., Greenstein, S., 2005. How did location affect adoption of the commercial Internet? Global village vs. urban leadership. Journal of Urban Economics 58 (3), 389–420.

Fort, F., Rastoin, J.L., Temri, L., 2005. Les determinants de l'innovation dans les petites et moyennes enterprises agroalimentaries. Revue Internationale PME: Économie et Gestiön de la Petite et Moyenne Enterprise 18 (1), 47–72.

Funke, M., Niebuhr, A., 2005. Regional geographic research and development spillovers and economic growth: evidence from West Germany. Regional Studies 39, 143–153.

Galliano, D., Roux, P., 2008. Organisational motives and spatial effects in Internet adoption and intensity of use: evidence from French industrial firms. Annals of Regional Science 42, 425–448.

García Alvarez-Coque, J.M., López-García Usach, T., Sanchez Garcia, M., 2013. Territory and innovation behaviour in agri-food firms: does rurality matter? New Medit (3), 2–10.

García, J.M., Alba, M., López-García, T., 2012. Innovation and sectoral linkages in the agri-food system in the Valencian Community. Spanish Journal of Agricultural Research 10 (1), 18–28.

Garcia-Martinez, M., Briz, J., 2000. Innovation in the Spanish food and drink industry. International Food and Agribusiness Management Review 3 (2), 155–176.

Garcia-Martinez, M., Burns, J., 1999. Sources of technological development in the Spanish food and drink industry. A supplier-dominant industry. Agribusiness: An International Journal 15 (4), 431–448.

Giunta, A., Trivieri, F., 2007. Understanding the determinants of information technology adoption: evidence from Italian manufacturing firms. Applied Economics 39 (10), 1325–1334.

Gorodnichenko, Y., Svejnar, J., Terrell, K., 2010. Globalization and innovation in emerging markets. American Economic Journal–Macroeconomics 2 (2), 194–226.

Harindranath, G., Dyerson, R., Barnes, D., 2008. Ict adoption and use in UK SMEs: a failure of initiatives? The Electronic Journal Information Systems Evaluation 11 (2), 91–96.

Hashmi, A.R., 2011. Competition and Innovation: The Inverted-U Relationship Revisited. Working Paper, No. 1101. Department of Economics; National University of Singapore.

Hauknes, J., Knell, M., 2009. Embodied knowledge and sectoral linkages: an input-output approach to the interaction of high- and low-tech industries. Research Policy 38, 459–469.

Hollenstein, H., 2004. The determinants of the adoption of ICT. Structural Change and Economic Dynamics 15, 315–342.

Hughes, K., 1986. Exports and innovation: a simultaneous model. European Economic Review 30 (2), 383–399.

Idris, A., Tey, L.S., 2011. Exploring the motives and determinants of innovation performance of Malaysian offshore international joint ventures. Management Decision 49, 1623–1641.

IGR/CGAAER, 2012. Une stratégie publique pour les industries alimentaires. In: Rapport, l'Nspection Générale des Finances – Conseil Général de l'Alimentation, de l'Agriculture et des Spaces Ruraux. Mai 2012.

Johanson, J., Mattsson, L.G., 1988. Internationalisation in industrial systems: a network approach. In: Hood, En.N., Vahlne, J.E. (Eds.), Strategies in Global Competition. Croom Helm, London, pp. 287–314.

Kafouros, M.I., Buckley, P.J., Sharp, J.A., Wang, C., 2008. The role of internationalization in explaining innovation performance. Technovation 28 (1–2), 63–74.

Kilponen, J., Santavirta, T., 2007. When Do R&D Subsidies Boost Innovation? Revisiting the Inverted U-shape. Bank of Finland Research Discussion Paper Number 10.

Kleinknecht, A., Oostendorp, R., 2002. R&D and export performance: taking account of simultaneity. In: Kleinknecht, A., Mohnen, P. (Eds.), Innovation and Firm Performance. Palgrave, London, pp. 310–320. 29.

Liao, J., Welsch, H., Stoica, M., 2003. Organizational absorptive capacity and responsiveness: an empirical investigation of growth-oriented SMEs'. Entrepreneurship: Theory and Practice 28 (1), 63–86.

Lopez, N., Montes-Peon, J.M., Vazquez-Ordas, C., 2003. Innovation in the Spanish food and beverage industry: an integrated approach. International Journal of Biotechnology 5 (3–4), 311–333.

López, E., Arcas, N., Alcón, F., 2014. Uso y calidad de los sitios web: evaluación de las empresas agroalimentarias murcianas. Revista Española de Estudios Agrosociales y Pesqueros 237 (1), 155–180.

López-García, T., García, J.M., Alba, M., 2013. Composición de la intensidad innovadora de la industria agroalimentaria: fuentes internas y externas al sector. Economía Industrial 353, 153–162.

Martins, M., Oliveira, T., 2008. Determinants of information technology diffusion: a study at the firm level for Portugal. The Electronic Journal Information Systems Evaluation 11 (1), 17–24.

Minbaeva, D., Pedersen, T., Björkman, I., Fey, C.F., Park, H.J., 2003. MNC knowledge transfer, subsidiary absorptive capacity, and HRM. Journal of International Business Studies 34 (6), 586–599.

Moulaert, F., Sekia, F., 2003. Territorial innovation models: a critical survey. Regional Studies 37 (3), 289–302.

Mowery, D.C., Oxley, J., Silverman, B.S., 1996. Strategic alliances and interfirm knowledge transfer. Strategic Management Journal 17, 77–91.

OECD, 1994. Creating Rural Indicators for Shaping Territorial Policy. OECD Publications, Paris, France.

OECD, 2009. OECD Rural Policy Reviews: Spain. OECD Publications, Paris, France.

ONTSI, 2012. Informe PYME 12. Análisis sectorial de la implantación de las TIC en las pyme españolas. Available at: http://www.ontsi.red.es/ontsi/sites/default/files/informe_epyme_2012_0.pdf.

Panayiotou, N., Katimertzoglou, P.K., 2015. Micro firms internet adoption patterns: the case of the Greek jewellery industry. Journal of Enterprise Information Management 28 (4), 508–530.

Prasad, A.N., 2008. Competition and Innovation – A Reexamination of Inverted-U Relationship. Ohio Northern University, mimeo. http://www.fma.org/Texas/Papers/RD_Competition_Inverted U.pdf.

Premkumar, G., Roberts, M., 1999. Adoption of new information technologies in rural small businesses. The International Journal of Management Science 27, 467–484.

Ritchie, B., Brindley, C., 2005. ICT adoption by SMEs: implications for relationships and management. New Technology, Work and Employment 20 (3), 205–217.

Roelandt, T.J.A., den Hertog, P., 1999. Cluster analysis and cluster-based policy making in OECD countries: an introduction to the theme. In: En Boosting Innovation. The Cluster Approach. OECD Proceedings 1999.

Rogers, M., 2004. Networks, firm size and innovation. Small Business Economics 22 (2), 141–153.

Sin Tan, K., Choy Chong, S., Lin, B., Cyril Eze, U., 2010. Internet-based ICT adoption: evidence from Malaysian SMEs. Industrial Management & Data Systems 109 (2), 224–244.

Tingvall, P.G., Poldahl, A., 2006. Is there an inverted U-shaped relation between competition and R&D? Economics of Innovation and New Technology 15 (2), 101–118.

Vinding, A., 2006. Absorptive capacity and innovative performance: a human capital approach. Economics of Innovation and New Technology 15 (4–5), 507–517.

Wymer, S., Regan, E., 2005. Factors influencing e-commerce adoption and use by small and medium businesses. Electronic Markets 15 (4), 438–453.

Zarka, M., Laroche, A., Juin 2015. De Nouveaux Modèles de Croissance Pour les Industries Agroalimentaires Françaises? Rapport La fabrique de l'industrie-SAF Agr'Idees.

CHAPTER

IMPLEMENTATION OF FOODOMICS IN THE FOOD INDUSTRY

13

J.-L. Sébédio[1,2], C. Malpuech-Brugère[1,2]

[1]INRA, Clermont Université, Unité de Nutrition Humaine, Clermont-Ferrand, France;
[2]CRNH Auvergne, Clermont-Ferrand, France

13.1 INTRODUCTION

Nowadays food is not only recognized as a source of energy but also for its potential in maintaining humans in better health for a longer time. Furthermore, minor components that are present in food products have a potential to prevent/delay the appearance of diseases (Capozzi and Bordini, 2013). Another goal is also to optimize nutritional recommendations to specific population groups (Brennan, 2015).

The word *foodomics* first appeared in scientific papers following a special issue of the *Journal of Chromatography A* in 2009, after the first meeting, which took place in Cesena, Italy (Capozzi and Bordini, 2013). It was defined (Cifuntes, 2009) as being "a discipline that studies the food and nutrition domains through the application of omics technologies." These technologies including genomics, transcriptomics, proteomics, and metabolomics generate data at different expression levels (gene, transcript [RNA], protein, and metabolite). Since then, special journal issues and international meetings (Bordoni et al., 2014) have been organized in order to demonstrate the progress made during the past 15 years in this field of research (Cifuentes et al., 2013; Cifuentes, 2014; Capozzi and Bordoni, 2013). The evolution in the approaches used in the research on food science and nutrition was not only due to advances in analytical techniques but also in data treatment, bioinformatics tools, and the development of new sample and fractionation techniques (Garcia-Canas et al., 2012).

The purpose of this chapter is to demonstrate the innovative characteristics that omics technologies provide in food science and nutrition, as well as to explore ways of implementing them in the food industry. As a first step, the different technologies that are being currently used in foodomics approaches are described by denoting their major advantages and their drawbacks. Thereafter, potential industrial applications are outlined and their implementation in the food industry is analyzed. Applications of metabolomics will, herein, concern firstly the food area and secondarily the nutrition area. More information on the latter applications can be found elsewhere (Sébédio and Brennan, 2015).

13.2 FOODOMICS TECHNOLOGIES AND TECHNIQUES

The recent development of high-throughput technologies ("the omics"), which allow generating large-scale molecular-level measurements, has increased attention of scientists in medical, biological,

Innovation Strategies in the Food Industry. http://dx.doi.org/10.1016/B978-0-12-803751-5.00013-1

nutritional, and food sciences. These tools offer the possibility of characterizing global alterations associated with disease conditions or nutritional exposition (food composition). They allow to move nutrition from a rather reductionist approach to a global one. Food scientists and nutritionists are not only focusing on the relationship between diseases and macro- and micronutrients consumption but also attempting to identify bioactive molecules that may be present in the diet in low quantities. They also try to understand the protective effects that these molecules may have and their mechanisms of action. These bioactive compounds interact in a number of metabolic pathways and could modulate health status after being metabolized by the gut microbiota. Dietary polyphenols are a typical example of the importance that gut microbiota may play (Moco et al., 2012).

The major "omics" technologies used in the field of foodomics, either to analyze/describe food products or samples collected during an intervention study, are mainly genomics, transcriptomics, proteomics, and metabolomics (Zheng and Chen, 2014). The utilization of such approaches generates data at different expression levels such as gene, transcript, protein, and finally metabolite. Among the different analytical approaches used, capillary electrophoresis (CE) coupled to mass spectrometry (CE-MS) has been playing an important role as it has been applied to proteomics, peptidomics, and metabolomics (Ibanez et al., 2013; Garcia-Canas et al., 2014). CE requires small sample and solvent volumes, as well as analytical time, but due to the difficulty to identify metabolites, it is frequently used in combination with a mass spectrometer. Furthermore, due to a low sensitivity, several sample preconcentration techniques have been developed (Oh et al., 2010).

In the field of human nutrition, genomics has been the basis for understanding the diversity of response of humans to foods. In genomics studies, modern DNA sequencing (so-called the "next generation of sequencing" [NGS] technologies) can quickly generate a huge amount of genomic information even on species having little previous genetic information (Pareek et al., 2011). The utilization of large-scale DNA sequencing is now being used, for example, in aquaculture, for agricultural products quality, and other applications (Cerda and Manchado, 2013).

Human health is not only determined by genes but also by the interaction of different factors (lifestyle/environment, gut microbiota). Among them, food intake is a crucial environmental factor, as food components may alter gene expression, posttraductional modification and consequently modify protein and metabolite compositions. The production and utilization of metabolites are more directly connected to the phenotype (sometimes called metabolic-phenotype) exhibited by an organism than the presence of mRNAs or proteins. The study of the transcriptome (characterization of all transcriptional activity: coding/noncoding) representing the whole set of mRNAs is essential, as it can vary with external conditions (eg, diet) and reflect the genes that are being expressed at any given time. As for genomics, transcriptomics technologies include two types of high-throughput analytical techniques, the microarray and the sequencing approaches (Valdes et al., 2013a; Mutz et al., 2013). While microarray analyses are relatively inexpensive compared to the other transcriptomic technologies, they usually display a background noise. The latest is a drawback for the detection of low-abundance transcripts (Valdes et al., 2013a), as cross-hybridization appeared introducing biases in measurements. However, until the advent of RNA sequencing, microarrays were the standard tool for gene expression quantification (Wolf, 2013). On the contrary, the competition between instrumental companies to develop sequencing platforms and the development of specific downstream analyses and software (Wolf, 2013) allow unlimited possibilities in biology and food science. However, NGS is still an expensive approach (Mutz et al., 2013) compared to the microarray approach.

Proteins are one of the essential components in many food products. In foodomics, proteomics and peptidomics are precious tools to define the detailed composition of food and its properties. They are a source of amino acids and components providing texture, functional, and sensory properties to foods. In this case, proteomics is utilized to define food quality and safety. It is also used in human cells, for example, after an intervention study to follow the impact of the diet. Mass spectrometry (MS) is a common procedure for proteomics and peptidomics analysis (Picariello et al., 2012). Proteomics approach usually includes three major steps:

1. isolation
2. fractionation of proteins followed by MS analysis
3. interpretation of the MS spectra by using database repositories

For example, a two-dimensional gel electrophoresis (2-DE) allows the separation and isolation of proteins that are then digested into peptides, which are further submitted to mass spectrometry (Gallardo et al., 2013). In another approach, the protein mixture is digested without previous fractionation and the peptides are analyzed by liquid chromatography coupled with mass spectrometry (LC-MS). For further details, the reader may want to look at the proteomics workflows commonly used in food research (Gallardo et al., 2013).

The methods previously described for protein analysis are complex, time-consuming, and need specially trained people to operate the different steps involved. Recently other simpler analytical processes (so-called "lab-on-chip") requiring less time and operating cost have been developed (Nazzaro et al., 2012). In this application, electrophoresis of proteins is carried out on a microchip taking a few seconds to minutes, and thus could be a possible alternative to the sophisticated conventional methods.

Metabolomics approaches that deal with the study of small molecules in a biofluid or tissue (Fiehn et al., 2000; Nicholson et al., 1999) are being used for several applications, including:

1. potential biomarker discovery in nutritional interventions (Brennan, 2013; Lodge, 2010; Rezzi et al., 2008)
2. risk prediction of evolution of a health status to a disease state (Wang et al., 2011)
3. potential food/nutrient biomarker consumption (Hedrick et al., 2012; Llorach et al., 2012; Pujos-Guillot et al., 2013)
4. to describe food composition in detail (Shepherd et al., 2007)

Obtaining accurate data on dietary intake in free living populations, and the study of the "food metabolome," which may be defined as the results of the digestion and metabolism, is a great challenge for epidemiological studies (Ismail et al., 2013).

The development of metabolomics in food science and nutrition is still relatively new (Chin and Slupsky, 2013), although this topic constitutes a growing domain of interest considering the shift observed in dietary habits (German et al., 2002; Gibney et al., 2005; Shepherd et al., 2011). The rapidly expending field of metabolomics has been driven by major advances in analytical tools such as nuclear magnetic resonance (NMR) (Eisenreich and Bacher, 2007) and mass spectrometry and the corresponding hyphenated methods such as LC-MS, and CE coupled with mass spectrometry (CE-MS). The latest may also be used in other omics (peptidomics and proteomics) technologies (Ibanez et al., 2013). Gas-liquid chromatography coupled with mass spectrometry (GC-MS) is another alternative (Dettmer et al., 2007; Pan and Raftery, 2007; Wilson et al., 2005; Gao et al., 2010; Shepherd et al., 2007). Advances in chemometric and bioinformatic technologies (Idle and Gonsalez, 2007; Moco et al., 2007; Wishart, 2007; Skov et al., 2014) also played an important role in foodomics analyses.

A typical metabolomic experiment (Fig. 13.1) includes six major steps:

1. design of the experiment and preparation of the sample
2. analysis of biofluids tissues or food product via LC-MS, NMR, or GC-MS, etc. or a multiplatform approach to maximize metabolite coverage
3. extraction and treatment of data set
4. statistical analyses
5. identification of key or discriminatory metabolites
6. interpretation of the data

The data acquisition and retrieval may be achieved by a targeted analysis when specific molecules are analyzed or by a hypothesis-driven approach (untargeted analysis). Both have advantages and disadvantages, and the approach taken depends ultimately on the study in question.

The choice of the analytical method is most of the time a critical step. While NMR does not require extensive sample preparation time, has high reproducibility and little interlaboratory variability, it is limited by its sensitivity compared to MS, which needs extensive sample treatments or prefractionation steps. On the other hand, NMR is a good method when large sample sets have to be analyzed as it is generally the case for clinical trials and analysis of food products (Capozzi, 2013) even if the number of detected metabolites is smaller compared to MS (Laghi et al., 2014). Furthermore, metabolite and biomarker identifications are easy to perform due to the existing databases. The possibility of working on intact tissues (high-resolution magic angle spinning nuclear magic resonance spectroscopy; HRMAS-NMR) is an essential advantage that avoids extraction of lipids or water-soluble metabolites prior to analysis. Samples of fruits, vegetables, raw meats, and processed foodstuffs were analyzed without any chemical or physical manipulation using only a few mg of tissue allowing determination of quality and safety of the products for human consumption (Mazzei and Piccolo, 2012; Ritota et al., 2012; Valentini et al., 2011). LC-MS is a very sensitive method, which most of the time requires extensive sample preparation steps but generally allows detecting many metabolites even at very low concentrations in complex samples. However, despite the major advances in the field over the last decade, metabolite identification is still one of the biggest challenges when using this technique. When metabolomics started to find applications in the field of nutrition research, many papers in the literature did not contain much structural information about the discriminant metabolites, except for data obtained by NMR, as existing spectral databases were of great help (Wishart et al., 2012). Nowadays mass spectrometry-based approaches, especially LC-MS, are being increasingly used in metabolomics explorations (Ferrara and Sébédio, 2015).

GC-MS has been proven to be a robust method for quantification of selected metabolites classes. Additionally, mass spectral databases such as commercially available NIST, Wiley, and "Agilent Fiehn GC/MS Metabolomics RTL Library" further facilitate the identification of metabolites. For comprehensive metabolome analysis based on GC-MS, derivatization is an effective method transforming highly polar materials into products sufficiently volatile and narrowing the boiling point window. Consequently they can be eluted at reasonable temperatures without thermal decomposition or molecular rearrangement. The active functional groups such as carboxylic, amide, amino, and hydroxyl are either alkylated, acylated, or silylated, and complex standardized procedures have been published in order to analyze complex mixtures (Kanani et al., 2008).

The silylation procedure needs nonaqueous environment for the reaction, while some nonsilylating derivatization techniques (eg, chloroformates derivatization) can be performed in the presence of water.

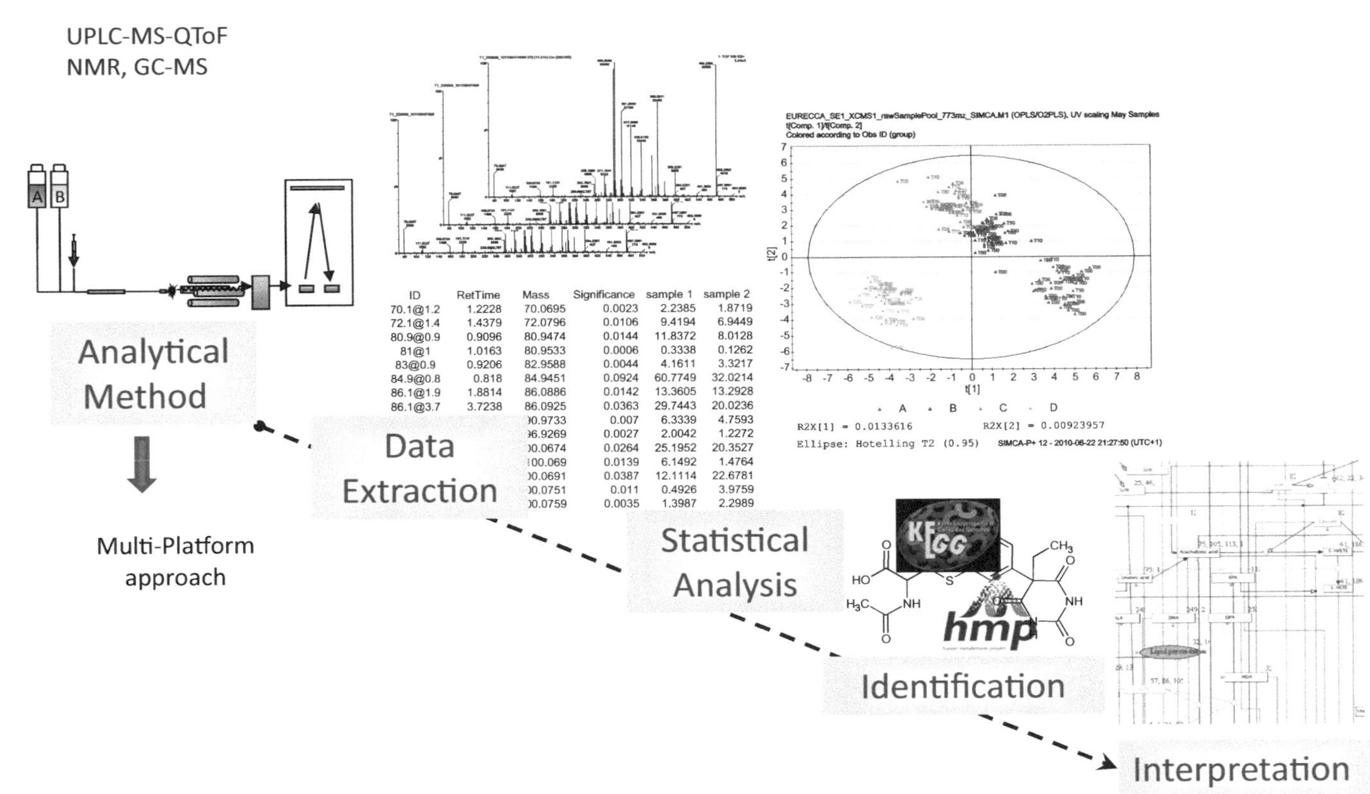

FIGURE 13.1

Workflow of a metabolomics analysis.

Reproduced by permission from Elsevier from Ferrara, M., Sébédio, J.L., 2015. Challenges in nutritional metabolomics: from experimental design to interpretation of data sets. In: Sebedio, J.L., Brennan, L. (Eds.), Metabolomics as a Tool in Nutrition Research, Woodhead Publishing, Kidlington, UK, pp. 3–35.

Chloroformates have been proven to be strong and rapid derivatizing reagents, and in contrast to trimethylsilylation, alkyl chloroformate derivatization reaction occurred directly in aqueous media without the requirement of heating. It simplifies sample pretreatment and derivatization procedure (Hušek, 1998). GC-MS may be used to analyze food products (Shepherd et al., 2007), and biofluids such as blood, urine, or fecal water (Gao et al., 2010; Tao et al., 2008). In order to improve the separation efficiency and peak capacity, two-dimensional gas chromatography–mass spectrometry may also be used as described by Koek et al. (2008).

13.3 APPLICATIONS OF FOODOMICS

Foodomics technologies have already been applied to different questions such as food safety, traceability, quality, and transgenic and functional foods. They have also been used as biomarkers of food intake and biomarkers of the evolution from a healthy state to disease and health effects of different food ingredients.

13.3.1 FOOD QUALITY, AUTHENTICITY, TRACEABILITY, AND SAFETY

Food authentication has been a major concern for the prevention of fraud but also to evaluate the safety of consuming food products that may be harmful for human health. Among the food products, authentication of animal species in milk has been of special interest to evaluate adulteration of mozzarella cheese with cow's milk. A new multiplex RT-PCR method was developed (Cottenet et al., 2011) for detecting simultaneously cow and buffalo mitochondrial DNA in milk samples. It was demonstrated that this method allows to detect cross-contamination with a detection limit of 1% and be a preliminary screening before using the official ISO 17678 procedure. This method showed possibilities of milk traceability and authenticity. Food proteomics using either mass spectrometry using matrix-assisted laser desorption/ionization (Maldi-TOF-MS) or LC-ESI-MS (electrospray ionization) analysis of casein-derived "proteotypic peptides" is also a powerful tool for milk speciation by determining the content of milk of four ruminant species in a single analysis (Cuollo et al., 2010).

Foodomics technologies such as metabolomics, protein, and DNA analyses are also used in aquaculture to authenticate commercial species and determine rearing conditions, as well as traceability of the product. For instance, a ^{1}H NMR lipid fingerprinting has been applied to differentiate wild or farmed gilthead sea bream and reared fish in different offshore sea cages (Melis et al., 2014). It is interesting to note that the n-3 fatty acids were the most discriminant molecules to differentiate the wild species from the farmed fish. In other research, Cerda and Manchado (2013) described European studies carried out using the new-generation sequencing technologies for flatfish aquaculture to produce high-quality and safe animals for the consumer. Aquaculture food products storage is also a problem as fish is often consumed far from the farm where it has been produced. NMR was proven to be an essential tool for monitoring the postmortem changes occurring during fish storage and to characterize fish metabolites responsible for fish quality (Shumilina et al., 2015). Authentication of seafood species (eg, shrimps, prawns), produced in different parts of the world and used in many food products, have interspecies phenotyping similarities that lead to mislabeling the food product. In order to better identify the species, methodologies using a combination of proteomics and the latest DNA-based approaches are being used (Ortea et al., 2012). However, major challenges such as multiple species identification in food samples, analysis of very complex matrices, and the final optimization of an inexpensive fast and robust method for routine analysis still have to be solved. In that sense, identification of new species-specific peptides by proteomics should offer the possibility to design fast detection analyses. According to Ortea et al. (2012), these species-specific peptides could be used as biomarkers in order to develop a fast and reliable method for multiple species identifications in complex food matrices. Potentially species-specific proteins were digested in order to investigate the nature of the peptides. Identification of peptides from arginine kinase (AK), a well-known allergen that was proposed as a biomarker of prawn species, was monitored by MS. Several selective peptides from AK were identified, allowing authentication of several prawn species.

Foodomics is also a useful tool to trace transformations applied to raw materials and screen safety of the food products. A food product is usually produced after a combination of physical and chemical treatments (heat treatment, extrusion, gel formation, cooking, etc.), which generally induce structural and chemical changes to the food. Proteomics of processed foods is still a great challenge (Picariello et al., 2012) due to the increase of protein complexity (protein–protein interaction, protein interaction with other molecules, protein oxidation, etc.). For example, a proteomic method by LC-MS/MS was developed to measure allergens and gluten markers in beverages such as beer (Weber et al., 2009). Another proteomic approach has also allowed detecting small quantities of harmful lectins in raw or processed foods in legume seeds. A common way to assure the safety of the raw or processed food is the destruction of these harmful lectins using boiling water at atmospheric pressure. The residual quantity of these molecules after heat treatment is determined using mass spectrometry (Nasi et al., 2009).

Considering that foods are nowadays produced and distributed throughout the world, detection of trace residues and contaminants has become a major issue. Among the different techniques, MS-based metabolomics has found applications in the study of chemical safety of food. However, some methodological questions such as robustness and repeatability of the methods still have to be confirmed as traces of compounds have to be analyzed (Antignac et al., 2011). One application is the utilization of food metabolomics (untargeted approach) to detect degradation and/or contamination of infant formula with components such as melamine or dicyandiamide (Inoue et al., 2015). Other applications include the screening of over 100 veterinary medicines in bovine muscle (Geis-Asteggiante et al., 2012), and of over 250 pesticides and veterinary drugs in animal feed (Aguillera-Luiz et al., 2013). Recently, a metabolomic approach using the performance of high-resolution accurate mass spectrometry operating in full scan MS (LC-LTQ-Orbitrap MS) was able to quantify amoxicillin (AMX) and detect one of its metabolites in several positive chicken muscle, liver, and kidney samples from animals medicated with AMX. This technique can discriminate medicated and nontreated animals (Hermo et al., 2014). This applied methodology was shown to be an excellent way to produce bioinformation in order to discriminate whether there is pharmacological adulteration in food. Poultry is, indeed, one of the most consumed foods by humans, and prior to consumption quality controls of poultry are mandatory to ensure protection of human health.

A method using GC-MS/MS for accurate quantification of bisphenol A (an industrial chemical used in the production of packaging materials) was also validated on different matrices, eg, more than 1000 food items (Deceuninck et al., 2014). However, improvements of the developed methods in the field of food contaminants still have to be done, especially for structural identification and matrix effects that may interfere in the identification and/or quantification procedures (Hird et al., 2014).

The detection of foodborne pathogens and their toxins is another important concern worldwide in spite of ensuring safe foods as well as concerning changes in production systems and global exchange increase. Methods for looking at the genome, proteome, and metabolome of pathogens have recently been reviewed by Giacometti and Josic (2013). Pathogenic prokaryotes (*Escherichia coli* species, the genus *Salmonella*, some *Bacillus*, and *Campylobacter*) are major concerns in fresh products. Among the genus *Listeria, Listeria monocytogenes* would impact processed products quality. The eukaryotic food pathogens such as *Aspergillus*, *Fusarium*, *Penicillium*, *Alternaria*, etc. may also produce metabolites highly toxic for humans (Giacometti et al., 2013). Application of the omic technologies using DNA microarray, GC-MS metabolomics, LC-MS/MS proteomics, and lipidomics were found more sensitive for the identification of microbial food contaminants and their toxins than using established methods based on immunochemical techniques. Mycotoxins, produced by molds are metabolites that can be found in a wide variety of products such as cereals, peanuts, fruits, and vegetables. Analytical methods for mycotoxin detection are ELISA techniques but reference methods using LC-MS are being developed especially for multitoxin determination. So far the genomes of 300 fungi species have also been sequenced, and a fungal secretome database (secreted proteins) of different fungal species has been produced (Giacometti et al., 2013). Similarly, the proteomics approaches have also shown a good potential to detect seafood toxins.

13.3.2 FOODOMICS AND TRANSGENIC FOODS

Genetic engineering is tied to the production of genetically modified crops having different traits, such as herbicide tolerance, resistance to insects, faster ripening, or nutrient quality (Dunwell, 2014). The authorization and commercialization of genetically modified organisms (GMOs) is strictly regulated in many countries, and a risk assessment is mandatory for introducing a new GMO product. This assessment involves the comparison of the GMO with the traditional variety. These evaluations are usually conducted using analyses of specific components (Simo et al., 2014). However, a report from the European Food Safety Authority (EFSA) published in 2006 recommended to use omics technologies, as the usual approach as stated before may not detect all the differences between the GMO and the original crop. In a recent review, Simo et al. (2014) summarized the recent metabolomics explorations that were carried out by the scientific community to compare GMOs with the wild-type original crops (eg, rice, maize, soybean, potato, etc.), their potential applications as well as their limitations. The used methodologies provided the ability to screen the changes induced in metabolites by genetic modifications. However, the method is not yet validated by regulatory agencies. Furthermore, even if other omics technologies such as transcriptomics and proteomics have been used for the same purpose (Valdés et al., 2013b), one should now report correlations between the different data sets in order to understand the biological significance of the changes induced by genetic engineering. However, if we consider the progress made in the past 10 years, the fast-evolving fields of omics technologies and bioinformatics should allow better peak annotation and multi-omics technologies integration in a reasonable time frame.

Recently a large-scale untargeted LC-MS-based metabolomics methodology was able to discriminate the conventional and organic growing conditions for the white cabbage culture (Mie et al., 2014). Although metabolomics has also been proposed as one of several approaches for organic food authentication (Capuano et al., 2013), Mie et al. (2014) suggest this technique only as a complement to other analytical techniques.

13.3.3 BIOMARKERS OF FOOD INTAKE

Metabolomics has been extensively used these past few years to characterize the composition of the "food metabolome." The latest has been described as a complex mixture of metabolites derived from the digestion of foods, including the molecules resulting from the biotransformation by the microbiota. The main goal of the investigations was to identify metabolites in blood or urine samples of specific food or nutrient intake. The purpose was to use these metabolites to estimate food intake in order to complement the food frequency questionnaires in nutritional epidemiology. Nutritional epidemiology aims to explore the relation between diet and health and to monitor the nutritional status of the different groups (young, elderly, etc.) in the population. A very comprehensive review on this subject was recently published by Manach et al. (2015). The Food Biomarker Alliance (FOODBALL project) that started in December 2014 is offering the possibility to carry out exploration of food components in different population groups within Europe in order to validate biomarkers of food intake. This project is supported by the European Union initiative "A Healthy Diet for a Healthy Life" and is coordinated by the Wageningen UR. The FOODBALL consortium includes 22 partners from 11 countries. The data so far acquired are promising but important technical problems such as metabolite identification and quantification still have to be solved before applying this approach to cohort studies. Moreover, the gut microbiota, a highly variable environmental factor that interacts with the digestive process could influence the overall outcome of ingested food. It should be taken into account, since many functional foods are in development, especially by adding living microorganisms (prebiotics, probiotics). The different omics methods used in metabolic profiling are tools that help to understand for instance the probiotic metabolism, as well as molecular mechanisms involved in the health aspect, while they can help to select the formula and the route of administration (Jiménez-Pranteda et al., 2015). As an example, in 2013 the first inulin-type fructans prebiotics intervention study using a metabolomics profiling approach on obese women demonstrated the link between plasmatic metabolic markers and bacterial content (Dewulf et al., 2013).

13.3.4 BIOMARKERS OF METABOLIC DISEASES

Many studies have been carried out during the past two decades using animal models, human intervention trials, and follow-up of cohorts. For a long period of time, analysis of blood and urine samples were compared before and after the diagnosis of a metabolic disease, and sometimes only samples from people having chronic diseases were compared to those of a control group. Only a few studies including kinetics during a short nutritional intervention (2 months) are available (Morio et al., 2015; Sébédio and Polakof, 2015). On the other hand, interesting data have been reported for acylcarnitine and branched-chain amino acids (BCAAs). Newgard et al. (2009) reported that obese individuals, when compared to lean subjects, showed specific increases in C3 and C5 acylcarnitine levels. This outcome was strongly related to the metabolism of BCAAs, whereas these changes reflect an overload of BCAAs catabolism in obese subjects and by changes in other metabolites, eg, decreased levels of α-ketoglutarate, as well as increased levels of glutamate, aromatic amino acids, phenylalanine, and tyrosine.

Among the studies carried out on cohorts to find predictive biomarkers, much data have been published using the Framingham Offspring Study (http://www.framinghamheartstudy.org for more details). Metabolite profiling from 189 individuals (negative for diabetes diagnosis at the time the blood was collected) (Wang et al., 2011) was performed at baseline using an LC-MS approach.

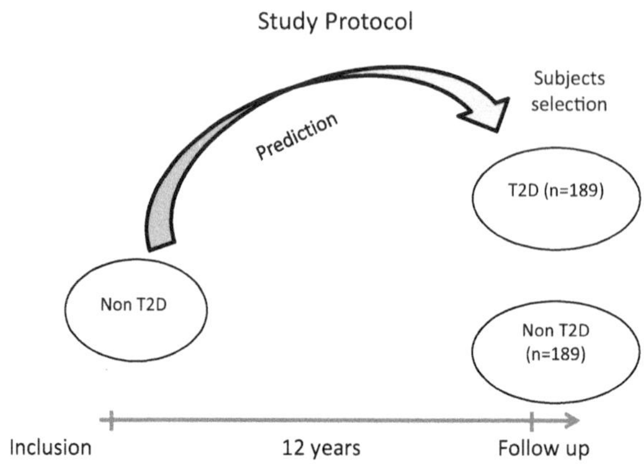

FIGURE 13.2

Protocol of the metabolic study carried out by Wang et al. (2011).

Reproduced by permission from Elsevier from Sébédio, J.L., Polakof, S., 2015. Using metabolomics to identify biomarkers for metabolic diseases: analytical methods and applications. In: Sébédio, J.L., Brennan, L. (Eds.), Metabolomics as a Tool in Nutrition Research, Woodhead Publishing, Kidlington, UK, pp. 145–166.

Controls (189 subjects, negative for diabetes diagnosis at the time the blood was collected and who did not develop diabetes during the follow-up) were matched for age, body mass index and fasting glucose. The experimental protocol is described in Fig. 13.2. Five branched chain (BCAAs) and aromatic amino acids—namely, isoleucine, leucine, valine, tyrosine, and phenylalanine— were found to be highly associated with the future incidence of diabetes. Further examination of the data revealed that the association of three molecules (isoleucine, tyrosine, and phenylalanine) could predict that individuals in the top quartile had a five- to sevenfold higher risk of developing diabetes, compared with subjects in the lowest quartile. These findings were validated using another cohort (Malmö Diet and Cancer study).

13.3.5 HEALTH EFFECTS OF FOOD INGREDIENTS

Many nutrients have shown some potential health benefits (polyphenols, peptides, n-3 polyunsaturated fatty acids, tocopherols, etc.), but the way they act for disease prevention is often poorly understood, and some issues are still controversial (Ibanez et al., 2012). Consequently, one of the major challenges of nutritionists is to unravel their mechanism of action at the molecular level such as their interactions with genes and their effects on proteins and metabolites. This is an important topic in order to provide sound data and scientific evidence on the claimed beneficial effects of new food and ingredients as requested by the EFSA.

In this research area, omics technologies allow to understand molecular mechanisms, but most studies so far published used one omic technology (Ibanez et al., 2012; Valdes et al., 2014) instead of a combination of the techniques described herein (see Section 13.2). In this sense, the combination of the information obtained from the different expression levels is of major importance to understand and sustain the health benefits of some food components. This section is herein illustrated with two studies using a multi-omics approach, one dealing with humans fed an anti-inflammatory dietary mix and one with cell culture, cancer, and polyphenols. In the first study (Bakker et al., 2010), a dietary mixture containing resveratrol, green tea extract, α-tocopherol, vitamin C, n-3 fatty acids, and tomato extract was fed versus a placebo for 5 weeks to overweight men with slightly increased C-reactive protein. Inflammatory and oxidative stress defense markers were analyzed in blood and urine, while plasma proteins, metabolites, and the transcriptomes of the blood mononuclear cells and adipose tissue were quantified. Using a combination of three omics technologies, this study allowed describing the molecular processes by which this dietary mix was able to influence oxidation, inflammation, and metabolism in human. Furthermore, an integrated analysis of the different omic data revealed subtle changes that indicated modulated inflammation of the adipose tissue, and improved endothelial function and increased fatty acid oxidation. One important limitation of such a study is the type of samples that could be collected. For example, some of the effects observed suggested possible benefit of the ingredients in other unanalyzed tissues such as liver and brain. An approach using the mini-pig as a model would be an alternative.

In the second study using a similar multiplatform approach, the chemopreventive effects of rosemary polyphenols extract were investigated on human HT29 colon cancer cells by Ibanez et al. (2012). According to the authors, this was the first time that three omic platforms were put together to study the health effects of dietary compounds against colon cancer cells. A total of 1308 genes, 17 proteins, and 210 metabolites were differently expressed (up- or downregulated) between control and treated cells. The major outcomes of the study were that these polyphenols induced modification of antioxidant activity inside the cell, induction of apoptosis, and cell cycle arrest. However, as outlined by the authors, sometimes no direct correlations between the different data sets could be found and work has still to be carried out for these hypotheses free approaches to reach their full potentials. This type of analysis faces two major problems:

- the different time scales of transcript production, protein expression, and metabolite production
- the different amount of information produced by the omics (1308 genes, only 17 proteins, and 65 identified metabolites).

Consequently, the integration of all the omic data using the Ingenuity Pathway Analysis software (see Fig. 13.3 in Ibanez et al., 2012) did not give all the information required to obtain a more global view. A manual approach was finally used and future studies will for certainly allow such a complex data integration. Solving this problem should be a great help to get health claims accepted at a European level.

13.4 CHALLENGES AND POTENTIAL STRATEGIES FOR THE IMPLEMENTATION OF FOODOMICS IN INDUSTRY

13.4.1 MAJOR CHALLENGES

In the last 20 years very sophisticated omics approaches have been introduced in food and nutrition sciences. Even if foodomics involves the utilization of many tools, mass spectrometry is now playing a crucial role due to its sensitivity and its ability to detect a number of molecular structures with different physical and chemical properties in a single run. However, mass spectrometry approaches still suffer in some cases from incomplete in-house databases for metabolomic analyses. Thereby, only a part of the biomarkers or discriminant metabolites is usually identified. Most of the techniques involved in foodomics approaches are complex, costly, and need highly trained people with complementary skills to carry out the analyses as well as to maintain the instruments in perfect running condition in order to ensure reliable and robust data. Even so, it is not always possible to integrate the data from the three omic technologies despite the availability of software as demonstrated by Ibanez et al. (2012). Such a team will be able to understand the action mechanism of bioactives and to evaluate their safety and authenticity characteristics. This should provide food products that will fulfill the consumer needs in terms of safety and health maintenance and the legal organizations (eg, such as the EFSA) in terms of effectiveness and safety.

This situation is even going to be more complex as new possibilities for the localization of molecules from small metabolites to large proteins (Kaspar et al., 2011) on a tissue section are now being developed to be used in food science (Taira et al., 2014), as well as in nutrition (Touboul et al., 2011). This method involves matrix-assisted laser desorption/ionization (MALDI)-imaging mass spectrometry, which provides both the spatial localization of metabolites and their relative abundance. It is being used for plant cultivation, food quality evaluation, localization of pesticide residue in plants, and of specific components such as lipids in animal organs like brain and gut. The development of such technology is opening new frontiers in food science and nutrition.

To our knowledge, only a few industries have invested in such analytical platforms. This moderate level of equipment investment made by the food industry to implement the foodomics technologies may arise from different reasons:

- The initial cost of the equipment and the recruitment of a team with complementary skills as previously described may represent a major investment for small companies. However, targeted analyses (quantification of one component or one family of molecules) are easier to develop and do not require as much expertise as, for example, needed for open metabolomics approaches, which may be the major activity of companies specialized in the omic technologies.
- The analytical techniques are progressing rapidly especially in the field of mass spectrometry, so that increased analytical capabilities are being offered, which necessitates more investments to use state-of-the-art methodologies. It seems that only companies able to carry out thousands of analyses a year would have the capabilities to stay on top of the technical advances. Consequently, it will be very hard for SMEs to innovate in this sector of research, except those that concentrate their efforts on targeted analyses.
- Omics technologies are still in their infancy and databases in some cases are still incomplete especially for mass spectrometry. Some databases are freely available such as the Human Metabolome Database, Kyoto Encyclopedia of Genes and Genomes, METLIN mass spectrometry database, etc. However, identification of molecules requires sometimes synthesizing standards, which is time- and money-consuming in order to obtain reference spectra. Consequently, some industrials feel that some of these technologies are not yet ready to be routinely used, considering that metabolomics is a young discipline.

13.4.2 POTENTIAL STRATEGIES

So far, industry, especially when working on large populations or sample sets, has subcontracted the analyses to the private "omics" sector, while keeping the intellectual property. However, research collaborations have also been carried out between industry and academia in order to create teams with complementary expertise. For example, the Metabolomics Society has initiated the creation of the "Industry Engagement Task Group" to enhance the relationship between the scientific community, the companies including software and hardware vendors, service providers, and industrial users (Viant et al., 2014). This is a new idea that should favor collaborations and stimulate the utilization of foodomics technologies by industry.

Different countries have built specialized platforms in life science research where these sophisticated instruments are being dedicated to research in food science, nutrition, toxicology, and pharmacology. These platforms are belonging to either large public research centers or a group of research centers from the same country, to private investigators associated with "omic technologies," or are joint ventures between large industries and the public sector.

The Netherlands Metabolomics Center (NMC) is a typical example of a public–private consortium including Unilever, TNO (Netherlands Organization for Applied Scientific Research), and several universities and research centers in the Netherlands. NMC has received a grant from the Netherlands Genomics Initiative and all partners together are invested in research studies. The NMC program started in the summer of 2008. Projects are focused on the development of novel technologies. In addition to the research program, NMC offers facilities (demonstration and competence laboratory, data support platform) to support the research community.

The public sector has also invested in creating centers of excellence in different countries. In Finland, the Biocenter Finland (BF) is a national research infrastructure of biocenters in six Finnish universities (Helsinki, Kuopio, Oulu, Tampere, Turku, and the Abo Akademi University). Funds were first received from the Ministry of Education and Culture. It is now supported by funds from the six host universities. BF offers services (bioinformatics, metabolomics, proteomics, biological imaging, translational research technology, etc.) to the entire Finnish research community from academia and industry and to a limited extent to users abroad. Investments in foodomics research were also realized in Wageningen in the Netherlands, in RIKILT, an independent organization that is part of Wageningen University and conducting research on food quality and safety, offering services in foodomics approaches.

In France, MetaboHUB, a project financed by the National Research Agency (ANR) as an "investissement d'avenir" (investment for the future) was created in 2013. MetaboHUB aims at providing state-of-the-art tools, services, and support in metabolomics and fluxomics to academic teams and industrial partners in the fields of nutrition, health, agriculture, and biotechnology. Three platforms constitute this national infrastructure, one in Bordeaux, one in Toulouse, and one in Clermont-Ferrand. It provides services to national as well as international organizations but it is also in charge of developing state-of-the-art technologies as requested by the different clients and researchers from different public laboratories such as INRA, INSERM, CNRS, and universities.

Outside Europe, over the past 12 years Canada has invested 10 million dollars in metabolomics. In Alberta, the Metabolomics Innovation Center, directed by David Wishart, offers metabolomics services for Canadian researchers and companies. However, the level of investment in metabolomics is low compared to what has been done for genomics, despite the growing interest in metabolomics coming from the agri-food sector as reported by Wishart. In the United States, the National Institutes of Health (NIH) has recently invested through the NIH common funds of about 14 million dollars to increase the capacity of research in metabolomics. These funds are used for technology development, training, standard synthesis, and data sharing.

Biocrates and Metabolon are two examples of private companies offering services in metabolomics. The first one is located in Austria while the second one has two laboratories in the United States. Biocrates, founded in 2002, mainly develops targeted approaches to identify potential biomarkers for diagnosis of multifactorial diseases in clinical research. It provides standardized diagnostic kits that give information about metabolites in cells, tissues, and biological fluids. It is also involved in many collaborative European-funded projects including partners from SMEs, universities, and academic clinical centers. Metabolon, a company founded in 2000, is also offering services in the same research area of life sciences. However, untargeted approaches are now proposed, allowing getting access to many chemical families. Targeted analyses are also possible in order to explore a particular metabolic class. The members of the company have also developed novel clinical diagnosis tests from their metabolic expertise. The scientific team has recently developed a novel lipidomic platform that should allow identification/quantification of a great number of complex lipid molecules. Metabolon is also involved in large long-term collaborations (Tachibana, 2014). We should also mention as a last example application notes published by some vendors of detection equipment and software for metabolomics.

The work published by Fournier et al. (2007) is a typical example of a targeted lipidomic method that was developed and validated as a result of collaboration between a food company and a research institute for quality control of ingredients. This was used to quantify small quantities of *trans* polyunsaturated fatty acids in refined commercially available oils, which contain eicosapentaenoic acid (EPA) and docosahexaenoic acid (DHA). These long-chain polyunsaturated n-3 fatty acids have *cis* ethylenic bonds and are essential components of biological membranes. In contrast, the *trans* isomers that can be formed during the refining process of oils have been shown to induce potential harmful effects (Fournier et al., 2007; Acar et al., 2002). It is therefore mandatory to have a robust method to quantify the major *trans* isomers that may be present in the oils. The limits of quantification of geometrical isomers of these two n-3 fatty acids were, respectively, 0.16 and 0.56 g/100 g of fish oil. The data obtained showed that the commercial oils analyzed contained low amounts (<1% of total fatty acids) of geometrical isomers of EPA and DHA. Consequently, these commercial fish oils could be used for supplementation of food products for human consumption.

13.5 CONCLUSIONS

Many studies have been published showing the power of omic technologies in the field of foodomics. However, many papers deal with one or two approaches like transcriptomics and metabolomics and a major challenge is to integrate and connect data sets from different expression levels such as gene, transcript, protein, and metabolite. Great progress already has been made (Skov et al., 2014), and collaboration between scientists from the public and private sectors as well as instruments' vendors has been initiated in different countries. Foodomics is now ready to provide data to be exploitable in food science, nutrition, and medicine. The next step will be their real implementation in the industry.

For metabolomics, especially when using LC-MS approaches, a number of technical challenges still have to be resolved, among them metabolite identification being the tedious one. Sharing of resources such as databases and chemical standards would be mandatory to carry out these identifications. Development of standardized data repositories would also help foodomics to reach its full potential.

REFERENCES

Acar, N., Chardigny, J.M., Bonhomme, B., Almanza, S., Doly, M., Sebedio, J.L., 2002. Long term intake of trans n-3 polyunsaturated fatty acids reduces the b-wave amplitude of electroretinograms in rats. Journal of Nutrition 132, 3151–3154.

Aguillera-Luiz, M., Romero-Gonzalez, R., Plaza-Bolanos, P., Martinez-Vidal, J.L., Garido Frenich, A., 2013. Wide-scope analysis of veterinary drug and pesticide residues in animal feed by liquid chromatography coupled to quadrupole-time-of-flight mass spectrometry. Analytical and Bioanalytical Chemistry 405, 6543–6553.

Antignac, J.-P., Courant, F., Pinel, G., Bichon, E., Monteau, F., Elliott, C., Le Bizec, B., 2011. Mass spectrometry-based metabolomics applied to the chemical safety of food. Trends in Analytical Chemistry 30, 292–301.

Bakker, G.C.M., van Erk, M.J., Pellis, L., Wopereis, S., Rubingh, C.M., Cnubben, N.H.P., Kooistra, T., van Ommen, B., Hendriks, H.F.J., 2010. An anti-inflammatory dietary mix modulates inflammation and oxidative and metabolic stress in overweight men: a nutrigenomics approach. American Journal of Clinical Nutrition 91, 1044–1059.

Bordoni, A., Capozzi, F., Ferranti, P., 2014. Preface: foodomics. The science for discovering. Food Research International 63, 125.

Brennan, L., 2013. Metabolomics in nutrition research: current status and perspectives. Biochemical Society Transactions 41, 670–673.

Brennan, L., 2015. Metabotyping: moving towards personalized nutrition. In: Sébeédio, J.L., Brennan, L. (Eds.), Metabolomics as a Tool in Nutrition Research. Woodhead Publishing, Kidlington, UK, pp. 137–144.

Capozzi, F., Bordoni, A., 2013. Foodomics: a comprehensive approach to food and nutrition. Genes & Nutrition 8, 1–4.

Capozzi, F., 2013. The new frontier in foodomics: the perspective of nuclear magnetic resonance spectrometry. Newfood 16, 44–48.

Capuano, E., Boerrigter-Eenling, R., van der Veer, G., van Ruth, S.M., 2013. Analytical authentication of organic products: an overview of markers. Journal of the Science of Food and Agriculture 93, 12–28.

Cerda, J., Manchado, M., 2013. Advances in genomics for flatfish aquaculture. Genes & Nutrition 8, 5–17.

Chin, E., Slupsky, C.M., 2013. Applications of metabolomics in food science: food composition and quality, sensory and nutritional attributes. In: Weimer, B.C., Slupsky, C. (Eds.), Metabolomics in Food and Nutrition. Woodhead Publishing, Kidlington, UK, pp. 217–230.

Cifuentes, A., Dugo, P., Fanali, S., 2013. Advances in food analysis. Journal of Chromatography A 1313, 1.

Cifuentes, A., 2009. Food analysis and foodomics. Journal of Chromatography A 1216, 7109.

Cifuentes, A., 2014. Foodomics: the necessary route to boost quality, safety, and bioactivity of foods. Electrophoresis 35, 1517–1518.

Cottenet, G., Blancpain, C., Golay, P.-A., 2011. Simultaneous detection of cow and buffalo species in milk from China, India, and Pakistan using multiplex real time PCR. Journal of Dairy Science 94, 3787–3793.

Cuollo, M., Caira, S., Fierro, O., Pinto, G., Picariello, Addeo, F., 2010. Toward milk speciation through the monitoring of casein proteotypic peptides. Rapid Communications in Mass Spectrometry 24, 1687–1696.

Deceuninck, Y., Bichon, E., Durand, S., Bemrah, N., Zendong, Z., Morvan, M.L., Marchand, P., Dervilly-Pinel, G., Antignac, J.P., Leblanc, J.C., Le Bizec, B., 2014. Development and validation of a specific and sensitive gas chromatography tandem mass spectrometry method for the determination of bisphenol A residues in a large set of food items. Journal of Chromatography A 1362, 241–249.

Dettmer, K., Aronov, P.A., Hammock, B.D., 2007. Mass spectrometry-based metabolomics. Mass Spectrometry Reviews 26, 51–78.

Dewulf, E.M., Cani, P.D., Claus, S.P., Fuentes, S., Puylaert, P.G., Neyrinck, A.M., Bindels, L.B., de Vos, W.M., Gibson, G.R., Thissen, J.P., Delzenne, N.M., 2013. Insight into the prebiotic concept: lessons from an exploratory, double blind intervention study with inulin-type fructans in obese women. Gut 62, 1112–1121.

Dunwell, J.M., 2014. Trangenic cereals: current status and future prospects. Journal of Cereal Science 59, 419–434.

Eisenreich, W., Bacher, A., 2007. Advances of high-resolution NMR techniques in the structural and metabolic analysis of plant biochemistry. Phytochemistry 68, 2799–2815.

Ferrara, M., Sébédio, J.L., 2015. Challenges in nutritional metabolomics: from experimental design to interpretation of data sets. In: Sebedio, J.L., Brennan, L. (Eds.), Metabolomics as a Tool in Nutrition Research. Woodhead Publishing, Kidlington, UK, pp. 3–35.

Fiehn, O., Kopka, J., Dormann, P., Altmann, T., Trethewey, R.N., Willmitzer, L., 2000. Metabolite profling for plant functional genomics. Nature Biotechnology 18, 1157–1161.

Fournier, V., Destaillats, F., Hug, B., Golay, P.-A., Joffre, F., Juanéda, P., Sémon, E., Dionisi, F., Lambelet, P., Sebedio, J.L., Berdeaux, O., 2007. Quantification of eicosapentaenoic and docosahexaenoic acid geometrical isomers formed during fish oil deodorization by gas-liquid chromatography. Journal of Chromatography A 1154, 353–359.

Gallardo, J.M., Ortea, I., Carrera, M., 2013. Proteomics and its applications for food authentification and food technology research. Trends in Analytical Chemistry 52, 135–141.

Gao, X., Pujos-Guillot, E., Sébédio, J.L., 2010. Development of a quantitative metabolomic approach to study clinical human fecal water metabolome based on trimethylsilylation derivatization and GC/MS analysis. Analytical Chemistry 82, 6447–6456.

Garcia-Canas, V., Simo, C., Herrero, M., Ibanez, E., Cifuentes, A., 2012. Present and future challenges in food analysis: foodomics. Analytical Chemistry 84, 10150–10159.

Garcia-Canas, V., Simo, C., Castro-Puyana, M., Cifuentes, A., 2014. Recent advances in the application of capillary electromigration methods for food analysis and foodomics. Electrophoresis 35, 147–169.

Geis-Asteggiante, L., Lehotay, S., Lightfield, A.R., Dutko, T., Ng, C., 2012. Ruggedness testing and validation of a practical analytical method for over 100 veterinary drug residues in bovine muscle by ultrahigh performance liquid chromatography-tandem mass spectrometry. Journal of Chromatography A 1258, 43–54.

German, J.B., Roberts, M.A., Fay, L., Watkins, S.M., 2002. Metabolomics and individual metabolic assessment: the next great challenge for nutrition. Journal of Nutrition 132, 2486–2487.

Giacometti, J., Josic, D., 2013. Foodomics in microbial safety. Trends in Analytical Chemistry 52, 16–22.

Giacometti, J., Buretic Tomljanovic, A., Josic, D., 2013. Application of proteomics and metabolomics for investigation of food toxins. Food Research International 54, 1042–1051.

Gibney, M.J., Walsh, M., Brennan, L., Roche, H.M., German, B., van Ommen, B., 2005. Metabolomics in human nutrition: opportunities and challenges. American Journal of Clinical Nutrition 82, 497–503.

Hedrick, V.E., Dietrich, A.M., Estabrooks, P.A., Savla, J., Serrano, E., Davy, B.M., 2012. Dietary biomarkers: advances, limitations and future directions. Nutrition Journal 11, 109.

Hermo, M.P., Saurina, J., Barbosa, J., Barrón, D., 2014. High-resolution mass spectrometry applied to the study of metabolome modifications in various chicken tissues after amoxicillin administration. Food Chemistry 153, 405–413.

Hird, S., Lau, B.P.-Y., Schuhmacher, R., Krska, R., 2014. Liquid chromatography-mass spectrometry for the determination of chemical contaminants in food. Trends in Analytical Chemistry 59, 59–72.

Hušek, P., 1998. Chloroformates in gas chromatography as general purpose derivatizing agents. Journal of Chromatography B 717, 57–91.

Ibanez, C., Valdes, A., Garcia-Canas, V., Simo, C., Celebier, M., Rocamora-Reverte, L., Gomez-Martinez, A., Herrero, M., Castro-Puyana, M., Segura-Carretero, A., Ibanez, E., Ferragut, J.A., Cifuentes, A., 2012. Global foodomics strategy to investigate the health benefits of dietary constituents. Journal of Chromatography A 1248, 139–153.

Ibanez, C., Simo, C., Garcia-Canas, V., Cifuentes, A., Castro-Puyana, M., 2013. Metabolomics, peptidomics and proteomics applications of capillary electrophoresis-mass spectrometry in foodomics: a review. Analytica Chimica Acta 802, 1–13.

Idle, J.R., Gonzalez, F.J., 2007. Metabolomics. Cell Metabolism 6, 348–351.

Inoue, K., Tanada, C., Sakamoto, T., Tsutsui, H., Akiba, T., Min, J.Z., Todoroki, K., Yamamo, Y., Toyo'oka, T., 2015. Metabolomics approach of infant formula for the evaluation of contamination and degradation using hydrophilic interaction liquid chromatography coupled with mass spectrometry. Food Chemistry 181, 318–324.

Ismail, N.A., Posma, J.M., Frost, G., Holmes, E., Garcia-Perez, I., 2013. The role of metabonomics as a tool for augmenting nutritional information in epidemiological studies. Electrophoresis 34, 2776–2786.

Jiménez-Pranteda, M.J., Pérez-Davó, A., Monteoliva-Sánchez, M., Ramos-Cormenzana, A., Aguilera, M., 2015. Food omics validation: towards understanding key features for gut microbiota, probiotics and human health. Food Analytical Methods 8, 272–289.

Kanani, H., Chrysanthopoulos, P.K., Klapa, M.I., 2008. Standardizing GC-MS metabolomics. Journal of Chromatography B 871, 191–201.

Kaspar, S., Peukert, M., Svatos, A., Matros, A., Mock, H.-P., 2011. Maldi-imaging mass spectrometry-an emerging technique in plant biology. Proteomics 11, 1840–1850.

Koek, M.M., Muilwijk, B., van Stee, L.L.P., Hankemeier, T., 2008. Higher mass loadability in comprehensive two-dimensional gas chromatography-mass spectrometry for improved analytical performance in metabolomics analysis. Journal of Chromatography A 1186, 420–429.

Laghi, L., Picone, G., Capozzi, F., 2014. Nuclear magnetic resonance for foodomics beyond food analysis. Trends in Analytical Chemistry 59, 93–102.

Llorach, R., Garcia-Aloy, M., Tulipani, S., Vazquez-Fresno, R., Andres-Lacueva, C., 2012. Nutrimetabolomic strategies to develop new biomarkers of intake and health effects. Journal of Agricultural and Food Chemistry 60, 8797–8808.

Lodge, J.K., 2010. Symposium 2: modern approaches to nutritional research challenges: targeted and non-targeted approaches for metabolite profiling in nutritional research. Proceedings of the Nutrition Society 69, 95–102.

Manach, C., Brennan, L., Dragsted, L.O., 2015. Using metabolomics to evaluate food intake: applications in nutritional epidemiology. In: Sébédio, J.L., Brennan, L. (Eds.), Metabolomics as a Tool in Nutrition Research. Woodhead Publishing, Kidlington, UK, pp. 167–202.

Mazzei, P., Piccolo, A., 2012. HRMAS-NMR metabolomics to assess quality and trceability of mozzarella cheese from Campania buffalo milk. Food Chemistry 132, 1620–1627.

Melis, R., Cappuccinelli, R., Roggio, T., Anedda, R., 2014. Addressing marketplace gilthead sea bream differentiation by H NMR-based lipid fingerprinting. Food Research International 63, 258–264.

Mie, A., Laursen, K.H., Åberg, K.M., Forshed, J., Lindahl, A., Thorup-Kristensen, K., Olsson, M., Knuthsen, P., Larsen, E.H., Husted, S., 2014. Discrimination of conventional and organic white cabbage from a long-term field trial study using untargeted LC-MS-based metabolomics. Analytical and Bioanalytical Chemistry 406, 2885–2897.

Moco, S., Vervoort, J., Moco, S., Bino, R.J., De Vos, R.C.H., Bino, R., 2007. Metabolomics technologies and metabolite identification. TRAC Trends in Analytical Chemistry 26, 855–866.

Moco, S., Martin, F.P., Rezzi, S., 2012. A metabolomics view on gut microbiome modulation by polyphenol-rich foods. Journal of Proteome Research 11, 4781–4790.

Morio, B., Comte, B., Martin, J.F., Chanseaume, E., Alliger, M., Junot, C., Lyan, B., Boirie, Y., Vidal, H., Laville, M., Pujos-Guillot, E., Sébédio, J.L., 2015. Metabolomics reveals differential metabolic adjustments of normal and overweight subjects during overfeeding. Metabolomics 11, 920–938.

Mutz, K.-O., Heilkenbrinker, A., Lonne, M., Walter, J.-G., Stahl, F., 2013. Transcriptome analysis using next-generation sequencing. Current Opinion in Biotechnology 24, 22–30.

Nasi, A., Picariello, G., Ferranti, P., 2009. Proteomic approaches to study structure, functions and toxicity of legume seeds lectins. Perspectives for the assessment of food quality and safety. Journal of Proteome Research 72, 527–538.

Nazzaro, F., Orlando, P., Fratianni, F., Di Lucia, A., Coppola, R., 2012. Protein analysis-on-chip systems in foodomics. Nutrients 4, 1475–1489.

Newgard, C.B., An, J., Bain, J.R., Muehlbauer, M.J., Stevens, R.D., Lien, L.F., Haqq, A.M., Shah, S.H., Arlotto, M., Slentz, C.A., Rochon, J., Gallup, D., Ilkayeva, O., Wenner, B.R., Yancy Jr., W.S., Eisenson, H., Musante, G., Surwit, R.S., Millington, D.S., Butler, M.D., Svetkey, L.P., 2009. A branched-chain amino acid-related metabolic signature that differentiates obese and lean humans and contributes to insulin resistance. Cell Metabolism 9, 311–326.

Nicholson, J.K., Lindon, J.C., Holmes, E., 1999. 'Metabonomics': understanding the metabolic responses of living systems to pathophysiological stimuli via multivariate statistical analysis of biological NMR spectroscopic data. Xenobiotica 1181–1189.

Oh, E., Hasan, M.D.N., Jamshed, M., Hong, H.-M., Song, E.J., Yoo, Y.S., 2010. Growing trend of CE at the omics level: the frontier of systems biology. Electrophoresis 31, 74–92.

Ortea, I., Pascoal, A., Canas, B., Gallardo, J.M., Barros-Velazquez, J., Calo-Mata, P., 2012. Food authentification of commercially-relevant shrimp and prawn species: from classical methods to Foodomics. Electrophoresis 33, 2201–2211.

Pan, Z., Raftery, D., 2007. Comparing and combining NMR spectroscopy and mass spectrometry in metabolomics. Analytical and Bioanalytical Chemistry 387, 525–527.

Pareek, C.S., Smoczynski, R., Tretyn, A., 2011. Sequencing technologies and genome sequencing. Journal of Applied Genetics 52, 413–435.

Picariello, G., Mamone, G., Addeo, F., Ferranti, P., 2012. Novel mass spectrometry-based applications of omic sciences in food technology and biotechnology. Food Technology and Biotechnology 50, 286–305.

Pujos-Guillot, E., Hubert, J., Martin, J.F., Lyan, B., Quintana, M., Claude, S., Chabanas, B., Rothwell, J.A., Bennetau-Pelissero, C., Scalbert, A., Comte, B., Hercberg, S., Morand, C., Galan, P., Manach, C., 2013. Mass spectrometry-based metabolomics for the discovery of biomarkers of fruit and vegetable intake: citrus fruit as a case study. Journal of Proteome Research 12, 1645–1659.

Rezzi, S., Martin, F.P., Kochhar, S., 2008. Defining personal nutrition and metabolic through metabonomics. Ernst Schering Foundation Symposium Proceedings 4, 251–264.

Ritota, M., Casciani, L., Failla, S., Valentini, M., 2012. HRMAS-NMR spectroscopy and multivariate analysis meat characterisation. Meat Science 92, 754–761.

Sebedio, J.L., Brennan, L., 2015. Metabolomics as a Tool in Nutrition Research. Woodhead Publishing, Kidlington, UK.

Sébédio, J.L., Polakof, S., 2015. Using metabolomics to identify biomarkers for metabolic diseases: analytical methods and applications. In: Sébédio, J.L., Brennan, L. (Eds.), Metabolomics as a Tool in Nutrition Research. Woodhead Publishing, Kidlington, UK, pp. 145–166.

Shepherd, L.V.T., Dobson, G., Verrall, S.R., Conner, S., Griffiths, D.W., McNicol, J.W., Davies, H., Stewart, D., 2007. Potato metabolomics by GC-MS: what are the limiting factors ? Metabolomics 3, 475–488.

Shepherd, L.V.T., Fraser, P., Stewart, D., 2011. Metabolomics: a second-generation platform for crop and food analysis. Bioanal 3, 1143–1159.

Shumila, E., Ciampa, A., Capozzi, F., Rustad, T., Dikiy, A., 2015. NMR approach for monitoring post-mortem changes in atlantic salmon fillets stored at 0 and 4°C. Food Chemistry 184, 12–22.

Simo, C., Ibanez, C., Valdés, A., Cifuentes, A., Garcia-Canas, V., 2014. Metabolomics of genetically modified crops. International Journal of Molecular Sciences 15, 18941–18966.

Skov, T., Honoré, A.H., Jensen, H.M., Naes, T., 2014. Chemometrics in foodomics: handling data structures from multiple analytical platforms. Trends in Analytical Chemistry 60, 71–79.

Tachibana, C., 2014. What's next in omics: the metabolome. Science 345, 1519–1521.

Taira, S., Uematsu, K., Kaneko, D., Katano, H., 2014. Mass spectrometry imaging: application in food science. Analytical Sciences 30, 197–203.

Tao, X., Liu, Y., Wang, Y., Qiu, Y., Lin, J., Zhao, A., Su, M., Jia, W., 2008. GC-MS with ethyl chloroformate derivatization for comprehensive analysis of metabolites in serum and its application to human uremia. Analytical and Bioanalytical Chemistry 391, 2881–2889.

Touboul, D., Brunelle, A., Laprévote, O., 2011. Mass spectrometry imaging: towards a lipid microscope? Biochimie 93, 113–119.

Valdes, A., Ibanez, C., Simo, C., Garcia-Canas, V., 2013a. Recent transcriptomics advances and emerging applications in food science. Trends in Analytical Chemistry 52, 142–154.

Valdés, A., Simo, C., Ibanez, C., Garcia-Canas, V., 2013b. Foodomics strategies for the analysis of transgenic foods. Trends in Analytical Chemistry 52, 2–15.

Valdes, A., Garcia-Canas, V., Simo, C., Ibanez, C., Miol, V., Ferragut, J.A., Cifuntes, A., 2014. Comprehensive foodomics study on the mechanisms operating at various molecular levels in cancer cells in response to individual rosemary polyphenols. Analytical Chemistry 86, 9807–9815.

Valentini, M., Ritota, M., Cafiero, C., Cozzolino, S., Leita, L., Sequi, P., 2011. The HRMAS-NMR tool in foodstuff characterisation. Magnetic Resonance in Chemistry 49, S121–S125.

Viant, M.R., Bearden, D.W., Creek, D.J., Fiehn, O., Robertson, D.G., 2014. Supporting the industry sector of the metabolomics community: the remit of the Metabolomics Society's Industry Engagement Task Group. Metabolomics 10, 541–542.

Wang, T.J., Larson, M.G., Vasan, R.S., Cheng, S., Rhee, E.P., Mccabe, E., Lewis, G.D., Fox, C.S., Jacques, P.F., Fernandez, C., 2011. Metabolite profiles and the risk of developing diabetes. Nature Medicine 17, 448–453.

Weber, D., Cléroux, Ben Rejeb, S., 2009. Emerging analytical methods to determine gluten markers in processed foods-method development in support of standard setting. Analytical and Bioanalytical Chemistry 395, 111–117.

Wilson, I.D., Plumb, R., Granger, J., Major, H., Williams, R., Lenz, E.M., 2005. HPLC-MS-based methods for the study of metabonomics. Journal of Chromatography. B, Analytical Technologies in the Biomedical and Life Sciences 817, 67–76.

Wishart, D.S., Jewison, T., Guo, A.C., Wilson, M., Knox, C., Liu, Y., et al., 2012. HMDB 3.0-The human metabolome database. Nucleic Acids Research 41, D801–D807.

Wishart, D.S., 2007. Current Progress in computational metabolomics. Briefings in Bioinformatics 8, 279–293.

Wolf, J., 2013. Principles of transcriptome analysis and gene expression quantification: an RNA-seq tutorial. Molecular Ecology Resources 13, 559–572.

Zheng, C., Chen, A., 2014. System biological research on food quality for personalized nutrition and health using foodomics techniques: a review. Journal of Food and Nutrition Research 2, 608–616.

CHAPTER

CONSUMER ACCEPTANCE OF NOVEL FOODS 14

A.R.H. Fischer[1], M.J. Reinders[2]

[1]Wageningen University, Wageningen, the Netherlands; [2]LEI Wageningen UR, Agricultural Economics Research Institute, the Hague, the Netherlands

The story so far: In the beginning the Universe was created. This has made a lot of people very angry and been widely regarded as a bad move.
Douglas Adams, *The Restaurant at the End of the Universe*

14.1 INTRODUCTION

Many innovations face some criticism when first launched. The ultimate success or failure of innovations depends to a considerable extent on a favorable overall consumer response to those innovations (see, eg, Gupta et al., 2012; Ronteltap et al., 2007; Siegrist, 2000). Novel foods fail due to the lack of consumer acceptance or even active consumer rejection of these products. Infamous examples of innovations that have suffered setbacks following negative consumer response are the introduction of genetically modified organisms in Europe (eg, Frewer et al., 2013a) and food irradiation (Bruhn, 1995; Diehl, 2002). Consumer responses to novel foods depend on how consumers perceive the benefits, risks, and costs of the novel foods and their underlying technologies (Ronteltap et al., 2007). Consumer perceptions and how they are combined into an overall evaluation of the novel food may systematically differ from expert assessments of the product (Lazo et al., 2000; Siegrist and Gutscher, 2006; Slovic et al., 1995). In addition, consumers may perceive and decide on the benefits of a product in different ways depending on their personal situation or the context of the food choice (Bagozzi et al., 2000).

To avoid unnecessary rejection of innovations, or to avoid investing in innovations that are inherently unacceptable to the public, it is therefore important to include consumer insights into the innovation process early on (Van Kleef et al., 2005). It is thus important to know: (1) which are the relevant perceptions of consumers in the context of food innovations and how they combine toward a final response, and (2) the products in which this innovation is applied (Gupta et al., 2012; Ronteltap et al., 2011). This also implies that we need to know how to measure consumer perceptions reliably (Churchill, 1979; Reinders et al., 2013). Both needs are complicated by the fact that perceptions and decisions are in the mind of the consumer and cannot be measured directly. It has been among the main tasks of (food-related) consumer scientists to develop theories to identify the determinants of consumer response to novel foods, eg, how these are related to each other and how we can measure them. This chapter will provide an overview of insights of these findings relevant to consumer acceptance of innovations in food.

Innovation Strategies in the Food Industry. http://dx.doi.org/10.1016/B978-0-12-803751-5.00014-3

Before we investigate consumer acceptance of food innovations, it is important to realize that the term *consumer acceptance* is used somewhat ambivalently across different disciplines (Ronteltap et al., 2011). In marketing and product innovation literature, the phrase "acceptance of a novel product" is generally used to indicate that a novel product has been taken up by sufficiently many users to be profitable. This implies that a product has been taken up by early adopters and subsequently has been adopted by larger groups of potential users (Rogers, 1962/1995). In the public understanding of science and risk analysis literature, "consumer acceptance" of a product or technology focuses on psychological acceptance of the product. In these approaches, the word *acceptance* indicates that people do not categorically reject a product or technology but are in principle willing to consider that product or technology, or may even be somewhat favorable toward it (eg, Bredahl, 2001; Siegrist, 2000). Acceptance defined as such does not imply the product is actually sold to anyone. It does predict that societal protest against the innovation is unlikely and that the product has a chance in the market. Arguable is that this latter type of psychological acceptance provides a necessary but not sufficient precondition for a product to be taken up by consumers, ie, to be accepted in the other meaning of the term.

In this chapter, the emphasis leans toward the latter definition, as actual societal adoption of novel products and products still under development cannot be fully studied. We start this chapter with a discussion of an important shift in the 20th century, where the dominance of expert and producer strategies became amended and even replaced by setting the consumer view as a central point. Then we introduce theories on consumer acceptance and communication commonly in use of innovations context. Subsequently, an overview of considerations when collecting consumer data is provided. Then we take a look at the more recent insights of consumer science and how these can be used to complement current theories, or where they show limitations of the current approaches. Finally, conclusions about the use of consumer science in food innovation are presented.

14.2 THE EMERGENCE OF CONSUMER OPINION

In the early 20th century Henry Ford could keep it simple. The T-model Ford was available in "...any color as long as it's black."

Ford's strategy worked since the early 20th century can be generally characterized as a seller's market in which demand was greater than supply. Consumers simply had to take whatever was available in the market. Around the mid-20th century, this perspective changed. The seller's market changed into a buyer's market as the number of new products that entered the market exceeded consumer demand. As a consequence, consumers turned to be more selective in the products they buy, and sales became more important. In these post-World War II years, marketing of new products was centered around the sales function (see Solomon and Banerjee, 2006). Even though the sales (instead of the production) department was now in charge, this view still relied on the dominance of the selling party. However, in the 1960s and 1970s it became clear that consumers do not like to be pushed and want their demands more visibly reflected in products. It was in this period that marketing adopted a consumer orientation and started to study consumers in an attempt to understand their needs and tailor products to different consumer groups. It became a widespread belief that companies that are able to uncover or even anticipate consumer demand, deliver against it, and communicate this effectively to consumers have a higher chance of survival and success in the marketplace (Costa-Font and Mossialos, 2006). Especially in the highly competitive food industry, it became critical to listen to consumers and communicate with them very carefully. Product innovation in the food industry should not merely reflect technological possibilities but also consumer needs and preferences (eg, Van Kleef et al., 2005).

Simultaneously, risk analysis experts realized that not only did public perceptions of risk and benefits systematically differ from technical assessments (Starr, 1969) but that this also started to pose societal problems for the acceptance of new technologies. In the 1970s it became clear that people judge risks and benefits to some extent by rules of thumb rather than making a full and deliberate assessment of the risk (Tversky and Kahneman, 1974). When measuring the perception of risks and benefits, monetary risks were seen as more negative than the same monetary benefits as positive, and with constant increases in monetary risks and benefits, the increments of perceived risks and benefits gradually declined (Kahneman and Tversky, 1979). Similarly, societal risks did not follow risk assessment. Protests in the 1970s focused around high perceived risks of environmental pollution and nuclear power (Fischhoff et al., 1978). This resulted in the psychometric paradigm of risk perceptions (Slovic, 1987). The psychometric paradigm explains which type of risks are systematically overestimated (ie, those leading to catastrophic and dreaded consequences as well as those that are uncontrollable and unobservable from a consumer point of view) and what risks are systematically underestimated (ie, leading to nondreaded consequences or those controllable and observable risks). In the debate of genetic modification, this approach shows that mainly the uncontrollable and unobservable nature of genetic modification contributed to increased risk perception (Fife-Schaw and Rowe, 1996, 2000). These findings opposed the then dominant world view that consumers are rational decision makers that calculate and optimize expected utility (that is the overall value of a product, objectively weighed), and that this is the best way to describe consumer behavior (Slovic et al., 2002; Von Neumann and Morgenstern, 1944).

During this debate the foundation for the theories on consumer behavior in general and consumer response to innovations in particular that are still in use today were laid.

14.3 MAJOR THEORIES ON CONSUMER ACCEPTANCE OF INNOVATIVE PRODUCTS

> Current state of consumer research is like an "exploded confetti factory"
>
> **Paraphrasing the metaphor popularized by Dutch psychologist Piet Vroon.**

A number of theories have been dominant in assessing consumer response to innovations (Ronteltap et al., 2011). These include the theory of planned behavior (Ajzen, 1991), the technology acceptance model (Davis, 1989), diffusion of innovations (Rogers, 1962/1995), the health belief model (Janz and Becker, 1984), and the protection motivation model (Prentice-Dunn and Rogers, 1986) as well as a number of studies focusing on risk–benefit perception, trust and knowledge (eg, Ronteltap and Van Trijp, 2007; Schenk et al., 2008; Siegrist, 2000).

The theory of planned behavior (TPB) (Ajzen, 1991), the successor of the theory of reasoned action (Fishbein and Ajzen, 1975), predicts consumer behavior from the intention of the consumer to conduct that behavior and the perceived control the consumer has about the behavior (Fig. 14.1). Intentions are in turn formed by attitudes, social norms, and perceived behavioral control. Attitudes are summary evaluations of the attitude object based on beliefs about how attributes of an object contribute to the overall valuation of the product and evaluations of the level of these attributes. Social norms are based

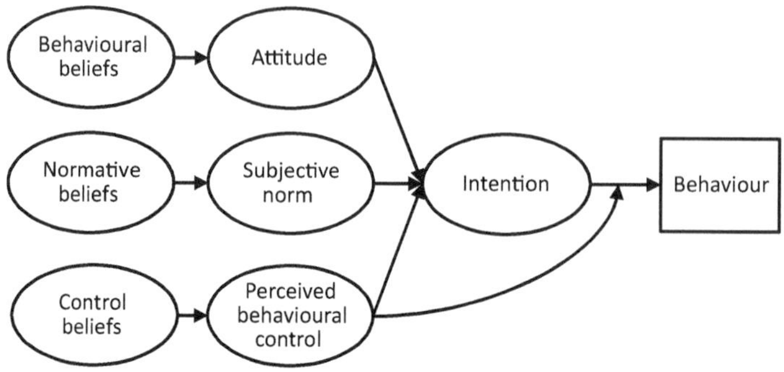

FIGURE 14.1

The theory of planned behavior, currently one of the most frequently applied models for predicting consumer preference formation.

Adapted from Ajzen, I., 1991. The theory of planned behavior. Organizational Behavior and Human Decision Processes 50, 179–211.

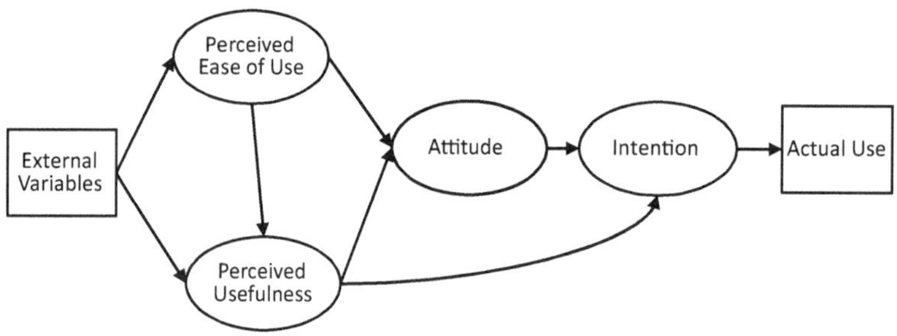

FIGURE 14.2

The technology acceptance model as originally proposed.

Adapted from Davis, F.D., 1993. User acceptance of information technology: system characteristics, user perceptions and behavioral impacts. International Journal of Man-Machine Studies 38, 475–487.

on beliefs and evaluations about social approval of conducting the behavior by relevant peers, and perceived behavioral control results from beliefs and evaluations of the likelihood that the behavior is under control (Ajzen, 1991).

The TPB is among the most frequently used theories in product choice (see, eg, Armitage and Conner, 2000) and is applied straightforwardly to novel foods (eg, Saba and Vassallo, 2002). The theory is versatile to some extent. Additional predictors to intention such as moral norm have been added and were shown to be relevant in the food context (Hübner and Kaiser, 2006; Spence and Townsend, 2006). Antecedents predicting attitude in more detail based on risk and benefit weighing and personality characteristics (Bredahl, 2001) can be included within the framework of TPB (Chen, 2008). Moderating variables such as cultural differences can be used to explain different weights between norms, control, and attitude in the TPB (Chen and Li, 2007; Frewer et al., 2008).

Another frequently used model is the technology acceptance model (TAM—Fig. 14.2) (Davis, 1989). The TAM shares many assumptions about how consumers process information and arrive at a decision with the TPB. Like the TPB it predicts the use of a technology from attitudes toward that technology. The attitudes themselves are in turn predicted by perceived usefulness and ease of technology usage. The central role of ease of use and usefulness in the technology acceptance model is related to its development in the use of electronics (in the 1980s) when more complicated consumer products entered the market with often highly technical user interfaces (eg, programmable video recorders and personal computers operating under MS-DOS). Since its original conception, the TAM has been extended to include elements from TPB and other models (Venkatesh and Bala, 2008; Venkatesh and Davis, 1996, 2000; Venkatesh et al., 2003, 2012).

The health belief model was introduced in the 1950s (Janz and Becker, 1984). It focuses on motivations why and when perceived threats to personal health result in consumer behavior aimed at avoiding or nullifying those threats. Based on ideas presented in the health belief model, a more generally applicable model related to avoiding risks was created in the protection motivation model (Rogers, 1983). An important distinction with the previous models is that it puts motivation (rather than attitude or intention) central in predicting consumer behavior. The protection motivation model introduces two ways of dealing with potentially harmful situations. The first route (labeled maladaptive) is based on avoiding threats, while the second route (adaptive) is about coping with threats. If a threat is appraised as substantial and coping is considered feasible, the motivation to engage in protective behavior is strong and action to deal with the threat is likely. From the perspective of innovations it should, however, be noted that this model focuses more on removing potentially negative consequences of a situation and how to deal with these than on specific properties of the product or innovation.

Several approaches consider both negative and positive outcomes by dealing with the role of risk–benefit perceptions in consumer preference (Fig. 14.3). These approaches do not usually follow a predefined theoretical model but specify a "theorette" for the problem at hand. Nevertheless, there are some general relations across these approaches. Perceived risks negatively influence, and perceived benefits positively influence, acceptance (eg, Bredahl, 2001; Fischer and Frewer, 2009; Frewer et al., 2015; Schenk et al., 2008; Siegrist, 2000; Siegrist et al., 2007). In addition, it is generally shown that perceptions of risk and benefit are negatively correlated (Alhakami and Slovic, 1994; Finucane et al., 2000; Siegrist et al., 2007), where one would expect a positive correlation as high risk–high benefit options may be viable, but high risk–low benefit options not (see Fig. 14.3A).

Different antecedents of risk and benefit perceptions have been studied. Trust in the agent, producer, or government that launches or communicates about an innovation plays an important role in reducing risk perception and increasing benefit perception (Frewer et al., 2003; Siegrist, 2000) (see Fig. 14.3B), although there is also evidence for a reversed causality, where we trust those who tell us what we want to hear (Eiser et al., 2002; Poortinga and Pidgeon, 2005). Furthermore, other antecedents of perceived risks and benefits, like knowledge, neophobia, and general attitudes toward nature and technology, have been conceptualized and empirically tested (see Fig. 14.3C).

All previous models take the psychological approach that information about a novel product or technology is processed in the mind leading to an opinion about that product. An alternative to the assumption that opinions are based on a systematic combination of arguments is to use already present opinions of similar products as indication for evaluation of the innovative product. This allows consumers to predict attributes of a product based on their knowledge of products in the same category. Categorization literature in marketing studies how people assign new products into existing categories (Loken, 2006), how these categories may need to be adjusted to accommodate the innovative product (Moreau et al., 2001), or when a new category has to be created to give the innovative product a place in the mind (Michaut, 2004).

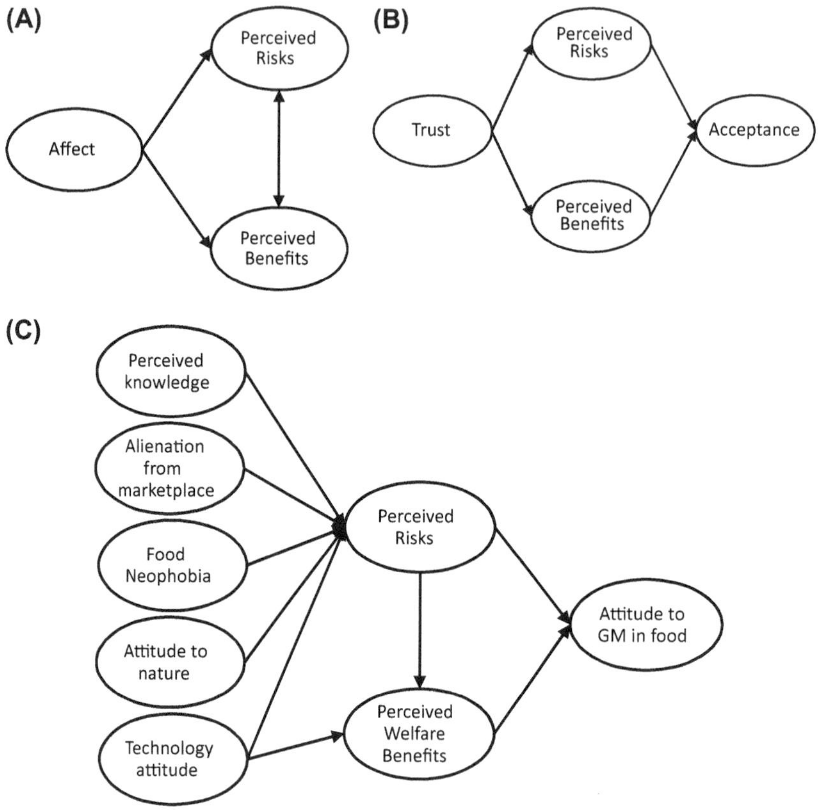

FIGURE 14.3

Examples of risk–benefit models. Panel (A) (Finucane et al., 2000), panel (B) (Siegrist, 2000), panel (C) (Bredahl, 2001). Note that these are only a few examples of this kind.

In addition, it has been studied how communication can facilitate placement of innovative products into relevant categories by creating analogies between the novel product and existing categories (Gregan-Paxton et al., 2002; Gregan-Paxton and John, 1997), and how this leads to opinions about and acceptance of the innovative product (Kardes et al., 2004; Sujan, 1985). Going one step further is the use of experiential analogies in which a novel product is compared to an emotional experience (Goode et al., 2010). Such an analogy has the power to focus consumers on the evaluative, emotional, and multisensory information associated with the product experience and could especially have potential for enhancing the acceptance of novel products within the food domain. However, in spite of the potential contribution to understanding consumer acceptance of innovations, the research into category-based inferences has not customarily been used in the literature on consumer response to innovation, and has developed in different scientific communities and with different jargon (Ranganath et al., 2010).

Besides the more psychologically based outlook considered in this chapter, sociological approaches such as the diffusion of innovations theory (Rogers, 1962/1995) are also used in the context of innovations in food. Diffusion of innovations has identified five characteristics of innovations that have been successfully adopted in the market. These are (1) relative advantage (ie, delivering a benefit over preceding technologies), (2) compatibility (ie, fitting with values, experiences, and needs of potential adopters), (3) complexity (ie, level of ease of use), (4) trialability (ie, the possibility to experiment with an innovation before actual adoption), and (5) observability by others. Innovation science investigates consumer response to innovative products (Smits and den Hertog, 2007) in an often more reflective way, describing specifics of an innovation trajectory. While diffusion of innovation has identified preconditions for acceptance and innovation science provides reflection on the launch of innovations, they provide no insight into deliberations leading consumers to object to or favor an innovation (Claudy et al., 2014).

In summary, there is a broad range of approaches to assess consumer response to innovations, both within consumer psychology and beyond. All presented models are models (sic) of consumer response. This means that they give a simplified reflection of reality and that many important determinants of behavior are left out. The specific question one has and the decision of which indicators of consumer response are most important to know at which stage of the innovation development process should be used to select, adapt, and apply the most relevant approach. This decision should be taken consciously and reevaluated frequently.

14.4 COMMUNICATION THEORIES

> The single biggest problem in communication is the illusion that it has taken place.
>
> **George Bernard Shaw**

Many approaches predicting consumer response to innovative products have in common that evaluations of the innovative product and/or its attributes lead to perceptions or attitudes. Positive attitudes in turn lead to positive response and more consumers buying the product. These models tend to be linear, as in that they start with the construction of perceptions and attitudes and end with behavior. Under this assumption, influencing the construction of consumer perceptions and attitudes will lead to predictable behavior change. The obvious choice is to influence attitude by providing additional information about the properties of the innovative products so that an attitude is formed based on more complete information on the innovation. This will only work, however, if we know to what extent the information is indeed communicated.

One of the leading models in this context is the elaboration likelihood model (ELM) (Petty and Cacioppo, 1986). The ELM posits that communication can be interpreted and processed in two different ways by consumers. From the communication perspective, the central route (as labeled by Petty and Cacioppo) is the most desirable one, leading to lasting change in attitudes in the direction of the provided information. The central route is, however, very demanding on consumers, as it requires that consumers are (Petty and Wegener, 1999):

1. motivated to actually use the information (eg, because it is personally important to them)
2. are able to process information at the time it is provided (are not distracted, information and jargon fits with their knowledge)
3. the information is given of sufficient quality and adds to existing knowledge
4. people take the time to repeat the information in their mind and memorize the information

If any of these conditions are not met, people will either keep their existing attitude or will temporarily change their opinion based on peripheral cues embedded in the message, ie, consumers will use the other, peripheral, route to process the communicated information. These peripheral cues may be specific words but may even be associations at the periphery of the message, eg, the layout, paper quality, and color scheme of a leaflet.

Other models of attitude change take this thought one step further and posit that as long as people think they have sufficient knowledge they will not consider new arguments at all (Heuristic Systematic Model Chaiken, 1980; Chen and Chaiken, 1999). The risk information seeking and processing model (RISP) follows this latter idea by positing that it is the perceived lack of information that is the consumers' main driver for seeking and using information about a product or situation (Griffin et al., 1999, 2002; Trumbo, 2002; Yang et al., 2014).

Regarding the implementation of communication, it is essential that the consumer considers the additional information as relevant, is willing to spend time and effort, as well as has sufficient knowledge and intelligence to process the information. If any of these preconditions are not present, no lasting predictable change in consumer opinion will result. Ideally we would therefore ensure consumers are highly motivated to choose the best food products available in order to make them use information to the fullest extent. Motivating consumers to think about food choice may, however, not be successful as food choices are very frequent (most of us make multiple food decisions every single day). Therefore considering each food decision in detail would quickly become overly demanding. If we consider the sheer amount of time needed to elaborate on each single food decision we take, it makes sense that many if not most of our food decisions are based on habit and past behavior (Honkanen et al., 2005). This is in line with the idea that "the more of the details of our daily life we can hand over to the effortless custody of automatism, the more our higher powers of mind will be set free for their own proper work" (James, 1890/1950). Considering the communication perspective, automatisms and habits pose barriers, as it is unlikely consumers will be always sufficiently motivated to process the provided information when they can already rely on a habit.

Consumers that make automatic and habitual choices do so based on a vast amount of prior experience and preference. If innovative products and technologies can be developed based on this vast amount of consumer knowledge and their use in daily life, the likelihood of successful introduction increases (Von Hippel, 1976), even without the need to change attitudes. This requires a reversal in communication, ie, rather than innovators communicating their plans to end users, end users now need to communicate and be involved as stakeholders early in the development of a new technology or product embodying it in order to capture a more usage-driven view into the development process itself (von Hippel, 1986). Bringing consumers into innovation teams as stakeholders is claimed to be a way to allow usage of end user evaluations early in development, which limits the number of adjustments that have to be made after the development stage (Nahuis et al., 2012). While early end user involvement seems promising, strong empirical evidence that this approach does indeed lead to fewer failed products and technologies is lacking (Reinders et al., 2013). In addition, best practices are difficult to identify, as this often involves a considerable change in the structure of the innovation process. In the consumer involvement approach the classical cascade method, where technology is developed first, prior to products and prior to market introduction, cannot be maintained. Instead, at each moment of technology development insights from markets and end consumers need to be fed back into the innovation process. This requires advanced methods to tap into consumer opinions during development of a product or technology and at the same time it requires that the innovation process is receptive to this information from consumers. This can only be achieved if the approaches are fully embedded in the organization including in the code of conduct, employee reward schemes, and top-level management support (Munksgaard and Freytag, 2011; Rossi, 2011).

14.5 METHODOLOGIES TO RECORD CONSUMER OPINIONS ON NOVEL FOODS

"Never used it myself," Ridcully said. "Wetting a finger and holding it up has always been good enough for me."

Terry Pratchett, *The Last Continent*

There are a number of difficulties associated with recording consumer opinions. First of all, measuring consumer responses requires indirect tools to identify noncorporeal, psychological ideas. Secondly, depending on the specific theoretical approach and research questions, we often need to measure several underlying psychological constructs for different problems. To do this rigorously, relevant knowledge of how each psychological construct is defined is necessary. There is, however, some ongoing disagreement about such definitions, even for central constructs like attitudes (Gawronski, 2007). A third practical problem is how to translate these psychological constructs into measurable indicators such as survey questions in such a way that they measure the same across language and cultural divides. Even a thoroughly designed, tested, and translated survey requires advanced statistics to provide the necessary checks and balances (Steenkamp and Baumgartner, 1998). These problems are amplified by the fact that there are a huge number of researchers who, independently, engage in survey research making it hard to identify a single best practice. Another problem is that it is very tempting to create survey questions that are close to the theoretical research question although this may not result in a reliable and valid instrument to measure consumer opinions. The large number of researchers and differences in operationalization of the research question in terms of survey questions leads to different and inconsistently used measures for the same psychological construct (Hendrick et al., 2013), but more seriously to problems with the reliability and validity of used methods (Churchill, 1979; Reinders et al., 2013). Together, in spite of the huge amount of published literature, this makes it nearly impossible to create a robust body of evidence based on meta-studies that go beyond the most mainstream models such as TPB (Reinders et al., 2013).

When several psychological constructs or elements need to be measured, different techniques may be required. Since there is a range of different psychological constructs, a broad range of techniques is used in practice as the research question requires (see, eg, Van Kleef et al., 2005). In our discussion of methodologies to record consumer response to novel foods, it is important to make a distinction between "methods" and "measures." With "methods" we mean the ways in which empirical data can be collected and how this relates to the research questions relevant to the specific stage of the innovation process. The term "measures" reflects psychological constructs that are operationalized in an instrument suitable for data recording. For example, it reflects how we ensure that survey questions on attitude truly measure attitude and not something else.

14.5.1 DATA COLLECTION METHODS

Generally, a distinction can be made between two families of methods: qualitative and quantitative ones. The choice of method depends on the specific research question. If a phenomenon is occurring in the world and is largely unknown, exploratory observations and the collection of data without a predefined theoretical approach can be helpful. Once sufficient data are collected, general rules can then be induced, following which predictions based on these rules can be deduced and tested. These results can be evaluated against real-world relevance to complete the empirical cycle (Fig. 14.4, de Groot, 1969). Incomplete cycles are, however, ubiquitous. The practice in postmodern science is that every

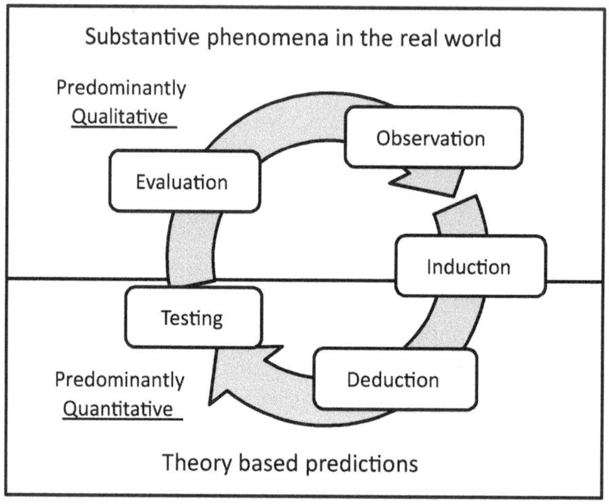

FIGURE 14.4

The empirical cycle (de Groot, 1969) in the context of relevant questions in consumer research (Lynch et al., 2012) and types of data collection.

situation is unique and hence theoretical predictions should be avoided (Ravetz, 1999). This results in a lock-in in the top half of the cycle and a stagnation of the number of generally applicable ideas that are generated. The fixation in consumer psychology on testing ever-more refined theories based on previous theories leads to a detachment with what matters in the real world (Lynch et al., 2012; Pham, 2013). This results in a lock-in in the lower half of the cycle and diminishes the relevance of consumer science to practitioners. Completing the cycle can overcome these lock-ins and is in our view an essential, yet currently underappreciated approach that the field of consumer science should strive for. To do so, the value and use of both qualitative and quantitative methods should be understood.

Qualitative methods are often used for exploratory or interpretative purposes that require a method capable of catching unexpected information from substantive real-world issues. These methods are suitable to identify which important issues exist and gain an understanding of underlying reasons, opinions, and motivations of consumers. Qualitative methods are supportive to the development of research ideas or hypotheses that tap into important, substantive real-world issues. Their aim is to bring as many new ideas to the table as possible. These methods tend to terminate when saturation is achieved, ie, when novel insights are no longer found. Qualitative methods therefore do not depend (nor should they) on a numerically representative sample of the population. Consequently, conclusions about overall opinions, preferences, effects of intervention, situations, and/or consumer segments cannot be drawn based on these methods. Consumer researchers mainly apply field observations, interviews, focus groups, and laddering interviews as qualitative methods.

Quantitative methods are used to obtain (numerical) data. Surveys, measuring self-reported judgements of end users, are often used as these are easy to apply and can often be filled out by participants at their leisure. Other methods that record numerical data, such as response time measures, can also be used. If sufficiently large data sets are collected in a sample that is representative for the population on relevant variables, quantitative data can be used to induce an overall population profile. The most well-known surveys of this type are probably election polls. In the context of food research, surveys are used to quantify attitudes, opinions, behaviors, and other defined variables. Another application of quantitative methods is the systematic testing of hypotheses in experimental settings. In the context of food innovations this can be used to compare the effectiveness of different communication types on consumer attitude. For experiments it is less important that the sample is representative of the population. At least this is of no major concern as long as differences between the sample and the population are not related to moderators. It is, however, essential that all experimental groups have comparable profiles. This can be ensured by randomization or more advanced techniques such as block randomization or matching.

Besides "pure" qualitative and quantitative methods, there is also a group of mixed methods such as Q-sort (eg, Cuppen et al., 2010), Delphi (eg, Frewer et al., 2011), or repertory grid (eg, Gupta et al., 2015). These methods can provide a bridge between identifying issues of importance and quantifying consumer opinions.

Among all data collection methods, surveys are by far the most often used to collect consumer information on acceptance of novel foods (Reinders et al., 2013). Surveys are a relevant method when a clear a priori idea of relevant attributes and contexts is known (Barrios and Costell, 2004; Iop et al., 2006). Reliance on a single, small-scale survey is an efficient type of data collection if specific research questions for target applications of a technology exist. Nevertheless, such isolated surveys have limited relevance to develop a more generally applicable resource for predicting consumer response.

The use of multiple methods can be relevant to triangulate toward real-world effects. If one method is not certain enough or leaves open questions, a second method may provide the relevant answers (see, eg, van Dijk et al., 2015). The more evidence is available from different approaches that point in the same direction, the more certain we can be that the identified findings are indicative of real-world consumer behavior. If qualitative, quantitative surveys and data from previous research point in the same direction, more confident conclusions can be drawn.

Merging data to form meta-data, or other systematic ways to aggregate data across multiple studies, is required to systematically investigate data from multiple sources in order to arrive at comprehensive conclusions relevant to the field in general (Moskowitz et al., 2005; Moskowitz and Hartmann, 2008). Recent efforts at creating systematic reviews and meta-analysis on genetic modification have adopted this approach (see, eg, Frewer et al., 2013b; Lusk et al., 2005), although the availability, comparability of data and the large differences in quality of published literature make many of the conclusions tentative. Thorough deliberation on which measure to use, and how to define it, will be required to consolidate the field.

14.5.2 MEASURES RELATED TO CONSUMER ACCEPTANCE OF NOVEL FOODS

An important step in collecting consumer data is to consider the operationalization of the constructs in the model—or, to put it differently, to decide what questions to ask to ensure all components in the model are validly and reliably measured. Although some constructs (eg, risk perception, benefit perception, trust, knowledge, attitude, willingness to pay) are relatively frequently used in this research area, different researchers may define these constructs in different ways. Hence the same term may have somewhat different meaning depending on research context (Reinders et al., 2013). Different theoretical approaches will almost unavoidably lead to different measures. But even if the construct is defined the same way, measures for that same construct are often wildly different (Reinders et al., 2013). This inconsistent use of measures has long been a problem to the dismay of consumer scientists (Churchill,

1979) but is sometimes even encouraged by leading practitioners (see, eg, Ajzen, 1991). More consistent use of measurement scales for these constructs would provide a much better baseline for future comparisons and limit the problems that meta-analyses encounter today (see, eg, Frewer et al., 2013b). To build toward a generally applicable body of research about consumer response to innovations, it is necessary to work toward situations where measures are (1) of the highest possible quality, ie, psychometrically validated, and (2) where possible using the same scale, to allow comparison of effects across different product samples and situations.

14.6 CRITICAL NOTES ON THEORIES AND MEASUREMENTS AND NOTABLE IDEAS FOR THE FUTURE

> History warns us ... that it is the customary fate of new truths to begin as heresies and to end as superstitions.
>
> **Thomas Henry Huxley, *Collected Essays of Thomas Henry Huxley***

Consumer science is constantly evolving, and it is important to sketch why current theories are criticized in order to decide whether to implement a specific approach or not. Indeed, while we criticize some of the current approaches, at the time of their development they provided essential, even heretical insights; and they still may be the best available tools depending on the job at hand (eg, Frewer et al., 2008). The bottom line is that the choice for an approach should be made consciously while being aware of the relevant alternatives. This section is therefore not intended to disqualify any approaches but to raise awareness about assumptions and weaknesses limiting the relevance of approaches in specific contexts. Thereby, alternative insights for theoretical approaches based on automatic behavior (Custers and Aarts, 2010; Verplanken and Orbell, 2003), implicit social cognition (Payne and Gawronski, 2010), construal level theory (Trope and Liberman, 2010), and choice architecture (Thaler and Sunstein, 2008) will be presented. Alternative, unobtrusive methods for tracing consumer decision-making using eye tracking (Kloosterman et al., 2015a; Kloosterman et al., 2015b; van Giesen et al., 2015), response time (Greenwald et al., 2009), and functional neuroimaging (fMRI Meuwese et al., 2014) will also be discussed.

14.6.1 CRITICAL NOTES ON CONSUMER ACCEPTANCE AND COMMUNICATION THEORIES

TPB, TAM, protection motivation, and risk–benefit approaches share several properties. All these approaches are based on the assumption that information about specific attributes of a product or technology are analyzed in terms of how positive they score and how they contribute to an overall judgment. These theories implicitly take the position that people create bottom-up judgments from attributes. It is, however, often argued that attitudes and other evaluations are stored after people have repeated a judgment several times (Fazio, 2001, 2007). These stored evaluations are activated immediately and determine choice, regardless of information (Eiser et al., 2003). The stronger such attitudes are, the less likely they are to change (Bizer and Krosnick, 2001), and attempts to change attitudes may only be successful after prolonged repetition (Tormala et al., 2006). In such cases, rational thought models such as TPB and information processing ones (eg, ELM) have limited relevance.

Strong attitudes may lead to a first unconscious gut response (Haidt, 2001). Such first implicit associations are often unconscious, and may contradict self-stated attitudes (Greenwald and Banaji, 1995; Fazio, 2001). In some situations, self-stated attitudes may be subjected to such implicit associations (Gawronski and Bodenhausen, 2006, 2011), which raises the question whether implicit associations, self-reported judgment, or a mix of these represent the most relevant consumer opinion (Gawronski, 2007; Gawronski and Bodenhausen, 2006). Generally it is assumed that the less time and effort available for a decision, the more the behavior is determined by implicit associations (Greenwald et al., 2009). In the context of consumer choice, implicit associations and self-stated attitudes tend to correlate positively (Hofmann et al., 2005) and self-stated attitudes tend to dominate behavior (Greenwald et al., 2009). Whether this is the same for innovations is, however, unclear. For example, there are indications that categorization of cultured meat as meat replacement can cause contrasting implicit associations and self-stated attitudes (Bekker et al., 2016a).

Predictive values of self-stated attitudes, a central construct in many mainstream consumer behavior models, on behavior are often very low (see, eg, Armitage and Conner, 2000; Van Dam and Fischer, 2015). This becomes even more problematic if the behavior is measured at a different level than the self-reported measure. For example, pro-environmental behavior is often measured as the purchase of specific products, while the measured consumer opinion is about ecological behavior in general. This is a problem that can be partially solved by measuring opinions at the same (preferably) concrete level (Kaiser, 1998). In general it is therefore essential to ensure that the specificity of behaviors and opinions closely correspond when designing the survey (Oppenheim, 2000).

Some models claim improvements by adding additional predictors to existing models. An example is the UTAUT (Venkatesh et al., 2012), which evolved from the TAM. The UTAUT is currently one of the most comprehensive models around. While this (necessarily) adds explained variance, it has also resulted in a complex model, which makes it hard to identify paths for improvement of an innovation. In addition, much of the additional explained variance from the additional predictors may introduce statistically significant, yet real-world irrelevant effects (Cohen, 1992; Lynch et al., 2012). Adding more variables may even distract us from the more important problem that many of the central mental processes in consumer decision-making remain largely unexplained (Bagozzi, 2007). Hence, most progress could probably be made by better understanding the core of consumer decision-making rather than adding ever-more detailed determinants to existing models.

14.6.2 ALTERNATIVE CONSIDERATIONS FOR CONSUMER SCIENCE RESEARCH

Besides potentially solving the problems encountered with the dominant approaches sketched so far, alternative insights may help to study these problems at hand. Many approaches to consumer behavior consider attitude and intention as the main predictor of behavior. Attitudes are often defined as a stable evaluation of an object (Fazio, 2001). Nevertheless, this ignores the fact that the evaluation of an object only makes sense in the context of the goal of consuming or buying that product. A difficult-to-prepare, rich, and high-end foodstuff may be evaluated positively in the context of a festive meal but negatively when the goal is to have a quick dinner between work and an evening meeting. Thus the evaluation of a product is influenced by motivations attached to the use of that product (Allen et al., 2008). In spite of the obvious relevance of motivations and goals, their importance is only sporadically raised in consumer sciences (Bagozzi and Dholakia, 1999; Fiske, 1992; James, 1890/1950; Kunda, 1990; Rogers, 1983). Currently there is renewed interest in goals and motivations as their unconscious pursuit (Custers and Aarts, 2010) may link better to implicit associations than the conscious attitudes generally measured in consumer research (Ferguson and Porter, 2009). One type of motivation that has received much attention is that of regulatory focus (Higgins, 2000). In this case, the motivation may either be to avoid negative consequences or to approach positive consequences (Carver and Scheier, 1998). While both motivations may push toward the same optimal state, it has been shown that people in a prevention mindset are more likely to choose products that facilitate avoiding negative consequences, while people in a promotion mindset choose products that help them get closer to their goal, ie, there is regulatory fit (Higgins and Scholer, 2009). For food innovations this may be of relevance when considering trying to communicate about avoiding risks or gaining novel benefits. Preliminary evidence has shown that when there is a fit between consumer regulatory focus (ie, promotion vs prevention mindset) and the framing of a message (ie, stressing the attainment of positive outcomes or the prevention of negative outcomes) more positive attitudes and greater willingness to buy GM foods can be accomplished (Fransen et al., 2010).

If we shift our attention from attitude as a comprehensive evaluation (eg, Fazio, 2007; Gawronski and Bodenhausen, 2006; Schwarz, 2007) to decision making, the option of out of hand rejection of an alternative based on a single attribute, regardless of other attributes makes sense. Such categorical rejection of non-compensatory attributes is an efficient way of limiting the number of alternatives before more effortful considerations are made (Brandstätter et al., 2006; Gigerenzer and Todd, 1999). In the context of innovation this may lead to the consequence that an innovation is rejected out of hand if the innovative attribute in itself is categorically rejected. This may have played a role in the rejection of Genetic Modification in Europe where perceived risks of Genetic Modification were deemed non-compensatory by consumer groups (Grunert et al., 2001); and there is evidence that such categorical rejection of Genetic Modification plays a role with a substantial number of consumers (Kaye-Blake et al., 2005).

Consumers are sometimes depicted as inconsistent. However, a person being a consumer may adopt a specific consumer role with different associated goals and motivations, and hence shows seemingly inconsistent behavior compared to the same person who acts in the role, and toward the goals of a responsible citizen (Johnston, 2008). This includes the case where people are negative, eg, about genetically modified food but still consume it, and in cases where people are positive but do not buy it (eg, organic produce Vermeir and Verbeke, 2006). Such seeming inconsistencies in behavior may be due to the fact that people can construe meaning to food products interchangeably at a high level or a low level. The high level is characterized by value relevance and is abstract. Long-term consequences, socially and geographically distant effects as well as hypothetical ones, and uncertainties all fit with a high construal (Trope and Liberman, 2010). A low construal level is more related to immediate, personally relevant, local effects that are certain and concrete. High construal levels may be more related to the citizen role of people and the way people fill out questionnaires, while consumers having to buy a food product right now are more likely to make these choices based on low construal level (Van Trijp and Fischer, 2011). For innovations, it is important to realize that discussing a future, uncertain technology as a whole (high construal) may result in answers more related to a citizen than a consumer. To consider innovation at a lower construal level, one could try to relate the technology to concrete and immediate effects.

As in communication and attitude change theories, it is becoming increasingly accepted that decision-making itself can also be conducted in two distinctly different ways (Kahneman, 2003). A first heuristic system takes care of most decisions and ensures that mind power remains available. Consider, for example, how incredibly tiring it would be to have to think through every step you take when walking down the corridor. This automatic, system 1, is somewhat susceptible to signals in the environment and may lead to suboptimal behavior from time to time. A second system is more reflective and is called upon when the first system cannot deal with the situation at hand, albeit at the cost of spending more mental effort (Kahneman, 2003). Which system is dominant is both context and situation dependent and is another reason for seemingly inconsistent consumer behavior.

Like goals that can be automatically activated and unconscious (Custers and Aarts, 2010), attitudes that can be implicit (Greenwald and Banaji, 1995), decisions can be made based on heuristics (Kahneman, 2003). Triggering the relevant goals, attitudes, and decisions by changing the environment in which the consumer behaves seems an obvious way of communicating and influencing consumer behavior. This is exactly the approach proposed in nudging (Thaler and Sunstein, 2008). More specifically, it was claimed that by taking specific decisions in the choice architecture (ie, the environmental context a consumer has to make a decision in), consumers can be unconsciously steered toward a choice that is optimal to their unconscious goals and demands. This takes a lot of the decisions out of the consumers' hands and puts them in the hands of the designer of the choice environment.

14.6.3 CRITICAL NOTES ON, AND DEVELOPMENT OF, MEASUREMENTS

Self-reported measures can only be filled out consciously. This implies that measures are sensitive to socially desirable or self-enhancing responses, especially for sensitive topics like new technologies or innovations. In social psychology, the past two decades have increasingly been devoted to developing instruments for measuring consumer behavior more indirectly, as implicit associations. Most empirical measures for implicit associations are based on participants' performance on computer-based, quick categorization tasks (Gawronski and Bodenhausen, 2006). Examples of implicit measurement instruments are the implicit association test (IAT; Greenwald et al., 2003), the affective priming task (Fazio, 2001), and the affect misattribution procedure (Payne et al., 2005). For a more complete overview of such measures, see Goodall (2011). The IAT is the most common way to measure implicit processes, although specific research questions may favor one of the other tests (Lebel and Paunonen, 2011). An IAT aims to measure individual differences in associative evaluations of specific objects. Conceptually, associative evaluations are defined as those associations that come to mind unintentionally, that are difficult to control once they are activated, and that may not necessarily be endorsed at a conscious level (Galdi et al., 2008). Associative measures are derived by computing the time that a respondent needs to pair two concepts (Karpinski and Steinman, 2006). Notice that results obtained from empirical attempts to implicitly measure consumer response within the domain of new technologies are often still very preliminary or nonconsistent (Bekker et al., 2016b; Spence and Townsend, 2006; Tenbült et al., 2008), leaving a lot of unresolved issues regarding their practical applicability.

Other sources of measuring implicit or intuitive reactions from consumers toward new technologies are often of psychophysiological nature, ie, measures that use physiological responses to a situation to measure psychological constructs such as preference. These measurement instruments include, eg, eye tracking, heart rate variability, galvanic skin response, electroencephalography, and functional magnetic resonance imaging (fMRI). While these methods and measures do pick up on bodily and neural functions and therefore do not rely on introspection, the measures are also indirect as the translation of these signals to mental constructs is not straightforward (Wang and Minor, 2008). For example, fMRI reliably measures brain activity intensity at specific locations at specific moments in time, but relating these active brain centers to specific thoughts has been and remains a challenge.

14.7 CONCLUSIONS AND DISCUSSION

Consumer science offers a wealth of theoretical and empirical approaches that can be of use in the development of food innovations. Each of these approaches is most suitable for the type of research question and within the context where they were intended. This has resulted in a large number of approaches, each of which has its own benefits and downsides. Choosing the more relevant approach and relevant measure among these alternatives is among the most important parts of conducting good consumer research. Even if this is achieved, consumer science has only modest predictive power, especially when studying the uncertain future of food innovation.

Current movements in consumer science are toward approaches where less of the consumer behavior is considered to be based on conscious and deliberate choice. Instead implicit, often unconscious, gut feelings and the influence of the environment through choice architecture are shown to influence consumer behavior even outside the awareness of the consumer. Cocreation aims to tap into the same unconscious knowledge base by involving consumers in the design phase. The usefulness and applicability of these methods, especially outside a lab setting, is however not yet well developed. While these innovations in consumer science are interesting developments in the field of consumer science itself, in applied work more traditional methods still dominate. When correctly, consistently, and relevantly applied, these methods can still contribute immensely to our understanding of consumer response to food innovations.

Raising and answering questions on what theory, methods, and measurements to use before data are collected lies at the heart of conducting relevant consumer research. By presenting many alternative theories, methods, and measurements, the current chapter may not have provided clear-cut answers. Instead, this chapter aimed to raise some of the relevant questions that need to be answered and present alternatives that maybe used in order to conduct high-quality consumer research relevant to innovations in food.

REFERENCES

Ajzen, I., 1991. The theory of planned behavior. Organizational Behavior and Human Decision Processes 50, 179–211.

Alhakami, A.S., Slovic, P., 1994. A psychological study of the inverse relationship between perceived risk and perceived benefit. Risk Analysis 14, 1085–1096.

Allen, M.W., Gupta, R., Monnier, A., 2008. The interactive effect of cultural symbols and human values on taste evaluation. Journal of Consumer Research 35, 294–308.

Armitage, C.J., Conner, M., 2000. Efficacy of the theory of planned behaviour: a meta-analytic review. British Journal of Social Psychology 40, 471–499.

Bagozzi, R.P., 2007. The legacy of the technology acceptance model and a proposal for a paradigm shift. Journal of the Association of Information Systems 8, 244–254.

Bagozzi, R.P., Dholakia, U., 1999. Goal setting and goal striving in consumer behavior. Journal of Marketing 63, 19–32.

Bagozzi, R.P., Wong, N., Abe, S., Bergami, M., 2000. Cultural and situational contingencies and the theory of reasoned action: application to fast food restaurant consumption. Journal of Consumer Psychology 9, 97–106.

Barrios, E.X., Costell, E., 2004. Review: use of methods of research into consumers' opinions and attitudes in food research. Food Science and Technology International 10, 359–371.

Bekker, G., Fischer, A.R.H., Tobi, H., Van Trijp, H.C.M., 2016a. Cultured meat: feels like fake, sound like real meat. Implicit and Explicit Attitudes Towards Cultured Meat (in preparation).
Bekker, G., Fischer, A.R.H., Tobi, H., van Trijp, H.C.M., 2016b. Explicit and implicit attitude toward a future food technology: the case of cultured meat (under review).
Bizer, G.Y., Krosnick, J.A., 2001. Exploring the structure of strength-related attitude features: the relation between attitude importance and attitude accessibility. Journal of Personality and Social Psychology 81, 566–586.
Brandstätter, E., Gigerenzer, G., Hertwig, R., 2006. The priority heuristic: making choices without trade-offs. Psychological Review 113, 409–432.
Bredahl, L., 2001. Determinants of consumer attitudes and purchase intentions with regard to genetically modified foods – results of a cross-national survey. Journal of Consumer Policy 24, 23–61.
Bruhn, C.M., 1995. Consumer attitudes and market response to irradiated food. Journal of Food Protection 58, 175–181.
Carver, C.S., Scheier, M.F., 1998. On the Self-Regulation of Behavior. Cambridge University Press, New York, NY, US.
Chaiken, S., 1980. Heuristic versus systematic information processing and the use of source versus message cues in persuasion. Journal of Personality and Social Psychology 39, 752–766.
Chen, M.F., 2008. An integrated research framework to understand consumer attitudes and purchase intentions toward genetically modified foods. British Food Journal 110, 559–579.
Chen, M.F., Li, H.L., 2007. The consumer's attitude toward genetically modified foods in Taiwan. Food Quality and Preference 18, 662–674.
Chen, S., Chaiken, S., 1999. The heuristic-systematic model in its broader context. In: Chaiken, S., Trope, Y. (Eds.), Dual Process Theories in Social Psychology. Guilford Press, New York, NY, pp. 73–96.
Churchill, G.A.J., 1979. A paradigm for developing better measures of marketing constructs. Journal of Marketing 16, 64–73.
Claudy, M.C., Garcia, R., O'Driscoll, A., 2014. Consumer resistance to innovation-a behavioral reasoning perspective. Journal of the Academy of Marketing Science (prepublished online).
Cohen, J., 1992. A power primer. Psychological Bulletin 112, 155–159.
Costa-Font, J., Mossialos, E., 2006. The public as a limit to technology transfer: the influence of knowledge and beliefs in attitudes towards biotechnology in the UK. Journal of Technology Transfer 31, 629–645.
Cuppen, E., Breukers, S., Hisschemöller, M., Bergsma, E., 2010. Q methodology to select participants for a stakeholder dialogue on energy options from biomass in the Netherlands. Ecological Economics 69, 579–591.
Custers, R., Aarts, H., 2010. The unconscious will: how the pursuit of goals operates outside of conscious awareness. Science 329, 47–50.
Davis, F.D., 1989. Perceived usefulness, perceived ease of use, and user acceptance of information technology. MIS Quarterly 13, 319–340.
Davis, F.D., 1993. User acceptance of information technology: system characteristics, user perceptions and behavioral impacts. International Journal of Man-Machine Studies 38, 475–487.
Diehl, J.F., 2002. Food irradiation – past, present and future. Radiation Physics and Chemistry 63, 211–215.
van Dijk, H., Fischer, A.R.H., Marvin, H.J.P., van Trijp, H.C.M., 2015. Determinants of stakeholders' attitudes towards a new technology: nanotechnology applications for food, water, energy and medicine. Journal of Risk Research (in press). http://dx.doi.org/10.1080/13669877.2015.1057198.
Eiser, J.R., Fazio, R.H., Stafford, T., Prescott, T.J., 2003. Connectionist simulation of attitude learning: asymmetries in the acquisition of positive and negative evaluations. Personality and Social Psychology Bulletin 29, 1221–1235.
Eiser, J.R., Miles, S., Frewer, L.J., 2002. Trust, perceived risk, and attitudes toward food technologies. Journal of Applied Social Psychology 32, 2423–2433.
Fazio, R.H., 2001. On the automatic activation of associated evaluations: an overview. Cognition and Emotion 15, 115–141.

Fazio, R.H., 2007. Attitudes as object-evaluation associations of varying strength. Social Cognition 25, 603–637.
Ferguson, M.J., Porter, S.C., 2009. Goals and (implicit) attitudes: a social-cognitive perspective. In: Moskowitz, G.B., Grant, H. (Eds.), The Psychology of Goals. Guilford, New York.
Fife-Schaw, C., Rowe, G., 1996. Public perceptions of everyday food hazards: a psychometric study. Risk Analysis 16, 487–500.
Fife-Schaw, C., Rowe, G., 2000. Extending the application of the psychometric approach for assessing public perceptions of food risks: some methodological considerations. Journal of Risk Research 3, 167–179.
Finucane, M.L., Alhakami, A.S., Slovic, P., Johnson, S.M., 2000. The affect heuristic in judgments of risks and benefits. Journal of Behavioral Decision Making 13, 1–17.
Fischer, A.R.H., Frewer, L.J., 2009. Consumer familiarity with foods and the perception of risks and benefits. Food Quality and Preference 20, 576–585.
Fischhoff, B., Slovic, P., Lichtenstein, S., 1978. How safe is safe enough? A psychometric study of attitudes towards technological risks and benefits. Policy Sciences 9, 127–152.
Fishbein, M., Ajzen, I., 1975. Belief, Attitude, Intention, and Behavior: An Introduction to Theory and Research. Addison-Wesley, Reading, MA.
Fiske, S.T., 1992. Thinking is for doing: portraits of social cognition from daguerreotype to laser photo. Journal of Personality and Social Psychology 63, 877–889.
Fransen, M.L., Reinders, M.J., Bartels, J., Maassen, R.L., 2010. The influence of regulatory fit on evaluation and intentions to buy genetically modified foods: the mediating role of social identification. Journal of Marketing Communications 16, 5–20.
Frewer, L.J., Fischer, A.R.H., Brennan, M., Bánáti, D., Lion, R., Meertens, R.M., Rowe, G., Siegrist, M., Verbeke, W., Vereijken, C., 2015. Risk/benefit communication about food – a systematic review of the literature. Critical Reviews in Food Science and Nutrition (in press). http://dx.doi.org/10.1080/10408398.2013.801337.
Frewer, L.J., Fischer, A.R.H., Brink, P.J.v.d., Byrne, P., Brock, T., Brown, C., Crocker, J., Goerlitz, G., Hart, A., Scholderer, J., Solomon, K., 2008. Potential for the adoptation of probabilistic risk assessments by end-users and decision–makers. Human and Ecological Risk Assessment 14, 166–178.
Frewer, L.J., Fischer, A.R.H., Wentholt, M.T.A., Marvin, H.J.P., Ooms, B.W., Coles, D., Rowe, G., 2011. The use of Delphi methodology in agrifood policy development: some lessons learned. Technological Forecasting and Social Change 78, 1514–1525.
Frewer, L.J., Kleter, G.A., Brennan, M., Coles, D., Fischer, A.R.H., Houdebine, L.M., Mora, C., Millar, K., Salter, B., 2013a. Genetically modified animals from life-science, socio-economic and ethical perspectives: examining issues in an EU policy context. New Biotechnology 30, 447–460.
Frewer, L.J., Scholderer, J., Bredahl, L., 2003. Communicating about the risks and benefits of genetically modified foods: the mediating role of trust. Risk Analysis 23, 1117–1133.
Frewer, L.J., van der Lans, I.A., Fischer, A.R.H., Reinders, M.J., Menozzi, D., Zhang, X., van den Berg, I., Zimmermann, K.L., 2013b. Public perceptions of agri-food applications of genetic modification – a systematic review and meta-analysis. Trends in Food Science & Technology 30, 142–152.
Galdi, S., Arcuri, L., Gawronski, B., 2008. Automatic mental associations predict future choices of undecided decision-makers. Science 321, 1100–1102.
Gawronski, B., 2007. Editorial: attitudes can be measured! but what is an attitude? Social Cognition 25, 573–581.
Gawronski, B., Bodenhausen, G.V., 2006. Associative and propositional processes in evaluation: an integrative review of implicit and explicit attitude change. Psychological Bulletin 132, 692–731.
Gawronski, B., Bodenhausen, G.V., 2011. The associative-propositional evaluation model. Theory, evidence, and open questions. Advances in Experimental Social Psychology 59–127.
van Giesen, R.I., Fischer, A.R.H., Van Dijk, H., Van Trijp, H.C.M., 2015. Tracing attitude expression: an eye tracker study. Journal of Behavioral Decision Making 29, 232–244.
Gigerenzer, G., Todd, P.M., 1999. Simple Heuristics That Make Us Smart. Oxford University Press, Oxford, UK.
Goodall, C.E., 2011. An overview of implicit measures of attitudes: methods, mechanisms, strengths, and limitations. Communication Methods and Measures 5, 203–222.

Goode, M.R., Dahl, D.W., Moreau, C.P., 2010. The effect of experiential analogies on consumer perceptions and attitudes. Journal of Marketing Research 47, 274–286.

Greenwald, A.G., Banaji, M.R., 1995. Implicit social cognition: attitudes, self-esteem, and stereotypes. Psychological Review 102, 4–27.

Greenwald, A.G., Nosek, B.A., Banaji, M.R., 2003. Understanding and using the implicit association test: I. An improved scoring algorithm. Journal of Personality and Social Psychology 85, 197–216.

Greenwald, A.G., Poehlman, T.A., Uhlmann, E.L., Banaji, M.R., 2009. Understanding and using the implicit association test: III. Meta-analysis of predictive validity. Journal of Personality and Social Psychology 97, 17–41.

Gregan-Paxton, J., Hibbard, J.D., Brunel, F.F., Azar, P., 2002. "So that's what that is": examining the impact of analogy on consumers' knowledge development for really new products. Psychology and Marketing 19, 533–550.

Gregan-Paxton, J., John, D.R., 1997. Consumer learning by analogy: a model of internal knowledge transfer. Journal of Consumer Research 24, 266–284.

Griffin, R.J., Dunwoody, S., Neuwirth, K., 1999. Proposed model of the relationship of risk information seeking and processing to the development of preventive behaviors. Environmental Research Section A 80, S230–S245.

Griffin, R.J., Neuwirth, K., Giese, J., Dunwoody, S., 2002. Linking the heuristic-systematic model and depth of processing. Communication Research 29, 705–732.

de Groot, A.D., 1969. Methodology. Foundations of Inference and Research in the Behavioral Sciences. Mouton & Co, The Hague, Paris.

Grunert, K.G., Lähteenmäki, L., Asger Nielsen, N., Poulsen, J.B., Ueland, O., Åström, A., 2001. Consumer perceptions of food products involving genetic modification – results from a qualitative study in four Nordic countries. Food Quality and Preference 12, 527–542.

Gupta, N., Fischer, A.R.H., Frewer, L.J., 2012. Socio-psychological determinants of public acceptance of technologies: a review. Public Understanding of Science 21, 782–795.

Gupta, N., Fischer, A.R.H., Frewer, L.J., 2015. Ethics, risk and benefits associated with different applications of nanotechnology: a comparison of expert and consumer perceptions of drivers of societal acceptance. Nanoethics 9, 93–108.

Haidt, J., 2001. The emotional dog and its rational tail: a social intuitionist approach to moral judgment. Psychological Review 108, 814–834.

Hendrick, T.A.M., Fischer, A.R.H., Tobi, H., Frewer, L.J., 2013. Self-reported attitude scales: current practice in adequate assessment of reliability, validity, and dimensionality. Journal of Applied Social Psychology 43, 1538–1552.

Higgins, E.T., 2000. Beyond pleasure & pain. In: Higgins, E.T., Kruglanski, A.W. (Eds.), Motivational Science: Social and Personality Perspectives. Key Reading in Social Psychology. Psychology Press, Philadelphia, PA, pp. 231–255.

Higgins, E.T., Scholer, A.A., 2009. Engaging the consumer: the science and art of the value creation process. Journal of Consumer Psychology 19, 100–114.

von Hippel, E., 1986. Lead users: a source of novel product concepts. Management Science 32, 791–805.

Hofmann, W., Gawronski, B., Gschwendner, T., Le, H., Schmitt, M., 2005. A meta-analysis on the correlation between the implicit association test and explicit self-report measures. Personality and Social Psychology Bulletin 31, 1369–1385.

Honkanen, P., Olsen, S.O., Verplanken, B., 2005. Intention to consume seafood–the importance of habit. Appetite 45, 161.

Hübner, G., Kaiser, F.G., 2006. The moderating role of the attitude-subjective norms conflict on the link between moral norms and intention. European Psychologist 11, 99–109.

Iop, S.C.F., Teixeira, E., Deliza, R., 2006. Consumer research: extrinsic variables in food studies. British Food Journal 108, 894–903.

James, W., 1890/1950. The Principles of Psychology, vol. I. Dover Publications, Oxford, England.

Janz, N.K., Becker, M.H., 1984. The health belief model: a decade later. Health Education Quarterly 11, 1–47.

Johnston, J., 2008. The citizen-consumer hybrid: ideological tensions and the case of whole foods market. Theory and Society 37, 229–270.
Kahneman, D., 2003. A perspective on judgment and choice: mapping bounded rationality. American Psychologist 58, 697–720.
Kahneman, D., Tversky, A., 1979. Prospect theory: an analysis of decision under risk. Econometrica 47, 263–292.
Kaiser, F.G., 1998. A general measure of ecological behavior. Journal of Applied Social Psychology 28, 395–422.
Kardes, F.R., Posavac, S.S., Cronley, M.L., 2004. Consumer inference: a review of processes, bases, and judgment contexts. Journal of Consumer Psychology 14, 230–256.
Karpinski, A., Steinman, R.B., 2006. The single category implicit association test as a measure of implicit social cognition. Journal of Personality and Social Psychology 91, 16–32.
Kaye-Blake, W., Bicknell, K., Saunders, C., 2005. Process versus product: which determines consumer demand for genetically modified apples? Australian Journal of Agricultural and Resource Economics 49, 413–427.
Kloosterman, N.A., Meindertsma, T., Hillebrand, A., van Dijk, B.W., Lamme, V.A.F., Donner, T.H., 2015a. Top-down modulation in human visual cortex predicts the stability of a perceptual illusion. Journal of Neurophysiology 113, 1063–1076.
Kloosterman, N.A., Meindertsma, T., van Loon, A.M., Lamme, V.A.F., Bonneh, Y.S., Donner, T.H., 2015b. Pupil size tracks perceptual content and surprise. European Journal of Neuroscience 41, 1068–1078.
Kunda, Z., 1990. The case for motivated reasoning. Psychological Bulletin 108, 480–498.
Lazo, J.K., Kinnell, J., Fisher, A., 2000. Expert and lay person perceptions of ecosystem risk. Risk Analysis 20, 179–193.
Lebel, E.P., Paunonen, S.V., 2011. Sexy but often unreliable: the impact of unreliability on the replicability of experimental findings with implicit measures. Personality and Social Psychology Bulletin 37, 570–583.
Loken, B., 2006. Consumer psychology: categorization, inferences, affect, and persuasion. Annual Review of Psychology 57, 453–485.
Lusk, J.L., Jamal, M., Kurlander, L., Roucan, M., Taulman, L., 2005. A meta-analysis of genetically modified food valuation studies. Journal of Agricultural and Resource Economics 30, 28–44.
Lynch Jr., J.G., Alba, J.W., Krishna, A., Morwitz, V.G., Gürhan-Canli, Z., 2012. Knowledge creation in consumer research: multiple routes, multiple criteria. Journal of Consumer Psychology 22, 473–485.
Meuwese, J.D.I., Scholte, H.S., Lamme, V.A.F., 2014. Latent memory of unattended stimuli reactivated by practice: an fMRI study on the role of consciousness and attention in learning. PLoS One 9.
Michaut, A.M.K., 2004. Consumer Acceptance of New Products With Applicaiton to Foods, Marketing and Consumer Behaviour Group. Wageningen University, Wageningen, NL.
Moreau, C.P., Markman, A.B., Lehmann, D.R., 2001. "What is it?" Categorization flexibility and consumers' responses to really new products. Journal of Consumer Research 27, 489–498.
Moskowitz, H.R., German, J.B., Saguy, I.S., 2005. Unveiling health attitudes and creating good-for-you foods: the genomics metaphor, consumer innovative web-based technologies. Critical Reviews in Food Science and Nutrition 45, 165–191.
Moskowitz, H.R., Hartmann, J., 2008. Consumer research: creating a solid base for innovative strategies. Trends in Food Science and Technology 19, 581–589.
Munksgaard, K.B., Freytag, P.V., 2011. Complementor involvement in product development. Journal of Business and Industrial Marketing 26, 286–298.
Nahuis, R., Moors, E.H.M., Smits, R.E.H.M., 2012. User producer interaction in context. Technological Forecasting and Social Change 79, 1121–1134.
Oppenheim, A.N., 2000. Questionnaire Design, Interviewing and Attitude Measurement. (Continuum Pages).
Payne, B.K., Cheng, C.M., Govorun, O., Stewart, B.D., 2005. An inkblot for attitudes: affect misattribution as implicit measurement. Journal of Personality and Social Psychology 89, 277–293.
Payne, B.K., Gawronski, B., 2010. A history of implicit social cognition: where is it coming from? Where is it now? Where is it going? In: Gawronski, B., Payne, B.K. (Eds.), Handbook of Implicit Social Cognition: Measurement, Theory, and Applications. Guilford, New York, pp. 1–15.

Petty, R.E., Cacioppo, J.T., 1986. Communication and Persuasion: Central and Peripheral Routes to Attitude Change. Springer-Verlag, New York, NY.
Petty, R.E., Wegener, D.T., 1999. The elaboration likelihood model: current status and controversies. In: Chaiken, S., Trope, Y. (Eds.), Dual Process Theories in Social Psychology. Guilford Press, New York, NY, pp. 37–72.
Pham, M.T., 2013. The seven sins of consumer psychology. Journal of Consumer Psychology 23, 411–423.
Poortinga, W., Pidgeon, N.F., 2005. Trust in risk regulation: cause or consequence of the acceptability of GM food? Risk Analysis 25, 199–209.
Prentice-Dunn, S., Rogers, R.W., 1986. Protection motivation theory and preventive health: beyond the Health Belief Model. Health Education Research 1, 153–161.
Ranganath, K.A., Spellman, B.A., Joy-Gaba, J.A., 2010. Cognitive "category-based induction" research and social "persuasion" research are each about what makes arguments believable: a tale of two literature. Perspectives on Psychological Science 5, 115–122.
Ravetz, J.R., 1999. What is post-normal science. Futures 31, 647–653.
Reinders, M.J., van der Lans, I.A., Fischer, A.R.H., van Trijp, H.C.M., 2013. A Review to Collate Information on External Communication as a Basis of Innovation Success. Connect4Action Deliverable, Wageningen.
Rogers, E.M., 1962/1995. Diffusion of Innovations. Free Press, New York.
Rogers, R.W., 1983. Cognitive and physiological processes in fear appeals and attitude change: a revised theory of protection motivation. In: Cacioppo, J.T., Petty, R.E. (Eds.), Social Psychophysiology: A Sourcebook. The Guilford Press, New York, pp. 153–176.
Ronteltap, A., Fischer, A.R.H., Tobi, H., 2011. Societal response to nanotechnology: converging technologies-converging societal response research? Journal of Nanoparticle Research 13, 4399–4410.
Ronteltap, A., Van Trijp, H., 2007. Consumer acceptance of personalised nutrition. Genes and Nutrition 2, 85–87.
Ronteltap, A., van Trijp, J.C.M., Renes, R.J., Frewer, L.J., 2007. Consumer acceptance of technology-based food innovations: lessons for the future of nutrigenomics. Appetite 49, 1–17.
Rossi, C., 2011. Online consumer communities, collaborative learning and innovation. Measuring Business Excellence 15, 46–62.
Saba, A., Vassallo, M., 2002. Consumer attitudes toward the use of gene technology in tomato production. Food Quality and Preference 13, 13–21.
Schenk, M.F., Fischer, A.R.H., Frewer, L.J., Gilissen, L.J.W.J., Jacobsen, E., Smulders, M.J.M., 2008. The influence of perceived benefits on acceptance of GM applications for allergy prevention. Health, Risk and Society 10, 263–282.
Schwarz, N., 2007. Attitude construction: evaluation in context. Social Cognition 25, 638–656.
Siegrist, M., 2000. The influence of trust and perceptions of risks and benefits on the acceptance of gene technology. Risk Analysis 20, 195–204.
Siegrist, M., Cousin, M.E., Kastenholz, H., Wiek, A., 2007. Public acceptance of nanotechnology foods and food packaging: the influence of affect and trust. Appetite 49, 459–466.
Siegrist, M., Gutscher, H., 2006. Flooding risks: a comparison of lay people's perceptions and expert's assessments in Switzerland. Risk Analysis 26, 971–979.
Slovic, P., 1987. Perception of risk. Science 236, 280–285.
Slovic, P., Finucane, M.L., Peters, E., MacGregor, D.G., 2002. Rational actors or rational fools: implications of the affect heuristic for behavioral economics. Journal of Socio Economics 31, 329–342.
Slovic, P., Malmfors, T., Krewski, D., Mertz, C.K., Neil, N., Purchase, I.F.H., 1995. Intuitive toxicology. II. Expert and lay judgments of chemical risks in Canada. Risk Analysis 15, 661–675.
Smits, R.E.H.M., den Hertog, P., 2007. TA and the management of innovation in economy and society. International Journal of Foresight and Innovation Policy 3, 28–52.
Solomon, B.D., Banerjee, A., 2006. A global survey of hydrogen energy research, development and policy. Energy Policy 34, 781.
Spence, A., Townsend, E., 2006. Implicit attitudes towards genetically modified (GM) foods: a comparison of context-free and context-dependent evaluations. Appetite 46, 67–74.

Starr, C., 1969. Social benefit versus technological risk. Science 165, 1232–1238.
Steenkamp, J.-B.E.M., Baumgartner, H., 1998. Assessing measurement equivalence in cross-national consumer research. Journal of Consumer Research 25, 78–90.
Sujan, M., 1985. Consumer knowledge – effects on evaluation strategies mediating consumer judgments. Journal of Consumer Research 12, 31–46.
Tenbült, P., De Vries, N.K., Dreezens, E., Martijn, C., 2008. Intuitive and explicit reactions towards "new" food technologies: attitude strength and familiarity. British Food Journal 110, 622–635.
Thaler, R.H., Sunstein, C.R., 2008. Nudge: Improving Decision About Health, Wealth and Happiness. Penguin, London.
Tormala, Z.L., Clarkson, J.J., Petty, R.E., 2006. Resisting persuasion by the skin of one's teeth: the hidden success of resisted persuasive messages. Journal of Personality and Social Psychology 91, 423–435.
Trope, Y., Liberman, N., 2010. Construal-level theory of psychological distance. Psychological Review 117, 440–463.
Trumbo, C.W., 2002. Information processing and risk perception: an adaptation of the heuristic-systematic model. Journal of Communication 52, 367–381.
Tversky, A., Kahneman, D., 1974. Judgment under uncertainty: heuristics and biases. Science 185, 1124–1131.
Van Dam, Y.K., Fischer, A.R.H., 2015. Buying green without being seen. Environment and Behavior 47, 328–356.
Van Kleef, E., Van Trijp, H., Luning, P.A., 2005. Consumer research in the early stages of new product development: a critical review of methods and techniques. Food Quality and Preference 16, 181–201.
Van Trijp, H.C.M., Fischer, A.R.H., 2011. Mobilizing consumer demand for sustainable development. In: van Latesteijn, H.C., Andeweg, K. (Eds.), Transforming Agro Innovation Toward Sustainable Development: The TransForum Model. Springer, Dordrecht, pp. 79–103.
Venkatesh, V., Bala, H., 2008. Technology acceptance model 3 and a research agenda on interventions. Decision Sciences 39, 273–315.
Venkatesh, V., Davis, F.D., 1996. A model of the antecedents of perceived ease of use: development and test. Decision Sciences 27, 451–477.
Venkatesh, V., Davis, F.D., 2000. Theoretical extension of the technology acceptance model: four longitudinal field studies. Management Science 46, 186–204.
Venkatesh, V., Morris, M.G., Davis, G.B., Davis, F.D., 2003. User acceptance of information technology: toward a unified view. MIS Quarterly: Management Information Systems 27, 425–478.
Venkatesh, V., Thong, J.Y.L., Xu, X., 2012. Consumer acceptance and use of information technology: extending the unified theory of acceptance and use of technology. MIS Quarterly: Management Information Systems 36, 157–178.
Vermeir, I., Verbeke, W., 2006. Sustainable food consumption: exploring the consumer "attitude – behavioral intention" gap. Journal of Agricultural and Environmental Ethics 19, 169–194.
Verplanken, B., Orbell, S., 2003. Reflections on past behavior: a self-report index of habit strength. Journal of Applied Social Psychology 33, 1313–1330.
Von Hippel, E., 1976. The dominant role of users in the scientific instrument innovation process. Research Policy 5, 212–239.
Von Neumann, J., Morgenstern, O., 1944. Theory of Games and Economic Behaviour. Princeton University Press.
Wang, Y.J., Minor, M.S., 2008. Validity, reliability, and applicability of psychophysiological techniques in marketing research. Psychology and Marketing 25, 197–232.
Yang, Z.J., Aloe, A.M., Feeley, T.H., 2014. Risk information seeking and processing model: a meta-analysis. Journal of Communication 64, 20–41.

CHAPTER

15

CHALLENGES AND OPPORTUNITIES

C.M. Galanakis

Galanakis Laboratories, Chania, Greece

15.1 INTRODUCTION

Change can be either a threat or an opportunity. Independently of how we choose to deal with it, change will require things to be done differently, with an ultimate goal to be improved. Following this consideration, companies pursue technological solutions that allow them to be more efficient or enable the creation of new products and services. However, the process of creating value out of new knowledge remains largely uncharted territory, with a high prospect of failure. The correct innovation management, from research and development to commercialization and demonstration, though not constituting enough guarantee of success, is nonetheless crucial in avoiding a debacle (Almeida Costa, 2008).

The term *innovation* has developed into a buzzword that is used whenever discussions focus on future developments. Nevertheless, despite its frequent reference, it is not easy to specify and identify. Innovation is both a process and an outcome, implying introduction of new products and processes, as well as opening of urgent markets and introduction of new organizational forms (Tepic et al., 2014). The food and beverage industry is typically described as a relatively mature and slow-growing area of business, quite conservative in the innovation types it introduces into the market. Despite its distinct character and difficulties such as protection of recipe secrecy and intellectual property rights in the food industry, its technological rate in spite of innovation is increasing. For a long time, the main changes of large, multinational, food processor companies have been directed to enlarging production capacity and reducing cost (Costa and Jongen, 2006; Geylani et al., 2008). However, the increased food demands, the food chain supply changes, and the increasing competition make innovation not only an unavoidable corporate activity but also one that is increasingly vital for overall agribusiness profitability (Sarkar and Costa, 2008; Tepic et al., 2014).

For instance, due to the fast development of technologies and the changes of customer demands, performance of the food sector has become increasingly dependent on continued improvement and introduction of new products and processes. Nowadays consumers demand unique flavors, convenience cooking, health-enhancing foods, and diets tailored to their individual needs and preferences (Costa et al., 2001, 2007). Besides, the food sector faces major challenges that arise from changes in the economic and noneconomic environments, increased food consumption, as well as by changing societal attitudes toward the consequences of the food industry activities, captured in the term *sustainability* (Fritz and Schiefer, 2008). In order to face these challenges, innovation has been increasingly

Innovation Strategies in the Food Industry. http://dx.doi.org/10.1016/B978-0-12-803751-5.00015-5

discussed, not only as an opportunity but also as a precondition for success. Indeed, innovation support has been identified as core requirement for assuring the sustainability and competitiveness of the food sector.

The larger role of technology in the food industry makes the innovation capabilities more important for the success of the implemented innovations. However, the customary lower levels of investment in research and development point out that the innovation process in the food and beverage field may have different preconditions and require different innovation capabilities (Tepic et al., 2014). Increasing innovation potential might require exploration of completely new and technologically complex innovative possibilities. At the same time, this complexity and uncertainty can complicate problem-solving (Atuahene-Gima and Evangelista, 2000). Innovation process quality is thus defined through proper planning, preparation, and continued assessment of new characteristics.

15.2 INNOVATION STRATEGIES AND LONG-TERM R&D FOR THE FOOD INDUSTRY

While in the past the food industry was traditionally focused on the minimization of production costs, thus devoting little attention to customer needs (Lienhardt, 2004), the food it has recently changed from a supply-based to a demand-based approach (chain reversal) in which the consumers tell producers what they want to eat (Bigliardi and Galati, 2013; Boland, 2008). Today food companies innovate for several reasons, eg, to improve the quality and increase the range of products and services, to develop flexible production processes, to increase market share, to reduce labor costs, to improve health and safety, etc. (Giannoulidis, 2013). Following these trends, the food industry faces challenges that cannot be dealt with by the sector's present state in organization, technology, cooperation, and communication. Major challenges include assuring food security and safety while respecting society's environmental and social concerns toward sustainability. These challenges must be addressed in a world with increasing population, increasing urbanization, increasing requirements on higher value diets but with limitations and reductions in the availability of resources. Many of them (eg, the delivery of sufficient food) become challenges because of responsibilities placed by society on the food sector.

While the challenges involve a variety of different needs, the magnitude of challenges is sometimes captured by the slogan "twice with half" where the near future requires a doubling of food production while production resources will decline. The progress of the sector will depend on innovative approaches that may change the present infrastructure, organization, production, distribution, or retail sales but also the sector's offering of consumer products. Dealing with the challenges, the sector is driven by market, consumer, or society requirements. These drivers are presently matched by a variety of enablers that could facilitate and support sector initiatives in meeting the challenges ahead. The diversity and creativity of small and medium enterprises (SMEs) may play an important role in implementing inventions, especially in case of being supported by appropriate network organizations that facilitate knowledge exchange and interaction with experts and research. However, moving from inventions to innovation depends on the relevance of drivers and needs and the availability of enablers that facilitate the uptake of inventions by the sector on a broad scale.

In this line, "open innovation" can enhance available options by accessing external assets and collaborations with unique opportunities, facilitating partnerships, and alleviating hurdles such as limited resources, R&D expertise, skills, etc. Open innovation presents a unique opportunity for all stakeholders, especially SMEs, to proactively engage in meeting future challenges and opportunities. The food industry is increasingly choosing to enhance internal idea development by pursuing an open approach to innovation, exploiting external knowledge and paths to market. Analysis of the extant literature (chapters: Open Innovation and Incorporation Between Academia and Food Industry and Open Innovation Opportunities Focusing on Food SMEs) leads to the identification of several challenges and needs for the successful implementation of the open innovation paradigm in the food industry. It is widely recognized that most food companies face a series of challenges when implementing open innovation (Bigliardi and Galati, 2012; Saguy, 2011; Sarkar and Costa, 2008). For instance, one challenge refers to the collaborations that a food company may establish. An open innovation approach usually involves more than one partner depending on the knowledge, technologies, and information required in the different stage of the innovation process. Therefore, a balance among numerous and often contrasting needs of the different partners has to be reached. Another challenge has to do with identifying and supporting a management process. Another one considers the tools (in particular the information and communications technologies [ICTs]) that a food company may adopt in order to implement an open innovation approach. Besides, adopting a sustainable approach to innovation could lead to a reduction of environmental impacts, healthier and safer food for an increasing population, which could also lead to a reduction of costs and time to market. Well-crafted innovation policy actions can accelerate the transition toward a sustainable agri-food model. Indeed, a change occurs when pressure exerted by landscape actors upon the regime, and technological pressure exerted by innovation niches ready to replace the incumbent technological regime, occur.

For activating the sector toward innovation, support of the creativity inherent in SMEs is of crucial relevance. Networking and the establishment of network organizations for supporting communication between enterprises is considered a core initiative for activating SME involvement in innovation development. In addition, a better understanding of boundary roles between academia and industry will help to decrease the perceived market demand uncertainty (Vauterin, 2012). When two or more project teams try to combine agile and open innovation principles, distinct problems may arise. Minimal documentation often reduces effective knowledge transfer while the use of short iterations, stand-up meetings, and presence of on-site customers reduce the amount of time for sharing ideas outside the team. Hence, a clear understanding of the inter- and intraorganizational applicability and implications of open innovation in agile systems development is required to address key challenges for research and practice (Conboy and Morgan, 2011). On the other hand, academia needs a revised intellectual property model to consider in addition to innovation's financial returns, the unique needs of SMEs, and its paramount role in social responsibility. To catch up, SME managements will have to be extremely flexible or face the consequences. Finally, more studies are still needed to investigate how to reduce the gap between academia and the food industry, as well as how such a collaboration can be developed and coordinated effectively.

15.3 DEVELOPMENT OF INNOVATIONS IN THE FOOD INDUSTRY

An important part of the European food industry produces traditional foods, the development of which is supported by SMEs. In addition to its economic and social importance, traditional foods are a significant part of culture, identity, and European gastronomic heritage (Ilbery and Kneafsey, 1999). In general, traditional foods are associated with a regional identity and a particular sensory quality, eg, marketed under different collective brands such as quality labels (Guerrero, 2001). The production of traditional foods still relies on low competitive and low-efficiency traditional production practices (Fito and Toldra, 2006). Thus, there appear to be a number of opportunities for traditional foods that have not yet been adequately exploited.

Traditional food producers face the challenge to implement different innovations to improve safety, health, and convenience of their products with an ultimate goal to maintain and even expand their current area of influence in a highly competitive and globalized market. Besides, the implementation of innovations in traditional foods generates in principle one main disadvantage: the innovation itself could make them lose their "traditional" nature (Caldentey and Gomez, 1997). In principle, there are certain possibilities of introducing innovations in traditional foods without diminishing their main competitive profile. Innovations that improve the nutritional quality of the products and make them more convenient are relatively well accepted as long as they do not involve changes in the sensory quality. Indeed, the existence of consumers with different beliefs and attitudes offers some opportunity of success to the implementation of innovations in the traditional food market. This information could help producers of traditional foods in making decisions when applying different innovations with respect to communication strategy, product placement, and new developments.

Another challenge of the food sector has to do with the fact that innovation is dealt with much less celebration than the innovation in other technology industries (eg, computers, mobile phones, or the Internet). For example, a new snack chip is a new snack chip. In contrast, a new computer chip or an application could lead to fortune for the inventors, patent holders, and a leap forward leading to a new generation of machine (chapter: Implementation of Innovation in the Food Industry). Besides, sustainability challenges occur at all stages in the food system from production through processing, distribution, and retailing to consumption and waste disposal (Galanakis and Schieber, 2014). Management of food processes in an adequate way can contribute to achieve a full sustainability concept with three dimensions implied: environmental, social, and economic. Integrated perspectives are particularly needed to improve understanding of the mutual interaction between technological change and these three contexts. There is a clear need for holistic solutions, combining adaptations in sociotechnical systems rather than single-minded technological responses (chapter: Sustainable Innovation in Food Science and Engineering).

The food industry has largely invested in processing facilities relying mostly on conventional thermal technologies, which show well-established reliability and efficacy (Mújica-Paz et al., 2011). However, these thermal processing approaches usually negatively impact important food quality attributes (eg, destruction of important nutrients, development of off-flavors, and color changes) (Galanakis, 2012, 2013; Deng et al., 2014). In recent years, the consumer is increasingly demanding fresh-like food products, with high organoleptic and nutritional quality. Consequently, novel processing technologies have been gaining interest among food researchers, due to their lower impact on nutritional and sensory properties of the products compared to the conventional thermal techniques. These methods include, eg, high-pressure processing (HPP), pulsed electric fields (PEF), ohmic heating, microwave, radiofrequency, and ultrasound. Most of them have only found niche applications in the food industry, replacing or complementing conventional preservation technologies.

The widest application of HPP within the food industry is in the context of food processing. These include solute diffusion processes (salting, sugaring), assisted freezing-thawing processes, and modification of functional properties of proteins and other macromolecules (San Martin et al., 2002). PEF is used for the inactivation of quality-related enzymes, as well as destruction of spoilage and pathogenic microorganisms, possibly due to loss of cell membrane functionality. PEF technology can also be used by the food industry to improve the functionality, extractability, and recovery of bioactive compounds by increasing cell permeability (Barba et al., 2015). Likewise, the achievement of high-voltage pulsed power systems led to PEF performance scales from laboratory scale to large production scale installations (Puértolas et al., 2010). However, this technology still shows some limitations in the context of food processing. For example, food products with large electrical conductivity are not suitable for PEF processing because the peak electric field across the chamber is reduced.

Ohmic heating is similar to PEF technique, allowing the rapid and uniform heating of liquids as well as maintaining the nutritional and sensory characteristics of foods (Sastry, 2005; Sun, 2012). However, there are still several limitations (eg, difficulties in controlling the heat rate during the processes) due to change of electric conductivity of the heating material. In contrast to PEF, the principal mechanisms of microbial inactivation in ohmic heating are thermal in nature. Further research is still needed to completely understand all the effects of ohmic heating in food products (Knirsch et al., 2010). Microwave heating has vast applications in the field of food processing, such as cooking, drying, pasteurization, and preservation of food materials (Chandrasekaran et al., 2013). Microwave pasteurization has been largely applied to fluid foods such as fruit juices and solid ones like eggs. Ultrasound is particularly suitable for preservation of fluid foods (Roselló-Soto et al., 2015 a,b; Zinoviadou et al., 2015). Its main advantage over the conventional thermal pasteurization methods include reduced water and energy consumption as well as minimal flavor and nutrient losses (Soria and Villamiel, 2010).

The great drawback for the implementation of these novel processing technologies in food industry seems to be the higher price of the final product when compared to products that are thermally processed. Subsequently, the consumer acceptance of products processed by emerging technologies is reduced, as most of them are not willing to pay more for these products. Thus, developers and sponsors of these technologies must take several measures to facilitate their implementation in the food industry (chapter: Implementation of Emerging Technologies). Additional corporative measures should also be considered as possible encouragements for the implementation of these novel technologies. One solution is the application of the open innovation model, where companies use external ideas, technologies, and solutions in addition to the internal ones to innovate, opening channels for knowledge access (Saguy and Sirotinskaya, 2014). This can be performed by the promotion of win–win collaborations between universities and industries. Finally, providing more information about the technologies seems to be the key to increase consumer acceptance of these (eg, PEF-processed) products (Sonne et al., 2012).

15.4 CUTTING-EDGE INNOVATION AREAS IN FOOD SCIENCE

15.4.1 FUNCTIONAL FOODS

Nowadays functional foods comprise the most important cutting-edge area in the food industry. This trend was forced by scientists supporting the idea that food can bring specific components with health benefits beyond those of basic nutrition. Subsequently, longer life expectancy has accelerated the consumption of certain foods that may help to improve the state of well-being and reduce the risk of diseases. The components of functional food health benefits are of diverse nature, including polyunsaturated fatty acids (PUFA, omega-3), antioxidants, vitamins, dietary fibers, probiotics, prebiotics (inulin, beta-glucans), peptides, etc. Examples of functional foods developed in the recent past years are foods with reduced content in salt and fat, or gluten-free, aimed at high-risk populations (eg, diabetics) (chapter: Development of Functional Foods).

The acceptance of functional foods is directly connected to attributes related to the product and its claim (eg, labeling simplicity or familiarity), as well as to the consumers themselves. In general, clear and transparent labeling upholds the confidence of consumers in the food industry (Lahteenmahteet al., 2010). Functional foods must also meet consumer expectations in terms of organoleptic and nutritional characteristics. With respect to the nutritional value, consumers perceive some nutrients as qualifying (fibers, vitamins, and minerals) or disqualifying (energy, fat, saturated fat, salt, sugars) (Hoefkens et al., 2011). An innovation field is to meet the nutrient profiling to obtain healthier products. Consumer-related attributes concern gender, age, and lifestyle. Generational differences in eating patterns, flavor preferences, and personal choices are creating new opportunities for the food and beverage companies (Verbeke, 2010). The millennials (born between 1981 and 2000), the children of baby boomers (born between 1946 and 1964), will redefine the food industry in the longer term. They are the most ethnically diverse and the most educated generation to date, who view their food choices as healthier, more expensive, less processed, and better tasting than those of their boomer parents. For instance, younger shoppers are likely to buy foods that address allergies, aging, appearance, digestive health, immunity, and mental clarity.

The actual framework for the development of functional foods is diverse in different countries, thus influencing the innovation policies of the food industry. Legal requirements for the commercialization of these products have profoundly affected the market of functional foods. Functional foods must be scientifically evaluated prior to commercialization. Once approved, the use of a health claim as a marketing tool significantly improves the competitive position of the food company. However, the design of a functional product means high development costs and important challenges. The building of consortia between research institutes, academia, and food industry aims to address these points and serves as a unique opportunity for food innovation.

15.4.2 FOODOMICS

Nowadays food is not only recognized as a source of energy but also a potential tool in maintaining humans in better health for a longer time. Furthermore, minor components that are present in food products have a potential to prevent the appearance of diseases (Capozzi and Bordini, 2013). Another goal is also to optimize nutritional recommendations to specific population groups (Brennan, 2015). In the past few years, research in food and nutrition sciences has moved from classical analytical methods to very sophisticated omics approaches that create an impressive amount of data. The latest can only be analyzed using bioinformatics. Foodomics is now ready to provide data to be exploitable in food science, nutrition, and medicine. The next step will be their real implementation in the industry. For metabolomics, especially when using LC-MS approaches, a number of technical challenges (eg, metabolite identification) still have to be resolved. Sharing of resources such as databases and chemical standards would be mandatory to carry out these identifications. However, developments still have to be done to transfer this approach used mainly in research laboratories to the industrial world. Most of the techniques involved in foodomics approaches are complex, costly, and need highly trained people to obtain robust data. Teams working in this field must be composed of people having complementary skills in chemistry, analysis, statistics, biology, and bioinformatics. Governments of different countries have developed large "omics" centers, sometimes using joint ventures with private industries. Other initiatives of sharing databases and bioinformatics tools have also been proposed. Consequently, one way to innovate in this field would be by developing closer collaborations between the academic world and industry, perhaps in an open innovation model.

15.4.3 FOOD WASTE RECOVERY

Another important and forthcoming innovation area in the food industry is food waste recovery. In the next 50 years, increasing population, urbanization, and rising incomes will bring a sharp growth in food processing industries and alter food supply chains worldwide. Pressures in natural resources to feed the world population and waste disposal costs are already global priorities, whereas the recovery of resources especially in the food chain will play a vital role for the management strategies in the years to come. Food processing industries generate a huge amount of liquid and solid wastes. The main methodologies for waste minimization and valorization include general methods such as anaerobic fermentation and composting. Indeed, the potential of food wastes to create new markets has been underestimated until very recent years. Today consumers' consciousness about environmental issues and legislative pressures increase the requirements of new methods for the recovery of food waste rather than its disposal. Food processing by-products can be used for the recovery of high added-value components with specific characteristics and their conversion to products with market value (Galanakis, 2011, 2012; Galanakis et al., 2014, 2013b; Patsioura et al., 2011; Rahmanian et al., 2014). For example, olive mill wastewater can be used as a source of phenols and dietary fibers, which could be recovered and reutilized in beverages and meat products, respectively (Galanakis et al., 2010a,b,c,d,e). Added value can be related to the shelf life or color of the final product, health benefits, formulation, or other product features. To prevent stifling of innovation (from the mounting safety concerns and strict regulations) and simultaneously address emerging wellness issues, a new direction for developing, acquiring, and implementing the vast potential of scientific breakthroughs is needed (Saguy and Sirotinskaya, 2014). The establishment of a new label (similar to organic foods) or reduced taxes on relevant products could reveal the potential of recovering high added-value ingredients from food by-products and reutilizing them in foods. In general, the key point for commercialization is to develop a recovery strategy that allows flexibility and provides alternative scenarios for each stage of processing (Galanakis, 2012).

What is strongly recommended is the implementation of nonthermal technologies, addition of green solvents, and safer materials (possessing generally recognized as safe status) (Galanakis, 2015b; Galanakis et al., 2013a,b; Tsakona et al., 2012). The development of tailor-made applications for the recovered products (crude or highly purified) is necessary, too, as it is difficult to survive competition in the market of functional foods. With regard to the recovery stages, cheaper methodologies are those containing fewer recovery steps. Product formation (the fifth one from the 5-Stages Universal Recovery Process) is the most essential step, as encapsulation enhances functionality and extends the shelf life of the products (Galanakis, 2015a; Galanakis et al., 2015). Researchers in the field will soon deal with the prospect of applying emerging technologies and particularly nanotechniques with an ultimate goal to optimize overall efficiency of suggested methodologies (Galanakis, 2012). It is also of particular interest that the companies active in the field and academia move on to a mutual commitment of resources and time by entering into a collaborative and long-term engagement. Selection of research topics should be guided not only by scientific interest but also by manufacturers and markets' specific needs and requests. This approach would positively influence application of obtained research results and use of scientific breakthroughs' potential. Last, but not least, new product development should aim at the fulfillment of consumer needs and the realization of consumer value rather than at the development of products or enabling technologies per se (Costa and Jongen, 2006). The food chain has become one of the primary sites of political resistance to globalization. To reinforce policy reforms and regulatory changes, all those involved in the chain have also sought to encourage consumers to buy "environmentally friendly" products. This in turn has stimulated "green" marketing and environmental quality assurance schemes, which generate multiple possibilities of innovations in the field.

15.4.4 BIOBASED MATERIALS FOR SUSTAINABLE PACKAGING

Sustainable packaging is a topic that attracts the attention of industry, government, and consumers. It concerns many different aspects such as the reduction of packaging material, the use of recycling packaging, or the use of materials with a lower ecological footprint. Biopolymers show some limitations in terms of performance, although they fulfill the environmental concerns. The use of nanoparticles exhibit great potential in the field of polymer science, whereas researchers and food professionals should focus on the printability of biodegradable films in order to make them cost-competitive with plastics. Besides, biobased packaging technology should not only focus on microstructure, physicochemical, and biodegradable properties of the films but also on consumer awareness and acceptance of such food packaging materials (chapter: Innovative Biobased Materials for Packaging Sustainability).

15.4.5 INFORMATION AND COMMUNICATIONS TECHNOLOGIES

The agri-food industry is a sector that is highly relevant in socioeconomic and territorial terms yet is less innovation intensive than other industrial sectors are. However, the food sector is more technology intensive than the primary agriculture sector is estimated to be and is indeed more technology intensive than some other industrial sectors (Fearne et al., 2013). The development of ICTs over the last two decades has considerably reduced the costs of processing and transferring information as well as diminished the relevance of the firm's physical location. Nevertheless, considering the perspective of territorial innovation models, the adoption of innovations such as ICTs may occur more quickly in urban areas because of the lack of infrastructure and qualified workers in rural areas (Moulaert and Sekia, 2003).

The adoption of ICTs represents a key innovation for firms activated in the food sector. Not only do ICTs directly affect the firm but they also lead to changes in the way the firm organizes itself and builds relationships with other entities. A prime example is the Internet, which offers a wide range of relationship opportunities, making the firm's location less relevant. The European agri-food industry has recently taken steps to adopt ICTs, which has allowed the industry to improve its innovation capacity (Alarcón et al., 2013). On the other hand, some studies show that innovation intensity has been low, as has the ICT adoption rate. The characteristics of the firm and business environment characteristics affect the adoption of ICTs. Thereby, firm's size and export activity are among the main determinants of the adoption of innovations. Large exporting firms are more likely to adopt ICTs. In terms of environmental variables, firms in rural environments adopt ICTs earlier than firms in urban environments, and rural firms are more likely to adopt ICTs. Besides, a moderate level of market concentration is positively related to a tendency to adopt ICTs (chapter: Adoption of ICT Innovations in the Agri-Food Sector: An Analysis of French and Spanish Industries).

15.5 CONSUMER ACCEPTANCE, AND CHAPTER CONCLUSIONS

The implementation of innovations in the food industry is governed by consumer acceptance. This is the critical parameter affecting to a considerable extent success or not. For instance, many innovations face some (or even severe) criticism when first launched. Genetically modified organisms (GMOs) in Europe is a classic example. Consumers' response to novel foods depends on how they perceive the benefits, risks, and costs from their consumption. In marketing, the phrase "acceptance of a novel product" is generally used to indicate that a novel product has been taken up by sufficiently many users to be profitable. Consumer perceptions and how they are combined into an overall evaluation of the novel food may systematically differ from expert product assessments (Siegrist and Gutscher, 2006). Besides, consumers may decide on the benefits of a product in different ways depending on their personal situation or the context of the food choice (Bagozzi et al., 2000).

To avoid investing in innovations that are inherently unacceptable to the public, it is important to include consumer insights into the innovation process early on (Van Kleef et al., 2005). Therefore it is important to know how to measure consumer perceptions reliably (Reinders et al., 2013). However, perceptions and decisions are in the mind of the consumer and cannot be measured directly. Numerous theories and approaches have been developed already to assess consumer response to innovations. Many of these approaches have in common that information about the innovative product and its attributes lead to perceptions. Modern insights into consumer behavior show that the picture is not as straightforward as perhaps previously thought. Implicit, often unconscious, gut feelings and the influence of the environment are also influencing consumer choice even without the awareness of the consumers themselves. Choosing the more relevant approach and relevant measure among these alternatives is among the most important parts of conducting good consumer research. Even if this is achieved, consumer science has only modest predictive power, especially when studying the uncertain future of food innovation. Current movements in consumer science are toward approaches where less of the consumer behavior is considered to be based on conscious and deliberate choice. The usefulness and applicability of these methods is however not yet well developed, and thus traditional methods are still applied.

Will this trend be established in consumer science concerning food products? We do not know yet. However, if this happens, it will definitely force implementation of innovations to a certain direction. In this case, consumers will hold the key to innovations, but their decisions would be unconscious. Although this situation is very tempting for strategy developers, it may stifle innovations in the field since there is a danger of generating a new skepticism toward consumers. If consumers somehow feel that they are out of the game or forced into a certain direction, their rejection would be immediate. The example of GMOs is an example of this situation. What we need more is extended collaborations that include all partners of interest (researchers, food industry, consumers, and policy makers) in a relevant open innovation model, as stated in chapters "Open Innovation and Incorporation Between Academia and Food Industry" and "Open Innovation Opportunities Focusing on Food SMEs." This process will definitely regenerate implementation of innovations in food science.

REFERENCES

Alarcón, S., González, L., Sánchez, M., 2013. Strategies for the development of new products in the Spanish agri-food industry. In: Zacharoula, A., Smarthrakis, V., Louca, S., Vlachopoulou, M. (Eds.), E-Innovation for Sustainable Development of Rural Resources During Global Economic Crisis. IGI Global, pp. 181–198.

Almeida Costa, A.A., 2008. Food innovation management. Trends in Food Science & Technology 19, 551–552.

Atuahene-Gima, K., Evangelista, F., 2000. Cross-functional influence in new product development: an exploratory study of marketing and R&D perspectives. Management Science 46, 1269–1284.

Bagozzi, R.P., Wong, N., Abe, S., Bergami, M., 2000. Cultural and situational contingencies and the theory of reasoned action: application to fast food restaurant consumption. Journal of Consumer Psychology 9, 97–106.

Barba, F.J., Galanakis, C.M., Esteve, M.J., Frigola, A., Vorobiev, E., 2015. Potential use of pulsed electric technologies and ultrasounds to improve the recovery of high-added value compounds from blackberries. Journal of Food Engineering 167, 38–44.

Bigliardi, B., Galati, F., 2013. Models of adoption of open innovation within the food industry. Trends in Food Science & Technology 30, 16–26.
Bigliardi, B., Dormio, A.I., Galati, F., 2012. The adoption of open innovation within the telecommunication industry. European Journal of Innovation Management 15, 27–54.
Boland, M.J., 2008. Innovation in the foods industry: personalisation and mass customisation. In: Marcure, J., Moughan, P., Bruhn, C. (Eds.), Innovation, Management Policy and Practice, vol. 10, pp. 1–132.
Brennan, L., 2015. Metabotyping: moving towards personalized nutrition. In: Sébeédio, J.l., Brennan, L. (Eds.), Metabolomics as a Tool in Nutrition Research. Woodhead Publishing, Kidlington, UK, pp. 137–144.
Caldentey, P., Gómez, A.C., 1997. Typical products, technical innovation and organizational innovations. In: Paper Presented at the Typical and Traditional Productions: Rural Effect and Agro-industrial Problems, 52nd EAAE Seminar (19–21 June 1997). Parma, Italia.
Capozzi, F., Bordoni, A., 2013. Foodomics: a comprehensive approach to food and nutrition. Genes & Nutrition 8, 1–4.
Chandrasekaran, S., Ramanathan, S., Basak, T., 2013. Microwave food processing – a review. Food Research International 52, 243–261.
Conboy, K., Morgan, L., 2011. Beyond the customer: opening the agile systems development process. Information and Software Technology 53, 535–542.
Costa, A.I.A., Jongen, W.M.F., 2006. New insights into consumer-led food product development. Trends in Food Science and Technology 8, 457–465.
Costa, A.I.A., Dekker, M., Beumer, R.R., Rombouts, F.M., Jongen, W.M.F., 2001. A consumer-oriented classification system for home meal replacements. Food Quality and Preference 12, 229–242.
Costa, A.I.A., Schoolmeester, D., Dekker, M., Jongen, W.M.F., 2007. To cook or not to cook: a means-end study of motives for choice of meal solutions. Food Quality and Preference 18, 77–88.
Deng, Q., Zinoviadou, K.G., Galanakis, C.M., Orlien, V., Grimi, N., Vorobiev, E., Lebovka, N., Barba, F.J., 2014. The effects of conventional and non-conventional processing on glucosinolates and its derived forms, isothiocyanates: extraction, degradation and applications. Food Engineering Reviews 7, 357–381.
Fearne, A., Álvarez-Coque, J.M., López-García Usach, T., Sánchez, M., 2013. Innovative firms and the urban/rural divide: the case of agro-food system. Management Decision 51, 1293–1310.
Fito, P., Toldra, F., 2006. Innovations in traditional foods. EFFOST 2005 conference. Trends in Food Science and Technology 17, 470.
Fritz, M., Schiefer, G., 2008. Food chain management for sustainable food system development: a European research agenda. Agribusiness 24, 440–452.
Galanakis, C.M., Schieber, A., 2014. Editorial. Special Issue on Recovery and utilization of valuable compounds from food processing by-products. Food Research International 65, 299–300.
Galanakis, C.M., Tornberg, E., Gekas, V., 2010a. A study of the recovery of the dietary fibres from olive mill wastewater and the gelling ability of the soluble fibre fraction. LWT-Food Science & Technology 43, 1009–1017.
Galanakis, C.M., Tornberg, E., Gekas, V., 2010b. Clarification of high-added value products from olive mill wastewater. Journal of Food Engineering 99, 190–197.
Galanakis, C.M., Tornberg, E., Gekas, V., 2010c. Dietary fiber suspensions from olive mill wastewater as potential fat replacements in meatballs. LWT-Food Science & Technology 43, 1018–1025.
Galanakis, C.M., Tornberg, E., Gekas, V., 2010d. Recovery and preservation of phenols from olive waste in ethanolic extracts. Journal of Chemical Technology & Biotechnology 85, 1148–1155.
Galanakis, C.M., Tornberg, E., Gekas, V., 2010e. The effect of heat processing on the functional properties of pectin contained in olive mill wastewater. LWT-Food Science & Technology 43, 1001–1008.
Galanakis, C.M., Goulas, V., Tsakona, S., Manganaris, G.A., Gekas, V., 2013a. A knowledge base for the recovery of natural phenols with different solvents. International Journal of Food Properties 16, 382–396.
Galanakis, C.M., Markouli, E., Gekas, V., 2013b. Fractionation and recovery of different phenolic classes from winery sludge via membrane filtration. Separation and Purification Technology 107, 245–251.

Galanakis, C.M., Chasiotis, S., Botsaris, G., Gekas, V., 2014. Separation and recovery of proteins and sugars from Halloumi cheese whey. Food Research International 65, 477–483.

Galanakis, C.M., Martínez-Saez, N., del Castillo, M.D., Barba, F.J., Mitropoulou, V.S., 2015. Chapter 15: Patented and commercialized applications. In: Galanakis, C.M. (Ed.), Food Waste Recovery: Processing Technologies and Industrial Techniques. Elsevier-Academic Press.

Galanakis, C.M., 2011. Olive fruit and dietary fibers: components, recovery and applications. Trends in Food Science and Technology 22, 175–184.

Galanakis, C.M., 2012. Recovery of high added-value components from food wastes: conventional, emerging technologies and commercialized applications. Trends in Food Science & Technology 26, 68–87.

Galanakis, C.M., 2013. Emerging technologies for the production of nutraceuticals from agricultural by-products: a viewpoint of opportunities and challenges. Food and Bioproducts Processing 91, 575–579.

Galanakis, C.M., 2015a. The universal recovery (Chapter 3). In: Galanakis, C.M. (Ed.), Food Waste Recovery: Processing Technologies and Industrial Techniques.

Galanakis, C.M., 2015b. Separation of functional macromolecules and micromolecules: from ultrafiltration to the border of nanofiltration. Trends in Food Science & Technology 42, 44–63.

Geylani, P.C., Pompelli, G.K., Stefanou, S.E., 2008. Productivity and innovation in the US food processing sector. In: Rama, R. (Ed.), Handbook of Innovation in the Food and Drink Industry. The Haworth Press, New York, NY.

Giannoulidis, N., 2013. Trends and Innovation Needs in the European Food and Drink Industry. http://www.innofoodsee.eu/downloads/trends_and_innovation.pdf (retrieved June 2015).

Guerrero, L., 2001. Marketing PDO (Products with Denominations of Origin) and PGI (Products with Geographical Identities). In: Frewer, L., Risvik, E., Shifferstein (Eds.), Food, People and Society. An European Perspective of Consumers' Food Choices. Springer Verlag, Berlin, pp. 281–296.

Hoefkens, C., Verbeke, W., Van Camp, J., 2011. European consumers' perceived importance of qualifying and disqualifying nutrients in food choices. Food Quality and Preference 22, 550–558.

Ilbery, B., Kneafsey, M., 1999. Niche markets and regional speciality food products in Europe: towards a research agenda. Environment and Planning A 31, 2207–2222.

Knirsch, M.C., Dos Santos, C.A., de Oliveira Soares, A.A.M., Penna, T.C.V., 2010. Ohmic heating – a review. Trends in Food Science & Technology 21, 436–441.

Lahteenmäki, L., Lampila, P., Grunert, K., Boztug, Y., Ueland, Ø., Aström, A., 2010. Impact of health-related claims on the perception of other product attributes. Food Policy 35, 230–239.

Lienhardt, J., 2004. The Food Industry in Europe, Statistics in Focus – Industry, Trade and Services, 39/2004. Eurostat, European Communities, Luxembourg.

Moulaert, F., Sekia, F., 2003. Territorial innovation models: a critical survey. Regional Studies 37, 289–302.

Mújica-Paz, H., Valdez-Fragoso, A., Samson, C.T., Welti-Chanes, J., Torres, J.A., 2011. High-pressure processing technologies for the pasteurization and sterilization of foods. Food and Bioprocess Technology 4, 969–985.

Patsioura, A., Galanakis, C.M., Gekas, V., 2011. Ultrafiltration optimization for the recovery of β-glucan from oat mill waste. Journal of Membrane Science 373, 53–63.

Puértolas, E., López, N., Condón, S., Álvarez, I., Raso, J., 2010. Potential applications of PEF to improve red wine quality. Trends in Food Science & Technology 21, 247–255.

Rahmanian, N., Jafari, S.M., Galanakis, C.M., 2014. Recovery and removal of phenolic compounds from olive mill wastewater. Journal of the American Oil Chemists' Society 91, 1–18.

Reinders, M.J., van der Lans, I.A., Fischer, A.R.H., van Trijp, H.C.M., 2013. A Review to Collate Information on External Communication as a Basis of Innovation Success. Connect4Action Deliverable, Wageningen.

Roselló-Soto, E., Barba, F.J., Parniakov, O., Galanakis, C.M., Grimi, N., Lebovka, N., Vorobiev, E., 2015a. High voltage electrical discharges, pulsed electric field and ultrasounds assisted extraction of protein and phenolic compounds from olive kernel. Food & Bioprocess Technology 8, 885–894.

Roselló-Soto, E., Galanakis, C.M., Brncic, M., Orlien, V., Trujillo, F.J., Mawson, R., Knoerzer, K., Tiwari, B.K., Francisco, J., Barba, F.J., 2015b. Clean recovery of antioxidant compounds from plant foods, byproducts and algae assisted by ultrasounds processing. Modeling approaches to optimize processing conditions. Trends in Food Science & Technology 42, 134–149.

Saguy, I.S., Sirotinskaya, V., 2014. Challenges in exploiting open innovation's full potential in the food industry with a focus on small and medium enterprises (SMEs). Trends in Food Science & Technology 38, 136–148.

Saguy, I.S., 2011. Paradigm shifts in academia and the food industry required to meet innovation challenges. Trends in Food Science & Technology 22, 467–475.

San Martin, M., Barbosa-Canovas, G., Swanson, B., 2002. Food processing by high hydrostatic pressure. Critical Reviews in Food Science and Nutrition 42, 627–645.

Sarkar, S., Costa, A.I.A., 2008. Dynamics of open innovation in the food industry. Trends in Food Science and Technology 19, 574–580.

Sastry, S.K., 2005. Advances in ohmic heating and moderate electric field (MEF) processing. In: Barbosa-Cánovas, G.V., Tapia, M.S., Cano, M.P. (Eds.), Novel Food Processing Technologies. CRC Press, Boca Raton, US, pp. 491–500.

Siegrist, M., Gutscher, H., 2006. Flooding risks: a comparison of lay people's perceptions and expert's assessments in Switzerland. Risk Analysis 26, 971–979.

Sonne, A.-M., Grunert, K.G., Veflen Olsen, N., Granli, B.-S., Szabó, E., Banati, D., 2012. Consumers' perceptions of HPP and PEF food products. British Food Journal 114, 85–107.

Soria, A.C., Villamiel, M., 2010. Effect of ultrasound on the technological properties and bioactivity of food: a review. Trends in Food Science & Technology 21, 323–331.

Sun, D.-W., 2012. Thermal Food Processing: New Technologies and Quality Issues. CRC Press, Boca Raton, US.

Tepic, M., Kemp, R.G.M., Omta, O., 2014. Innovation capabilities in food and beverages and technology-based innovation projects. British Food Journal 116, 228–250.

Tsakona, S., Galanakis, C.M., Gekas, V., 2012. Hydro-ethanolic mixtures for the recovery of phenols from Mediterranean plant materials. Food & Bioprocess Technology 5, 1384–1393.

Van Kleef, E., Van Trijp, H., Luning, P.A., 2005. Consumer research in the early stages of new product development: a critical review of methods and techniques. Food Quality and Preference 16, 181–201.

Vauterin, J.J., 2012. The Demand for Global Student Talent: Capitalizing on the Value of University-industry Collaboration. Lappeenranta University of Technology.

Verbeke, W., 2010. Consumer reactions to foods with nutrition and health claims. Agro Food Industry Hi-Tech 21, 5–8.

Zinoviadou, K.G., Barba, F.J., Galanakis, C.M., Brnčić, M., Trujillo, F., Mawson, R., Knoerzer, K., 2015. Fruit juice sonication: implications on food safety and physicochemical and nutritional properties. Food Research International 77 (4), 743–752.

Index

Note: Page numbers followed by "f" indicate figures, "t" indicate tables.

F

P

Q

R

Kolibri 360 GmbH
Ferdinand-Jühlke-Straße 7
99095 Erfurt, Deutschland
produktsicherheit@kolibri360.de